Lattices and Ordered Sets

Steven Roman

Lattices and Ordered Sets

Dr. Steven Roman
Irvine, CA 92603
USA

ISBN: 978-1-4419-2704-0 e-ISBN: 978-0-387-78901-9
DOI: 10.1007/978-0-387-78901-9

Mathematics Subject Classification (2000): 06-xx, 03-xx, 03Exx, 04-xx

Printed on acid-free paper

9 8 7 6 5 4 3 2 1

springer.com

To Donna
and to my friend
Gian-Carlo Rota,
who is missed

Preface

This book is intended to be a thorough introduction to the subject of order and lattices, with an emphasis on the latter. It can be used for a course at the graduate or advanced undergraduate level or for independent study. Prerequisites are kept to a minimum, but an introductory course in abstract algebra is highly recommended, since many of the examples are drawn from this area. This is a book on pure mathematics: I do not discuss the applications of lattice theory to physics, computer science or other disciplines.

Lattice theory began in the early 1890s, when Richard Dedekind wanted to know the answer to the following question: Given three subgroups A, B and C of an abelian group G, what is the largest number of distinct subgroups that can be formed using these subgroups and the operations of intersection and sum (join), as in

$$A + B, (A + B) \cap C, A + (B \cap C)$$

and so on? In lattice-theoretic terms, this is the number of elements in the relatively free modular lattice on three generators. Dedekind [15] answered this question (the answer is 28) and wrote two papers on the subject of lattice theory, but then the subject lay relatively dormant until Garrett Birkhoff, Oystein Ore and others picked it up in the 1930s. Since then, many noted mathematicians have contributed to the subject, including Garrett Birkhoff, Richard Dedekind, Israel Gelfand, George Grätzer, Aleksandr Kurosh, Anatoly Malcev, Oystein Ore, Gian-Carlo Rota, Alfred Tarski and Johnny von Neumann.

This book is divided into two parts. The first part (Chapters 1–9) outlines the basic notions of the theory and the second part (Chapters 10–12) contains independent topics.

Specifically, Chapter 1 describes the basic theory of partially ordered sets, including duality, chain conditions and Dilworth's theorem. Chapter 2 is devoted to the basics of well-ordered sets, especially ordinal and cardinal numbers, a topic that is usually presented only in classes on set theory. However, I have included it here for two reasons. First, we will have a few occasions to use ordinal numbers, most notably in describing conditions that characterize the incompleteness of a lattice and in the discussion of fixed points. Second, it is my experience that many students of algebra do not take a class in set theory and so this seems a reasonable place to get a first exposure to this important subject. In any case, the chapter may be omitted upon first reading without severe consequences: Only a few theorems will be inaccessible to readers with no background in the subject. However, I have also included nonordinal versions (for countable lattices) of these theorems.

Chapter 3 begins the study of lattices, introducing the key properties such as join-irreducibility, completeness, denseness, lattice homomorphisms, the B-down map, ideals and filters, prime and maximal ideals and the Dedekind-MacNeille completion.

Chapter 4 is devoted to the basics of modular and distributive lattices, including the many characterizations of these important types of lattices and the effect of distributivity on join-irreducibility and prime ideals.

Chapter 5 is devoted to Boolean algebras. This is a vast field, with many applications, but we concentrate on the lattice-theoretic concepts, such as characterizing Boolean algebras as Boolean rings, or in terms of weak forms of modularity or de Morgan's laws. We also discuss complete and infinite distributivity.

Chapter 6 concerns the representation theory of distributive lattices. We describe the representation of a distributive lattice with the DCC as a power set sublattice, the representation of complete atomic Boolean algebras as power sets and the representation of an arbitrary distributive lattice as a sublattice of down-sets of prime ideals.

Chapter 7 is devoted to algebraic lattices, a topic that should be of special interest to all those interested in abstract algebra. Here we prove that a lattice is algebraic if and only if it is isomorphic to the subalgebra lattice of some algebra.

Chapter 8 concerns the existence of maximal and prime ideals in lattices. Here we discuss various types of separation theorems and their relationship with distributivity, including the famous Boolean prime ideal theorem. This material is key to an understanding of the Stone and Priestley duality theory, described in a later chapter. Chapter 9 is devoted to congruence relations on lattices, quotient lattices, standard congruences and sectionally complemented lattices.

Chapters 10–12 are topics chapters. They can be covered in any order. Chapter 10 is devoted to the duality theory for bounded distributive lattices and Boolean algebras. We cover the finite case first, where topological notions are not required. Then we discuss the necessary topics concerning Boolean and ordered Boolean spaces. Although very little point-set topology is used, a brief appendix is included for readers who are not familiar with the subject. In both the finite and nonfinite cases, the dualities are most elegantly described using the language of category theory. For readers who are not familiar with this language, an appendix is provided with just enough category theory for our discussion in the text. (It is not so much the *results* of category theory that we need, but rather the *language* of category theory.)

Chapter 11 is an introduction to free lattices and Chapter 12 covers fixed-point theorems for monotone and for inflationary functions on complete partially ordered sets and complete lattices.

Contents

Part I
Basic Theory

Chapter 1
Partially Ordered Sets

Basic Definitions

We begin with some basic definitions.

System of Distinct Representatives

We will have occasions to use the following basic concept.

Definition *Let $\mathcal{F} = \{A_i \mid i \in I\}$ be a family of sets. A* **system of distinct representatives** *or* **SDR** *for $\mathcal{F}$ is a set $\{a_i \mid i \in I\}$ of distinct elements with the property that $a_i \in A_i$ for all i.* $\square$

Cartesian Products

The cartesian product of two sets S and T is generally defined as the set

$$S \times T = \{(s,t) \mid s \in S, t \in T\}$$

of ordered pairs. However, this does not easily generalize to the cartesian product of an arbitrary family of sets. For this, we use functions.

Definition *The* **cartesian product** *of a family $\mathcal{F} = \{X_i \mid i \in I\}$ of sets is the set*

$$\prod \mathcal{F} = \prod_{i \in I} X_i = \left\{ f\colon I \to \bigcup X_i \,\middle|\, f(i) \in A_i \right\} \qquad \square$$

Binary Relations

Definition *Let A be a nonempty set. A* **binary relation** *on A is a subset R of the cartesian product $A \times A$. We write $(a,b) \in R$ as $a \sim b$. A binary relation is*

S. Roman (ed.), *Lattices and Ordered Sets*, doi: 10.1007/978-0-387-78901-9_1,

1) **reflexive** *if for all* $a \in A$,

$$a \sim a$$

2) **irreflexive** *if for all* $a \in A$,

$$a \not\sim a$$

3) **symmetric** *if for all* $a, b \in A$,

$$a \sim b \quad \Rightarrow \quad b \sim a$$

4) **asymmetric** *if for all* $a, b \in A$,

$$a \sim b \quad \Rightarrow \quad b \not\sim a$$

5) **antisymmetric** *if for all* $a, b \in A$,

$$a \sim b, \quad b \sim a \quad \Rightarrow \quad a = b$$

6) **transitive** *if for all* $a, b \in A$,

$$a \sim b, \quad b \sim c \quad \Rightarrow \quad a \sim c$$

□

By far the two most important combinations of these axioms are found in the definition of an equivalence relation and a partial order.

Definition *An* **equivalence relation** *on a nonempty set* A *is a binary relation on* A *that is reflexive, symmetric and transitive. If* $a \in A$, *the set* $[a]$ *of all elements of* A *that are equivalent to* a *is called the* **equivalence class** *containing* a. *If* $\mathcal{E}$ *is an equivalence relation on a set* X, *then a system of distinct representatives for the equivalence classes of* $\mathcal{E}$ *is also called a* **system of distinct representatives** *for* $\mathcal{E}$.□

Partial Orders and Posets

We can now define one of the principal objects of study of this book.

Definition *A* **partial order** (*or just an* **order**) *on a nonempty set* P *is a binary relation* $\leq$ *on* P *that is reflexive, antisymmetric and transitive, specifically, for all* $x, y, z \in P$:

1) **(reflexive)**

$$x \leq x$$

2) **(antisymmetric)**

$$x \leq y, \quad y \leq x \quad \Rightarrow \quad x = y$$

3) **(transitive)**

$$x \leq y, \quad y \leq z \quad \Rightarrow \quad x \leq z$$

The pair $(P, \leq)$ *is called a* **partially ordered set** *or* **poset**, *although it is often said that* P *is a poset, when the order relation is understood. If* $x \leq y$, *then* x *is*

less than or equal to y *or* y *is* **greater than or equal to** x. *We also say that* x *is* **contained in** y *or that* y **contains** x. *If* $x \leq y$ *but* $x \neq y$, *we write* $x < y$ *or* $y > x$. *If* $x \leq y$ *or* $y \leq x$, *then* x *and* y *are said to be* **comparable**. *Otherwise,* x *and* y *are* **incomparable** *or* **parallel**, *denoted by* $a \parallel b$.□

If S and T are subsets of a poset P, then $S \leq T$ means that $s \leq t$ for all $s \in S$, $t \in T$. If $T = \{t\}$, then $S \leq \{t\}$ is written $S \leq t$ and similarly for $s \leq T$.

Definition *A poset* $(P, \leq)$ *is* **totally ordered** *if every* $x, y \in P$ *are comparable, that is,*

$$x \leq y \quad \text{or} \quad y \leq x$$

In this case, the order is said to be **total** *or* **linear**.□

We will assume unless otherwise noted that a nonempty subset S of a poset P inherits the order from P, making S a poset as well.

We will have some use for the following concept in the exercises.

Definition *A* **preorder** *or* **quasiorder** *on a nonempty set* P *is a binary relation that is reflexive and transitive.*□

Example 1.1

1) Let $\mathbb{N} = \{0, 1, \ldots\}$ be the set of natural numbers. Then $(\mathbb{N}, \leq)$ is a poset under the ordinary order. Also, $(\mathbb{N}, \mid)$ is a poset under division, that is, where $x \mid y$ means that $y = kx$ for some $k \in \mathbb{N}$, that is, x divides y.
2) If X is a nonempty set, then the **power set** $\wp(X)$ of X is the set of all subsets of X. It is well known that $\wp(X)$ is a poset under set inclusion.
3) The set $\mathcal{S}(G)$ of all subgroups of a group G is a poset under set inclusion. The set $\mathcal{N}(G)$ of all normal subgroups of G is a subset of $\mathcal{S}(G)$, inheriting the order relation from $\mathcal{S}(G)$. Many other such examples are possible, involving other algebraic structures.
4) The set $\mathcal{P}$ of all partitions of a nonempty set X is a poset, where $\lambda \leq \sigma$ if λ is a **refinement** of σ, that is, if every block of σ is a union of blocks of λ. Put another way, the blocks of λ are constructed by further partitioning some (or none) of the blocks of σ.
5) Let P be the collection of **partial functions** on a nonempty set X, that is, functions on X whose domain is a subset of X. Order P by saying that $f \leq g$ if f is an extension of g, that is, f and g agree on the dom(f). The P is a poset under this order.
6) The **product order** on the cartesian product $P \times Q$ is defined by

$$(p_1, q_1) \leq (p_2, q_2) \quad \text{if} \quad p_1 \leq p_2 \quad \text{and} \quad q_1 \leq q_2$$

The set $P \times Q$ with this order is called the **product** of P and Q.

7) The **lexicographic order** on the cartesian product $P \times Q$ is defined by

$$(p_1, q_1) \leq (p_2, q_2) \quad \text{if} \quad p_1 < p_2 \quad \text{or} \quad (p_1 = p_2 \quad \text{and} \quad q_1 \leq q_2)$$

This is also a partial order on $P \times Q$.□

Strict Orders

To every partial order on a set P there corresponds a strict partial order.

Definition *A binary relation* $<$ *on a nonempty set* P *is called a* **strict partial order** (*or* **strict order**) *if it is asymmetric and transitive.*□

Given a partially ordered set $(P, \leq)$, we may define a strict order $<$ by

$$x < y \quad \text{if} \quad x \leq y \text{ and } x \neq y$$

Conversely, if $<$ is a strict order on a nonempty set P, then the binary relation defined by

$$x \leq y \quad \text{if} \quad x < y \text{ or } x = y$$

is a partial order and so $(P, \leq)$ is a poset. Thus, there is a one-to-one correspondence between partial orders and strict partial orders on a nonempty set P and so a partially ordered set can be defined as a nonempty set with a strict order relation.

The Covering Relation

We next define the covering relation on a poset.

Definition *Let* $(P, \leq)$ *be a poset. Then* y **covers** x *in* P*, denoted by* $x \sqsubset y$*, if* $x < y$ *and no element in* P *lies strictly between* x *and* y*, that is,*

$$x \leq z \leq y \quad \Rightarrow \quad z = x \quad \text{or} \quad z = y$$

If $a \sqsubset b$ *or* $a = b$*, we write* $a \sqsubseteq b$.□

For a finite poset P, the covering relation uniquely determines the order on P, since $a \leq b$ if and only if there is a finite sequence of elements of P such that

$$a \sqsubset p_1 \sqsubset p_2 \sqsubset \cdots \sqsubset p_n \sqsubset b$$

Small finite posets are often described with a diagram called a **Hasse diagram**, which is a graph whose nodes are labeled with the elements of the poset and whose edges indicate the *covering* relation. This is illustrated in the following example.

Example 1.2 Figure 1.1 shows the Hasse diagram of the poset $\{1, 2, 3, 6, 12\}$ under division.

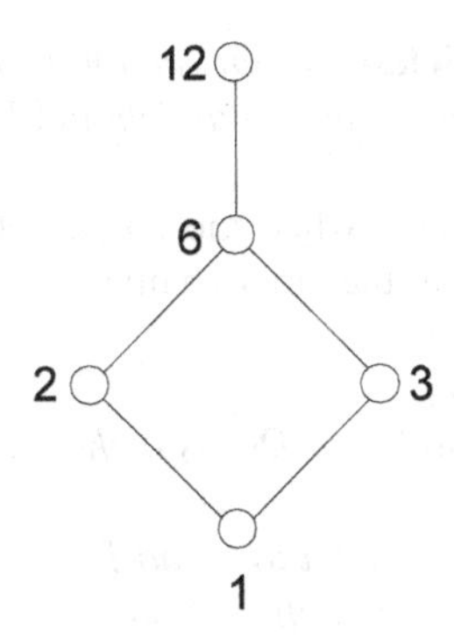

Figure 1.1

Figure 1.2 shows the Hasse diagrams of two posets

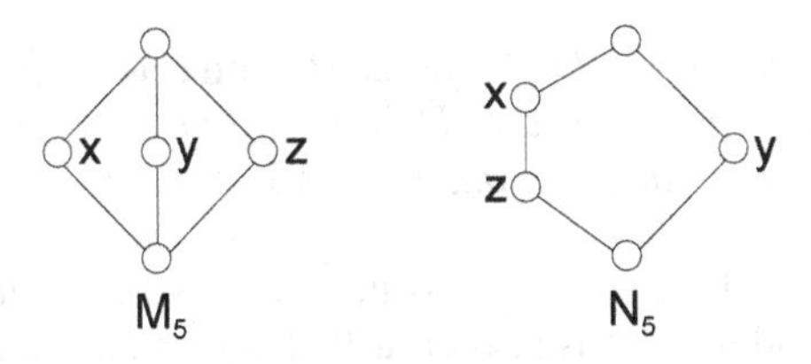

Figure 1.2

that will play key roles in later discussions.□

Chains and Directed Sets

Totally ordered subsets of a poset play an important role in the theory of partial orders.

Definition *Let $(P, \leq)$ be a poset.*

1) *A nonempty subset S of P is a* **chain** *in P if S is totally ordered by $\leq$. A finite chain with n elements can be written in the form*

$$c_1 < c_2 < \cdots < c_n$$

Such a chain said to have **length** $n-1$. *If the set of lengths of all chains contained in P is bounded, then P has* **finite length** *and the maximum length is called the* **length** *of P. Otherwise, P has* **infinite length**. *If $a < b$, then a* **chain from** *a to b in P is a chain in P whose smallest element is a and whose largest element is b. A* **maximal chain** *from a to b is a chain from a to b that is not contained in a larger (in the sense of set inclusion) chain from a to b.*

2) *A nonempty subset S of P is an* **antichain** *in P if every two elements of S are incomparable. An antichain with n elements is said to have* **width** n. *If the set of widths of all antichains in P is bounded, then P has* **finite width** *and the maximum width is called the* **width** *of P. Otherwise, P has* **infinite**

width. *A* **maximal antichain** *is an antichain that is not contained in a larger (in the sense of set inclusion) antichain.*□

It is possible to combine chains when the largest element of one chain is the same as the smallest element of the other chain.

Theorem 1.3 *Let P be a poset.*

1) *If $\mathcal{C}_{a,b}$ is a chain from a to b and $\mathcal{D}_{b,c}$ is a chain from b to c, then $\mathcal{C}_{a,b} \cup \mathcal{D}_{b,c}$ is a chain from a to c.*
2) *If $\mathcal{C}_{a,b}$ is a maximal chain from a to b and $\mathcal{D}_{b,c}$ is a maximal chain from b to c, then $\mathcal{C}_{a,b} \cup \mathcal{D}_{b,c}$ is a maximal chain from a to c.*□

In some circumstances, the property of being a chain is stronger than is required and the following property is sufficient.

Definition *A nonempty subset D of a poset P is* **directed** *if every pair $\{a, b\}$ of elements of D has an upper bound in D, that is, if for every $a, b \in D$, there is a $c \in D$ with the property that $a \leq c$ and $b \leq c$.*□

It is easy to see that a directed set P has the property that every *finite* subset of P has an upper bound in P. It is also clear that any chain is a directed set. Here is an example where a directed set is a more suitable choice than a chain: The union of every *directed family* of subspaces of a vector space V is a subspace of V. Similar statements can be made about subgroups of a group, ideals in a ring and so on.

Maximal and Minimal Elements

Maximal and maximum elements can be defined in posets.

Definition *Let $(P, \leq)$ be a partially ordered set.*

1) *A* **maximal** *element is an element $m \in P$ that is not contained in any other element of P, that is,*

$$p \in P, \quad m \leq p \quad \Rightarrow \quad m = p$$

A **maximum** **(largest** *or* **greatest)** *element m in P is an element that contains every element of P, that is,*

$$p \in P \quad \Rightarrow \quad p \leq m$$

We will generally denote the largest element by 1 and call it the **unit element**.

2) *A* **minimal element** *is an element $n \in P$ that does not contain any other element of P, that is,*

$$p \in P, \quad p \leq n \quad \Rightarrow \quad p = n$$

A **minimum** **(smallest** *or* **least)** *element n in P is an element contained in* *all other elements of P, that is,*

$$p \in P \quad \Rightarrow \quad n \leq p$$

We will generally denote the smallest element by 0 *and call it the* **zero element**.

A partially ordered set is **bounded** *if it has both a* 0 *and a* 1.□

If a poset P does not have a smallest element, then we can adjoin one as follows. Let 0 be a symbol that does not represent any element of P and let

$$P_{\perp} = P \sqcup \{0\}$$

be the disjoint union of P and $\{0\}$. Extend the order in P to $P_{\perp}$ by including the condition that $0 < x$, for all $x \in P$. Similarly, a 1 can be adjoined by taking

$$P_{\top} = P \sqcup \{1\}$$

and specifying that $x < 1$ for all $x \in P$.

Definition *If a poset* P *has a smallest element* 0*, then any cover of* 0 *is called an* **atom** *or* **point** *of* P*. The set of all atoms of a poset* P *is denoted by* $\mathcal{A}(P)$*. A poset with* 0 *is* **atomic** *if every nonzero element contains an atom. If* P *has a* 1*, then any element covered by* 1 *is called a* **coatom** *or* **copoint** *of* P.□

Upper and Lower Bounds

Upper and lower bounds can be defined in a poset.

Definition *Let* $(P, \leq)$ *be a partially ordered set and let* $S \subseteq P$.

1) *An* **upper bound** *for* S *is an element* $x \in P$ *for which*

$$S \leq x$$

The set of all upper bounds for S *is denoted by* S^u*. We abbreviate* $\{s\}^u$ *by* s^u*. If* S^u *has a least element, it is called the* **join** *or* **least upper bound** *or* **supremum** *of* S *and is denoted by* $\bigvee S$*. The join of a finite set* $S = \{a_1, \ldots, a_n\}$ *is denoted by*

$$a_1 \vee \cdots \vee a_n$$

2) *A* **lower bound** *for* S *is an element* $x \in P$ *for which*

$$x \leq S$$

The set of all lower bounds for S *is denoted by* S^ℓ*. We abbreviate* $\{s\}^\ell$ *by* s^ℓ*. If* S^ℓ *has a greatest element, it is called the* **meet** *or* **greatest lower bound** *or* **infimum** *of* S *and is denoted by* $\bigwedge S$*. The meet of a finite set* $S = \{a_1, \ldots, a_n\}$ *is denoted by*

$$a_1 \wedge \cdots \wedge a_n$$

□

Note that the join of the empty set $\emptyset$ is, by definition, the least upper bound of the elements of $\emptyset$. Since every element of P is an upper bound for $\emptyset$, we have $\bigvee\emptyset = 0$ if P has a smallest element 0; otherwise $\emptyset$ has no join. Similarly, $\bigwedge\emptyset = 1$ if 1 exists; otherwise $\bigwedge\emptyset$ does not exist.

Zorn's Lemma and Friends

We will assume various equivalent forms of the axiom of choice.

The Axiom of Choice

Theorem 1.4 *The following are equivalent:*

1) **(Axiom of choice)** *Let $\mathcal{F} = \{S_i \mid i \in I\}$ be a nonempty family of nonempty sets. Then there is a function*

$$f: \mathcal{F} \to \bigcup_{i\in I} S_i$$

for which $f(S_i) \in S_i$ for all $i \in I$. Such a function is called a **choice function**, *since it chooses a single element $f(S_i)$ from each set S_i.*

2) *Every partition $\mathcal{P}$ of a set S has a system of distinct representatives.*
3) *The cartesian product ΠX_i of a set $\mathcal{F} = \{X_i \mid i \in I\}$ of nonempty sets is nonempty.*

Proof. To see that 1) implies 2), if $\mathcal{P} = \{B_i \mid i \in I\}$, then $\mathcal{P}$ is a set of sets and a choice function f gives a system $\{f(B_i)\}$ of distinct representatives. To see that 2) implies 3), the set

$$S = \{(i,x) \mid i \in I, x \in X_i\}$$

can be partitioned using the first coordinate, that is, into the blocks

$$B_i = \{(i,x) \mid x \in X_i\}$$

for each $i \in I$. A system of distinct representatives $f = \{(i,x_i) \mid i \in I\}$ for this partition belongs to the cartesian product $\prod X_i$, which is therefore nonempty.

To see that 3) implies 1), let S be a set of sets. We may assume that no set in S is empty. Index each set using itself, that is, let $X_s = s$ for all $s \in S$. Then the indexed family

$$\{X_s \mid s \in S, s \neq \emptyset\}$$

has a nonempty cartesian product, that is, there is a function $f: S \to \bigcup X_s$ such that $f(s) \in X_s = s$. This is a choice function for S.□

Zorn's Lemma

One of the most commonly used equivalences of the axiom of choice is *Zorn's lemma.*

Theorem 1.5 *The following are equivalent.*

1) (**Zorn's lemma**) *If a poset P has the property that every chain in P has an upper bound, then P has a maximal element.*
2) *If a poset P has the property that every chain in P has an upper bound, then for every $p \in P$, there is a maximal element $q \in P$ such that $q \geq p$.*

Proof. It is clear that 2) implies 1), since any maximal element $q \in P$ such that $q \geq p$ is a maximal element of P. Conversely, if 1) holds and if P has the property that every chain in P has an upper bound then p^u also has this property and so 1) implies that p^u has a maximal element q, which is also maximal in P and satisfies $q \geq p$.□

The Well-Ordering Principle

The well-ordering principle is also a commonly used equivalence of the axiom of choice.

Definition *A poset $(W, \leq)$ is a* **well-ordered set** *if every nonempty subset of S has a least element. In this case, the order $\leq$ is called a* **well-ordering** *of W. A nonempty set X for which a well-ordering exists is called a* **well-orderable** *set.*□

Note that a well-ordering is a total ordering, since if $a \neq b$, then the set $\{a, b\} \subseteq P$ has a least element. However, the usual ordering on $\mathbb{Z}$ is a total ordering that is not a well-ordering.

Definition (The well-ordering principle) *Every nonempty set P is well-orderable.*□

Tukey's Lemma and the Hausdorff Maximality Principle

We will have relatively few occasions to use the following equivalences of the axiom of choice, so the reader may skip this now and return to it later when needed.

Definition *Let X be a nonempty set. A family $\mathcal{F}$ of subsets of X has* **finite character** *if it has the property that $A \in \mathcal{F}$ if and only if every finite subset of A is in $\mathcal{F}$.*□

An important example of a family of finite character is the family of all linearly independent subsets of a vector space.

Definition (Tukey's lemma) *Every nonempty family $\mathcal{F}$ of sets of finite character has maximal element.*□

Definition (Hausdorff maximality principle) *Let P be a poset. Any chain in P is contained in a maximal chain.*□

Equivalences

Let us now state formally that the previous statements are equivalent.

Theorem 1.6 *The following are equivalent:*

1) **(Zorn's lemma)** *If P is a partially ordered set in which every chain has an upper bound, then P has a maximal element.*
2) **(Hausdorff maximality principle)** *Let P be a poset. Any chain in P is contained in a maximal chain.*
3) **(Tukey's lemma)** *Every nonempty family $\mathcal{F}$ of sets that has finite character has a maximal element.*
4) **(Axiom of choice)** *Every set of sets has a choice function.*
5) **(Well-ordering principle)** *Every nonempty set can be well-ordered.* □

Miscellaneous

The following concepts are often useful.

Definition *Let $(P, \leq)$ be a poset. For $a, b \in P$ the* **closed interval** *from a to b is the set*

$$[a, b] = \{x \in P \mid a \leq x \leq b\}$$

The **initial segment** *up to a is the set*

$$(-\infty, a) = \{x \in P \mid x < a\}$$

and the **closed initial segment** *up to a is the set*

$$(-\infty, a] = \{x \in P \mid x \leq a\}$$

A nonempty subset S of a poset P is **convex** *if whenever $a, b \in S$, then $[a, b] \subseteq S$.* □

Duality

We now turn to a discussion of duality.

Definition *Let $(P, \leq)$ be a poset. The* **dual poset** *$(P^\partial, \preceq)$ is the poset with the same underlying set but whose order relation is the opposite of $\leq$, that is, $x \preceq y$ if and only if $x \geq y$.* □

Of course, $(P^\partial)^\partial = P$.

Statements and Their Duals

A **statement about posets** or a **property of posets** is just a statement that may or may not be true in any given poset. If s is a statement about posets, we write $P(s)$ to denote the statement s as applied to the poset P.

If s is true in P, we say that P **satisfies, possesses** or **has** property s and write $P \models s$ (also read P *models* s). We can also view a property of posets as a subfamily of the family of all posets, namely, the subfamily consisting of all posets in which s is true. If Π is a set of properties, we say that a poset P **satisfies** Π and write $P \models \Pi$ if P has every property in Π.

Two statements s and t are **logically equivalent** if they are true in exactly the same posets, that is, if

$$P \models s \quad \Leftrightarrow \quad P \models t$$

for all posets P. Let s be a statement about a poset P. One often defines the *dual statement* to s to be the statement formed by replacing all references (direct or indirect) to $\leq$ with $\geq$. Thus, least upper bound is replaced by greatest lower bound, maximal by minimal and so on. This is a bit vague, so we make the following definition.

Definition *Let s be a statement about posets. A* **dual statement** s^∂ *is a statement about posets that satisfies the following property:*

$$P \models s \quad \Leftrightarrow \quad P^\partial \models s^\partial$$

for all posets P. If $\Pi = \{p_i \mid i \in I\}$, then Π^∂ denotes any set of statements $\{p_i^\partial \mid i \in I\}$, where p_i^∂ is a dual to p_i.□

Note that if t and u are dual statements to s, then

$$P \models t \quad \Leftrightarrow \quad P^\partial \models s \quad \Leftrightarrow \quad P \models u$$

for all posets P and so t and u hold in exactly the same posets, that is, t and u are logically equivalent. In fact, if s and s^∂ are dual, then t is dual to s if and only if t is logically equivalent to s^∂.

As mentioned earlier, one way to find a dual statement to s is to translate $P^\partial(s)$ into a logically equivalent statement about P, where P denotes an arbitrary poset. For example, if $P(s)$ is the statement

$P(s)$: If the join of a nonempty subset of P exists, it is unique

then

$P^\partial(s)$: If the join of a nonempty subset of P^∂ exists, it is unique

Since a join in P *is* a meet in P^∂ and vice versa, an equivalent dual statement is

$P(s)^\partial$: If the meet of a nonempty subset of P exists, it is unique

Definition *A statement s about posets is* **self-dual** *if it is logically equivalent to any (and therefore all) of its dual statements s^∂, that is, if*

$$P \models s \quad \Leftrightarrow \quad P \models s^\partial$$

A family $\Pi = \{p_i \mid i \in I\}$ of properties of posets is **self-dual** *if*

$$P \models \Pi \quad \Leftrightarrow \quad P \models \Pi^\partial$$

for all posets P.□

If a statement s holds in all posets satisfying Π, that is, if

$$P \models \Pi \quad \Rightarrow \quad P \models s$$

for all P, we write $\Pi \Rightarrow s$. Now we can state the duality principle for posets.

Theorem 1.7 (The duality principle for posets) *Let* Π *be a set of properties of posets and let* s *be a property of posets. Then*

$$\Pi \Rightarrow s \quad \textit{implies} \quad \Pi^\partial \Rightarrow s^\partial$$

and if Π *is self-dual, then*

$$\Pi \Rightarrow s \quad \textit{implies} \quad \Pi \Rightarrow s^\partial$$

that is, if s *holds in all posets satisfying* Π*, then so does* s^∂*. In particular,*

$$\emptyset \Rightarrow s \quad \textit{implies} \quad \emptyset \Rightarrow s^\partial$$

that is, if s *holds in all posets, then so does* s^∂*.*□

Let us give a simple example of the use of the duality principle.

Theorem 1.8
1) *Every finite poset with* 0 *and with at least two elements has an atom.*
2) *Dually, every finite poset with* 1 *and with at least two elements has a coatom.*

Proof. For part 1), let P be a finite poset with 0 and let $a_1 \in P$ be nonzero. If a_1 is not an atom, there is an element a_2 for which $0 < a_2 < a_1$. If a_2 is not an atom, we may repeat the process to get $0 < a_3 < a_2 < a_1$. Since P is finite, this process must stop to reveal an atom.

Part 2) follows by duality. To illustrate how duality is applied in this case, let $\Pi = \{b, f, t\}$ consist of the properties

$$b\text{: has } 0, \quad f\text{: finite} \quad \text{and} \quad w\text{: at least two elements}$$

Since P has 0 if and only if P^∂ has 1, we have

$$b^\partial = t\text{: has } 1$$

Of course, f and w are self-dual. Hence, $\Pi^\partial = \{t, f, w\}$ represents the family of all finite posets with 1. Also, if

$$a\text{: has an atom}$$

then

$$a^\partial\text{: has a coatom}$$

Thus, the principle of duality gives

$$\Pi \Rightarrow a \quad \text{implies} \quad \Pi^\partial \Rightarrow a^\partial$$

which is part 2).□

Monotone Maps

We now define order-preserving maps.

Definition *Let P and Q be posets and let $f\colon P \to Q$.*

1) *f is* **order-preserving** *or* **monotone** *or* **isotone** *if*

$$x \le y \quad \Rightarrow \quad f(x) \le f(y)$$

and **strictly monotone** *if*

$$x < y \quad \Rightarrow \quad f(x) < f(y)$$

2) *f is an* **order embedding** *if*

$$x \le y \quad \Leftrightarrow \quad f(x) \le f(y)$$

(Note that such a map must be injective.) If $f\colon P \to Q$ is an order embedding, we say that P can be **order embedded**, *or simply* **embedded** *in Q and write $f\colon P \hookrightarrow Q$ or just $P \hookrightarrow Q$.*

3) *An order embedding f is an* **order isomorphism** *if it is also surjective. If $f\colon P \to Q$ is an order isomorphism, we write $f\colon P \approx Q$ or just $P \approx Q$.*□

Note that if $f\colon P \to Q$ is a monotone bijection, then f^{-1} need not be monotone, that is, f need not be an order isomorphism. (Map two incomparable elements to two comparable elements.)

Down-Sets and the Down Map

The following types of sets will occur frequently.

Definition *A subset D of a poset P is a* **down-set** *or* **order ideal** *if*

$$x \in D, \quad y \le x \quad \Rightarrow \quad y \in D$$

We denote the down-sets in P by $\mathcal{O}(P)$. Dually, a subset U of P is an **up-set** *or* **order filter** *if*

$$x \in U, \quad y \ge x \quad \Rightarrow \quad y \in U$$

We denote the up-sets in P by $\mathcal{U}(P)$.□

It is easy to see that D is a down-set in P if and only if its complement $P \setminus D$ is an up-set in P. Note also that the down-set property is transitive, that is,

$$D \in \mathcal{O}(E), \quad E \in \mathcal{O}(F) \quad \Rightarrow \quad D \in \mathcal{O}(F)$$

Theorem 1.9 *Let P be a poset.*
1) The union or intersection of any family of down-sets is a down-set.
2) The union or intersection of any family of up-sets is an up-set.
That is, $\mathcal{O}(P)$ and $\mathcal{U}(P)$ are closed under arbitrary union and arbitrary intersection. □

Theorem 1.10 *Let S be a subset of a poset P.*
1) The **down closure** *of S is the smallest down-set containing S. This set is*

$$\downarrow S = \{x \in P \mid x \leq s \text{ for some } s \in S\}$$

and is read "down S." The down-set $\downarrow p = \downarrow\{p\}$ is called the **principal ideal generated by** *p.*
2) Dually, the **up closure** *of S is the smallest up-set of P containing S. This set is*

$$\uparrow S = \{x \in P \mid x \geq s \text{ for some } s \in S\}$$

and is read "up S." The up-set $\uparrow p = \uparrow\{p\}$ is called the **principal filter generated by** *p.* □

The following maps are among the most important in the theory of order.

Theorem 1.11 *Let P be a poset.*
1) The map $\phi_\downarrow \colon P \to \wp(P)$ defined by

$$\phi_\downarrow(a) = \downarrow a = \{x \in P \mid x \leq a\}$$

is called the **down map**. *The down map $\phi_\downarrow$ is an order embedding of P into $\wp(P)$. Thus, any poset P can be order embedded in a power set $\wp(P)$.*
2) The map $\phi_\uparrow \colon P \to \wp(P)$ defined by

$$\phi_\uparrow(a) = \uparrow a = \{x \in P \mid x \geq a\}$$

is called the **up map**. *The up map $\phi_\uparrow$ is an order* **anti-embedding** *of P into $\wp(P)$, that is,*

$$a \leq b \quad \Leftrightarrow \quad \phi_\uparrow(a) \supseteq \phi_\uparrow(b)$$

□

In loose terms, Theorem 1.11 tells us that every poset P can be *represented* as a family of subsets of some set, under set inclusion. This is the first of several *representation theorems* for describing abstract order structures in terms of more familiar order structures. (One of the most famous representation theorems is the *Cayley representation theorem* of group theory, which says that every group can be represented as a subgroup of the group of permutations of a set.)

We will also have frequent use for the following generalization of the down map.

Definition *Let B be a nonempty subset of a poset P.*

1) *For $a \in P$, we write*

$$B_{\downarrow}(a) := B \cap \downarrow a = \{b \in B \mid b \leq a\}$$

*read "B-down of a." The B-**down map** $\phi_{B,\downarrow} : P \to \wp(B)$ is defined by*

$$\phi_{B,\downarrow}(a) = B_{\downarrow}(a)$$

2) *Dually, the B-up of $a \in P$ is*

$$B_{\uparrow}(a) := B \cap \uparrow a = \{b \in B \mid b \geq a\}$$

*and the corresponding B-**up map** $\phi_{B,\uparrow} : P \to \wp(B)$ is defined by*

$$\phi_{B,\uparrow}(a) = B_{\uparrow}(a)$$ □

We observe that the B-down map is order preserving.

Height and Graded Posets

In a poset P with no infinite chains, we can define the distance $d(a, b)$ from a to b, where $a \leq b$ are elements of P.

Definition *Let P be a poset with no infinite chains.*

1) *If $a \leq b$, then the* **distance** *$d(a, b)$ from a to b is the maximum length of all chains from a to b, if this maximum is finite. Otherwise, the distance is infinite.*
2) *If P has 0, then the* **height** *of $a \in P$ is the distance $d(0, a)$.*□

(Note that d is not a distance function in the sense of a metric on a set, since it is not symmetric.) It is clear that the distance $d(a, b)$ is equal to the maximum length among all *maximal* chains from a to b. Of course, not all maximal chains from a to b need have the same length, which prompts the following definition.

Definition *The* **Jordan–Dedekind chain condition** *on a poset P is the condition that for every $a < b$ in P, all maximal chains from a to b have the same finite length.*□

If a poset P has the Jordan–Dedekind chain condition, then $d(a, b)$ is the length of *any* maximal chain from a to b and so, for $a \leq b \leq c$, we have

$$d(a, b) + d(b, c) = d(a, c)$$

and in particular,

$$h(b) = h(a) + d(a, b)$$

and if $a \sqsubset b$, then

$$h(b) = h(a) + 1$$

Definition *A poset P is* **graded** *if there is a function $g: P \to \mathbb{Z}$ for which*
1) $a < b \Rightarrow g(a) < g(b)$
2) if $a \sqsubset b$ then $g(b) = g(a) + 1$
We refer to such a function as a **grading function.**□

Loosely speaking, a graded poset is a poset in which we can assign a well-defined "level" to each element. We have just shown that if P has the Jordan–Dedekind chain condition, then P is graded by its height function. The converse is also true.

Theorem 1.12 *Let P be a poset with 0 and with no infinite chains. Then P satisfies the Jordan–Dedekind chain condition if and only if P is graded by its height function.*
Proof. Suppose P is graded by its height function h. If

$$a = c_0 < c_1 < c_2 < \cdots < c_n = b$$

is a maximal chain from a to b then $c_i \sqsubset c_{i+1}$ and so $h(c_{i+1}) = h(c_i) + 1$. Thus, $h(c_i) = h(a) + i$ and so

$$h(b) - h(a) = n$$

Hence, all maximal chains from a to b have length $h(b) - h(a)$ and so P has the Jordan–Dedekind chain condition. The converse has already been established, since $a < b$ implies that $d(a, b) \geq 1$.□

Chain Conditions

The chain conditions are a form of finiteness condition.

Definition *Let P be a poset.*
1) *P has the* **ascending chain condition** *(ACC) if it has no infinite strictly ascending sequences, that is, for any ascending sequence*

$$p_1 \leq p_2 \leq p_3 \leq \cdots$$

there is an index n such that $p_{n+k} = p_n$ for all $k \geq 0$.
2) *Dually, P has the* **descending chain condition** *(DCC) if it has no infinite strictly descending sequences, that is, for any descending sequence*

$$p_1 \geq p_2 \geq p_3 \geq \cdots$$

there is an index n such that $p_{n+k} = p_n$ for all $k \geq 0$.□

The following characterizations of ACC and DCC are very useful.

Definition

1) A poset P has the **maximal condition** *if every nonempty subset of P contains a maximal element.*

2) Dually, a poset P has the **minimal condition** *if every nonempty subset of P contains a minimal element.* □

Theorem 1.13

1) A poset P has the ACC if and only if it has the maximal condition.

2) Dually, a poset P has the DCC if and only if it has the minimal condition.

Proof. Suppose P has the ACC and let $S \subseteq P$ be nonempty. Let $s_1 \in S$. If s_1 is maximal, we are done. If not, then we can pick $s_2 \in S$ such that $s_2 > s_1$. Continuing in this way, we either arrive at a maximal element in S or we get an ascending sequence that does not become constant, which contradicts the ACC. Hence, P has the maximal condition. Conversely, if P has the maximal condition, then any ascending sequence in P has a maximal element, at which point the sequence becomes constant. □

Chain Conditions and Finiteness

It seems intuitively clear that an infinite poset must have either an infinite chain or an infinite antichain, although a careful proof is a bit more subtle than one might think at first. It is also true that if a poset has an infinite chain, then it must have either an infinite *ascending* chain or an infinite *descending* chain.

Theorem 1.14

1) An infinite poset must have one of the following:

a) an infinite antichain

b) an infinite strictly ascending chain

c) an infinite strictly descending chain

2) A poset with no infinite chains or infinite antichains must be finite.

3) A poset with an infinite chain has an infinite strictly ascending chain or an infinite strictly descending chain.

Proof. For part 1), assume that a poset P has no infinite antichains and has the ACC. We show that P fails to have the DCC. Since P has the maximal condition, it has a maximal element. Let

$$\mathcal{M} = \{m_i \mid i \in I\}$$

be the nonempty set of all maximal elements of P. Since $\mathcal{M}$ is an antichain, it must be finite. Moreover, $P = \bigcup(\downarrow m_i)$ and so one of the sets, say $\downarrow m_{i_1}$, must be infinite. Let $A_1 = \{m_{i_1}\}$ and consider the poset

$$P_1 = (\downarrow m_{i_1}) \setminus \{m_{i_1}\}$$

of all elements of P strictly below m_{i_1}. This poset is infinite and also has no infinite antichains and no infinite ascending chains. Thus, we may repeat the above argument and select an element $m_{i_2} \in P_1$ such that $P_2 = (\downarrow m_{i_2}) \setminus \{m_{i_2}\}$ is infinite. Note that $m_{i_1} > m_{i_2}$. Continuing in this way, we get an infinite descending chain. Thus, P fails to have the DCC. Parts 2) and 3) follow from part 1).□

Dilworth's Theorem

In 1950, Robert Dilworth proved the following very interesting result.

Definition *Let P be a finite poset. A* **chain cover** *of P is a collection of chains in P whose union is P. We say that a chain cover is* **disjoint** *if the chains in the cover are pairwise disjoint.*□

If P has an antichain of size n, then any chain cover will clearly have size at least n, since no two elements in an antichain can be in the same chain of a chain cover. Thus, if P has width w, then all chain covers have size at least w. Dilworth proved that there always is a disjoint chain cover of size w.

Theorem 1.15 (Dilworth's theorem) *A finite poset has a disjoint chain cover of size $w = \text{width}(P)$.*
Proof. The proof is by induction on the size n of P. The result is true if $n = 1$. Assume it is true for posets of size less than $|P|$. We show that P has a chain cover of size w.

The idea is to use a maximum-sized antichain A to split P into two proper subsets, to which we can apply the inductive hypothesis to get two disjoint chain covers that can be pieced together at the members of A to produce a disjoint chain cover of P.

A maximum-sized antichain $A = \{a_1, \ldots, a_w\}$ splits P into two subsets $\uparrow A$ and $\downarrow A$, that is,

$$\uparrow A \cup \downarrow A = P$$

whose intersection is A. Also, A is a maximum-sized antichain of each subset $\uparrow A$ and $\downarrow A$. To ensure that we choose an A for which each of these subsets is proper, we proceed as follows. Since P is finite, there exist elements $m \leq M$ such that M is maximal in P and m is minimal in P. Let

$$Q = P \setminus \{m, M\}$$

If $\text{width}(Q) = k < w$, then the inductive hypothesis implies that Q has a disjoint chain cover $\mathcal{C}$ of size k and so $\mathcal{C} \cup \{m, M\}$ is a disjoint chain cover of P of size w (since it cannot have size less than w). In this case, we are done.

If $\text{width}(Q) = w$, then a maximum-sized antichain A of Q is also a maximum-sized antichain of P, for which $\downarrow A$ and $\uparrow A$ are proper in P. Let $\mathcal{C}$ be a disjoint chain cover of size w for $\uparrow A$ and let $\mathcal{D}$ be a disjoint chain cover of size w for $\downarrow A$. Then for each $a \in A$, there is a $C_a \in \mathcal{C}$ and a $D_a \in \mathcal{D}$ for which $C_a \cap D_a = \{a\}$. Since a is the smallest element of C_a and the largest element of D_a, it follows that $C_a \cup D_a$ is a chain in P. The family $\{C_a \cup D_a \mid a \in A\}$ is a disjoint chain cover of P of size w.□

The following important theorem on the existence of SDRs for finite families is a consequence of Dilworth's theorem.

Theorem 1.16 (Philip Hall's theorem) *A family $\mathcal{F}$ of subsets of a finite set X has an SDR if and only if it satisfies* **Hall's condition**: *For any $1 \leq k \leq |\mathcal{F}|$, the union of any k members of $\mathcal{F}$ has size at least k, that is,*

$$|A_{i_1} \cup \cdots \cup A_{i_k}| \geq k$$

Proof. Let $X = \{a_1, \ldots, a_m\}$ and $\mathcal{F} = \{A_1, \ldots, A_n\}$. If $\mathcal{F}$ has an SDR, then any k members of $\mathcal{F}$ have an SDR and so their union must have size at least k. For the converse, suppose that $\mathcal{F}$ satisfies Hall's condition. Consider the "membership" partial order on the set $P = X \cup \mathcal{F}$ defined by

$$a_i \leq A_j \quad \text{if} \quad a_i \in A_j$$

The chains of P are the singletons and the two-element chains $a < A_i$ where $a \in A_i$. Hence, all disjoint chain covers in P have the form (after reindexing if necessary)

$$\mathcal{C} = \{a_1 < A_1, \ldots, a_r < A_r, a_{r+1}, \ldots, a_m, A_{r+1}, \ldots, A_n\}$$

Thus, since X is an antichain in P, a chain cover of size m (if it exists) must have the form

$$\mathcal{C} = \{a_1 < A_1, \ldots, a_n < A_n, a_{n+1}, \ldots, a_m\}$$

and so $\{a_1, \ldots, a_n\}$ is an SDR for $\mathcal{F}$. Hence, it is sufficient to show that P has a chain cover of size m, and Dilworth's theorem implies that it is sufficient to show that P has width m. We have just noted that P has width at least m, since X is an antichain.

Suppose that, after possible reindexing, the set

$$\mathcal{A} = \{a_1, \ldots, a_u, A_1, \ldots, A_v\}$$

is an antichain in P. We count the number of elements in

$$S = \{a_1, \ldots, a_u\} \cup A_1 \cup \cdots \cup A_v$$

which, of course, is at most m. None of $a_1, \ldots, a_u$ lie in $A_1 \cup \cdots \cup A_v$ and Hall's condition implies that $A_1 \cup \cdots \cup A_v$ has size at least v. Hence,

$$|\mathcal{A}| = u + v \leq |S| \leq m$$

and so $\text{width}(P) \leq m$.□

Symmetric and Transitive Closures

If a binary relation fails to be symmetric or transitive, the relation can be augmented to fix the problem.

Definition *Let R be a binary relation on a nonempty set X.*
1) *The* **symmetric closure** R^{sc} *of R is the smallest symmetric binary relation on X that contains R.*
2) *The* **transitive closure** R^{tc} *of R is the smallest transitive binary relation on X that contains R.*□

Theorem 1.17 *Let R be a binary relation on a nonempty set X.*
1) *The symmetric closure R^{sc} of R is*

$$R^{sc} = R \cup R^{-1}$$

where $R^{-1} = \{(a,b) \mid (b,a) \in R\}$.
2) *The transitive closure R^{tc} of R is the relation defined as follows: $(a,b) \in R^{tc}$ if and only if there is a finite sequence $a_0, \ldots, a_n \in X$ with $n \geq 1$ for which $a_0 = a, a_n = b$ and*

$$a_0 R a_1 R a_2 \cdots a_{n-2} R a_{n-1} R a_n$$

that is, $a_i R a_{i+1}$ for $i = 0, \ldots, n-1$.
3) a) *If R is reflexive, then so is R^{tc}.*
   b) *If R is symmetric, then so is R^{tc}*
   c) *If R is reflexive and symmetric, then R^{tc} is an equivalence relation.*
   d) *R antisymmetric does not imply that R^{tc} is antisymmetric.*

Proof. For 3b), if $aR^{tc}b$, then there is a finite sequence $a_0, \ldots, a_n \in X$ for which $a_0 = a, a_n = b$ and

$$a_0 R a_1 R a_2 \cdots a_{n-2} R a_{n-1} R a_n$$

Since R is symmetric, we may reverse the direction of the sequence to get

$$a_n R a_{n-1} R a_{n-2} \cdots a_2 R a_1 R a_0$$

and so $bR^{tc}a$. For 3d), let $X = \{a,b,c\}$ have three distinct elements and consider the antisymmetric relation

$$R = \{(a,b),(b,c),(c,a)\}$$

Then R^{tc} contains (a,c) as well as (c,a), where $a \neq c$.□

The covering relation of a poset can be characterized as follows.

Definition *A binary relation R is* **totally intransitive** *if whenever*

$$a_1 R a_2 R a_3 R \cdots R a_n$$

for $n \geq 3$, it follows that $a_1 \not R a_n$ and $a_1 \neq a_n$.□

Theorem 1.18 *A binary relation R on a set P is the covering relation $\sqsubset$ for a partial order $\leq$ on P if and only if R is totally intransitive.*

Proof. If $\sqsubset$ is a covering relation on P and if $a_1 \sqsubset a_2 \sqsubset a_3 \sqsubset \cdots \sqsubset a_n$ for $n \geq 3$, then $a_1 < a_n$ and so $a_1 \neq a_n$. Also, $a_1 \not\sqsubset a_n$. Hence, $\sqsubset$ is totally intransitive. Conversely, suppose that R is totally intransitive. Let $\leq$ be the transitive closure of the set

$$R \cup \{(a, a) \mid a \in P\}$$

Then $\leq$ is reflexive and transitive. To see that it is also antisymmetric, suppose that $a \leq b$ and $b \leq a$. If $a \neq b$, then $aR \cdots Rb$ and $bR \cdots Ra$ and so $aR \cdots Ra$, which implies that $a \neq a$, a contradiction. Thus, $\leq$ is a partial order.

To see that R is the covering relation $\sqsubset$ for $\leq$, first note that aRa implies $aRaRa$ which implies that $a \not R a$. Therefore, if aRb then $a \neq b$ and $a \leq b$, which implies that $a < b$. But if $a < c < b$, then total intransitivity implies that $a \not R b$, a contradiction. Hence, aRb implies $a \sqsubset b$. Also, if $a \sqsubset b$ then $a < b$ and so $aR \cdots Rb$. But if $aRcR \cdots Rb$ then $a \leq c \leq b$ and so $a \sqsubset b$ implies that $c = a$ or $c = b$, which is false.□

Compatible Total Orders

If $(P, \leq)$ is a poset, then a total order $\preceq$ on P is **compatible** with the partial order $\leq$ if

$$a \leq b \quad \Rightarrow \quad a \preceq b$$

The problem of finding a compatible total order for a given partial order is a practical one. For example, if $(P, \leq)$ is finite, it may be desirable to input the data P into a computer using a (linear) input order that is compatible with the partial order.

For a finite poset, it is easy to find a compatible total order using a simple algorithm called **topological sorting**, implemented by simply taking a maximal (or minimal) element at each stage. In this way, larger (or smaller) elements are input first.

We wish to prove that for any poset $(P, \leq)$ there is a compatible total order. In fact, for a given pair of incomparable elements a and b in $(P, \leq)$, we can find a compatible total order $\preceq$ for which $a \preceq b$.

Theorem 1.19 *Let $(P, \leq)$ be a poset and let $a \parallel b$ in P. The transitive closure of the relation*

$$R = \ \leq \cup \{(a, b)\}$$

is the relation

$$S = \ \leq \cup \{(x, y) \mid x \leq a, b \leq y\}$$

Moreover, $S = R^{tc}$ *is a partial order on* P.

Proof. It is clear that S is contained in R^{tc}. Thus, if we show that S is a partial order on P, then it must be the transitive closure of R. Of course, S is reflexive since $\leq$ is reflexive. As to antisymmetry, if xSy and ySx, then there are cases to consider. Let us write

$$S' = S \setminus \leq \ = \{(x, y) \mid x \leq a, b \leq y\}$$

1) If $x \leq y$ and $y \leq x$, then $x = y$.
2) If $xS'y$ and $y \leq x$, then $x \leq a, b \leq y$ and so
$$b \leq y \leq x \leq a$$
which is false since $a \parallel b$.
3) If $x \leq y$ and $yS'x$, then $y \leq a$, $b \leq x$ and so
$$b \leq x \leq y \leq a$$
which is false since $a \parallel b$.
4) If $xS'y$ and $yS'x$, then $x \leq a, b \leq y$ and $y \leq a, b \leq x$ and so
$$b \leq x \leq a$$
which is false since $a \parallel b$.

Finally, as to transitivity, suppose that xSy and ySz. Then
1) If $x \leq y$ and $y \leq x$, then $x \leq z$ and so xSz.
2) If $xS'y$ and $y \leq z$, then $x \leq a, b \leq y$ and so
$$x \leq a, b \leq y \leq z$$
which implies that xSz.
3) If $xS'y$ and $yS'z$, then $y \leq a, b \leq z$ and so
$$x \leq a, b \leq z$$
which implies that xSz.
4) If $xS'y$ and $yS'z$, then $x \leq a, b \leq y$ and $y \leq a, b \leq z$ and so
$$x \leq a, b \leq z$$
which implies that xSz.□

Theorem 1.20 *Let* $(P, \leq)$ *be a poset and let* $a \parallel b$ *in* P. *Then* $\leq$ *is contained in a total order* $\preceq$ *on* P *for which* $a \preceq b$.

Proof. Let $\mathcal{S}$ be the collection of all partial orders on P that contain $\leq$ and (a, b). Then $\mathcal{S}$ is not empty since the transitive closure of $R = \leq \cup \{(a, b)\}$ is in $\mathcal{S}$. Order $\mathcal{S}$ by set inclusion. Let

$$\mathcal{C} = \{P_i \mid i \in I\}$$

be a chain in $\mathcal{S}$. The union $P = \bigcup P_i$ is a partial order on P containing $\leq$ and (a, b). Thus, $\mathcal{C}$ has an upper bound and so by Zorn's lemma, there is a maximal order M containing $\leq$ and (a, b). Then M must be linear, for if not, there are elements $x, y \in P$ such that $x \parallel y$ in M and we may extend M to a strictly larger partial order on X containing $\leq$ and (x, y), contradicting the maximality of M.□

The Poset of Partial Orders

If P is a nonempty set, then the family $\mathcal{F}$ of all partial orders on P is itself a partial order, where $\theta \leq \mu$ in $\mathcal{F}$ if

$$a\theta b \quad \Rightarrow \quad a\mu b$$

for all $a, b \in P$. The smallest partial order on P is equality, since

$$a = b \quad \Rightarrow \quad a\mu b$$

for all partial orders μ on P. We leave proof of the following as an exercise.

Theorem 1.21 *Let P be a nonempty set.*
1) The smallest partial order on P is equality.
2) The maximal partial orders on P are the total orders.
3) If θ is a partial order on P, then θ is the intersection of all total orders on P that contain θ.□

Exercises

1. A **strict order** $<$ on a nonempty set P is an irreflexive, transitive binary relation on P. Prove that there is a one-to-one correspondence between strict orders on P and partial orders on P.
2. Let P and Q be disjoint posets. Show that the union $P \cup Q$ is a poset, with the inherited order on P and Q and where $p \parallel q$ for all $p \in P$ and $q \in Q$.
3. Let P and Q be disjoint posets. Let $P \cup Q$ be the union with the inherited order on P and Q and such that $p < q$ for all $p \in P$ and $q \in Q$. Prove that this is a poset. It is called the **linear sum** of P and Q and denoted by $P \oplus Q$.
4. Find a poset P that contains an antichain $A = \{x, y, z\}$ with the property that no pair of elements from A has a join but A has a join.
5. Let $A = \{a_1, a_2, \ldots\}$ and $B = \{b_1, b_2, \ldots\}$ be disjoint countably infinite sets. Define a relation R on the union $A \cup B$ by

$$a_{i+1} R a_i, \quad b_{i+1} R b_i \quad \text{and} \quad b_i R a_i$$

for $i = 1, 2, \ldots$. Show that R is a covering relation. Draw a picture of the corresponding poset.

6. Let P and Q be posets.
   a) Describe the covering relation in the product $P \times Q$.
   b) If P and Q are chains, when is $P \times Q$ a chain?
7. Let P be a poset. Let $\mathcal{C}_{a,b}$ be a chain from a to b and let $\mathcal{D}_{b,c}$ be a chain from b to c.
   a) Show that $\mathcal{C}_{a,b} \cup \mathcal{D}_{b,c}$ is a chain from a to c.
   b) Show that if $\mathcal{C}_{a,b}$ and $\mathcal{D}_{b,c}$ are maximal chains, then so is $\mathcal{C}_{a,b} \cup \mathcal{D}_{b,c}$.
8. Let P be a poset. Prove that the set $\mathcal{O}(P)$ of down-sets is closed under arbitrary union and arbitrary intersection.
9. Let X be a nonempty set and let $P = \wp(X)$. Define a relation $\ll$ by $A \ll B$ if for every $a \in A$, there is a $b \in B$ for which $a \leq b$. Prove that if $A \ll B$ and $B \ll A$, then A and B have the same set of maximal elements.
10. Let P and Q be posets. Prove that a monotone map $f \colon P \to Q$ is an order isomorphism if and only if f has a monotone inverse map $f^{-1} \colon Q \to P$.
11. Let P and Q be finite posets and let $f \colon P \to Q$ be a function. Prove that $f \colon P \to Q$ is an order embedding if and only if

$$f(x) < f(y) \quad \Leftrightarrow \quad x < y$$

for all $x, y \in P$.

12. Let P and Q be posets and let $f \colon P \to Q$.
    a) Prove that f is monotone if and only if the induced inverse map satisfies $f^{-1}(D) \in \mathcal{O}(P)$ for all $D \in \mathcal{O}(Q)$.
    b) Assume that f is monotone and let $f^{-1} \colon \mathcal{O}(Q) \to \mathcal{O}(P)$ be the induced inverse map from part a). Show that f is an order embedding if and only if f^{-1} is surjective.
    c) Assume that f is monotone and let $f^{-1} \colon \mathcal{O}(Q) \to \mathcal{O}(P)$ be the induced map from part a). Show that f is surjective if and only if f^{-1} is injective.
13. Let P be a poset. Prove that the down map is an order embedding from P into $\wp(P)$.
14. Let $I_n = \{1, \ldots, n\}$ and let $f \colon \wp(I_n) \to \{0, 1\}^n$ be defined by

$$fE = (e_1, \ldots, e_n)$$

where $e_i = 1$ if and only if $i \in E$. Prove that f is an order isomorphism.

15. Let P be a poset and let P^∂ be the dual poset. Prove that

$$\mathcal{O}(P) \approx (\mathcal{O}(P^\partial))^\partial$$

16. Let P and Q be posets. Prove that $\mathcal{O}(P \sqcup Q) \approx \mathcal{O}(P) \times \mathcal{O}(Q)$.
17. Let P be a poset of size at least $ab + 1$. Prove that P contains either an antichain of size at least $a + 1$ or a chain of size at least $b + 1$.
18. Let P be a finite poset.
    a) Show that $|\mathcal{O}(P)|$ is equal to the number of antichains in P.

b) Show that for $x \in P$,

$$|\mathcal{O}(P)| = |\mathcal{O}(P \setminus \{x\})| + |\mathcal{O}(P \setminus (\downarrow x \cup \uparrow x))|$$

19. (Dual to Dilworth's theorem) An **antichain cover** of P is a collection of antichains in P whose union is P. Prove the dual to Dilworth's theorem: A poset P with finite height h has an antichain cover of size h.
20. Let $\preceq$ be a preorder on X. Define a binary relation $\sim$ on X by

$$a \sim b \quad \text{if} \quad a \preceq b \quad \text{and} \quad b \preceq a$$

a) Show that $\sim$ is an equivalence relation on X. Denote the equivalence class containing $a \in X$ by $[a]$.
b) Let $a, b \in X$. Show that if $a \preceq b$, then $x \preceq y$ for all $x \in [a]$ and $y \in [b]$.
c) Show that the binary relation on the set $X/\sim$ of equivalence classes of X under $\sim$ defined by

$$[a] \leq [b] \quad \text{if} \quad a \preceq b$$

is a partial order on $X/\sim$.
d) Given a partial order $\leq$ on P, let $\preceq$ be defined on the power set $\wp(P)$ by $S \preceq T$ if for each $s \in S$, there is a $t \in T$ for which $s \leq t$. Prove that $\preceq$ is a quasiorder.
21. Let $(P, \leq)$ be a poset and let $a \not\leq b$ in P. Show that $\leq$ is contained in a total order $\preceq$ on P for which $b \preceq a$.
22. Let P be a nonempty set. Prove the following:
a) The smallest partial order on P is equality.
b) The maximal partial orders on P are the total orders.
c) (Szpilrajn, 1930) If θ is a partial order on P, then θ is the intersection of all total orders on P that contain θ.
23. Let X be a topological space. Let $\text{cl}(A)$ denote the closure of A in X.
a) Show that $x \leq y$ if $x \in \text{cl}(\{y\})$ is a *preorder* on X.
b) Show that $\leq$ is a partial order if and only if X is T_0, that is, given any $x \neq y \in X$ there is an open set containing one of x or y but not both.
24. The **linear kernel** $K(P)$ of a poset P is the set of all elements of P that are comparable to every element of P. Prove that $K(P)$ is the intersection of all maximal chains of P.
25. Prove that Zorn's lemma is equivalent to the Hausdorff maximality principle.
26. Prove that the Hausdorff maximality principle implies Tukey's lemma.
27. Prove that Tukey's lemma implies the axiom of choice.

Chapter 2
Well-Ordered Sets

This chapter is an introduction to the subject of well-ordered sets and ordinal and cardinal numbers. While this material is more typically found in a textbook on set theory, we will have a few occasions to use the basic theory of ordinals, most notably in describing conditions that characterize the incompleteness of a lattice and in the discussion of fixed points. However, the chapter may be omitted upon first reading without severe consequences.

Well-Ordered Sets

We recall some earlier facts.

Definition *A poset* $(W, \leq)$ *is a* **well-ordered set** *if every nonempty subset of* W *contains a least element. In this case, the order* $\leq$ *is called a* **well-ordering** *of* W*. A nonempty set* X *for which a well-ordering exists is called a* **well-orderable** *set.* □

A well-ordered set is totally ordered, since any doubleton set $\{a, b\}$ has a least element. Also, every nonmaximal element $a \in W$ has a unique cover, since the set

$$U_a = \{x \in W \mid x > a\}$$

is nonempty and so has a least element z, which covers a. Moreover, if $a \sqsubset y$, then $y \in U_a$ and so $a < z \leq y$. Hence, $y = z$.

Initial Segments

In a well-ordered set, the following down-sets are special.

Definition *Let* P *be a poset. For* $a \in P$*, the set*

$$P[a] := \{x \in P \mid x < a\}$$

is called the **initial segment** *of* P *at* a*. (Note that* $a \notin P[a]$*.) We denote the*

S. Roman (ed.), *Lattices and Ordered Sets*, doi: 10.1007/978-0-387-78901-9_2,

family of initial segments of P by $\mathrm{seg}(P)$. *When we write $P[a]$, it is with the tacit understanding that $a \in P$.*□

If P is *totally* ordered, then

$$a \leq b \quad \Leftrightarrow \quad P[a] \subseteq P[b]$$

and so the *initial segment map* $I: P \to \mathrm{seg}(P)$ defined by $I(a) = P[a]$ is an order isomorphism. It follows that $\mathrm{seg}(P) \approx P$ and so $\mathrm{seg}(P)$ is totally ordered by set inclusion. We can say more when P is well-ordered.

Theorem 2.1 *Let W be a well-ordered set.*

1) The initial segments in W are precisely the proper *down-sets of W, that is,*

$$\mathrm{seg}(W) = \mathcal{O}^*(W) := \mathcal{O}(W) \setminus \{W\}$$

2) The function $I: W \approx \mathcal{O}^*(W)$ defined by

$$I(a) = W[a]$$

is an order isomorphism and so $\mathcal{O}(W)$ is well-ordered by set inclusion.

Proof. For part 1), the sets $W[a]$ are clearly proper down-sets. If D is a proper down-set of W, then $W \setminus D$ has a least element a and $D = W[a]$. To see this, it is clear that $W[a] \subseteq D$, since a is the least element not in D. Also, if $d \in D$ but $d \geq a$, then $a \leq d \in D$ and since D is a down-set, we get $a \in D$, which is false.

For part 2), if $a \leq b$, then certainly $W[a] \subseteq W[b]$. Conversely, if $W[a] \subseteq W[b]$ but $b < a$, then $b \in W[a] \subseteq W[b]$ and so $b \in W[b]$, which is false. Hence, $a \leq b$ and so I is an order embedding Finally, I is surjective by part 1) and so I is an order isomorphism.□

Isomorphisms of Well-Ordered Sets

We will have use for the following concept.

Definition *Let P be a poset. A function $f: P \to P$ is* **inflationary** *if*

$$x \leq f(x)$$

for all $x \in P$.□

The term *increasing* is also used for inflationary functions but unfortunately, some authors use *increasing* to mean monotone, so we have decided to avoid the term altogether.

Strictly monotone functions need not be inflationary. However, the situation is different for well-ordered sets and this has some important consequences.

Theorem 2.2 *Let W be a well-ordered set.*

1) *A strictly monotone function* $f: W \to W$ *is inflationary.*
2) W *is not isomorphic to any subset of any initial segment* W*, nor are any two distinct initial segments of* W *isomorphic.*
3) *There is at most one isomorphism between any two well-ordered sets. In particular, the identity is the only automorphism of* W.

Proof. For part 1), if there is an $x \in W$ such that $f(x) < x$, then the set

$$S = \{x \in W \mid f(x) < x\}$$

is nonempty and so has a least element a. But applying f to the inequality $f(a) < a$ gives

$$f(f(a)) < f(a)$$

which shows that $f(a) \in S$, a contradiction to the minimality of a.

For part 2), if $f: W \approx S \subseteq W[a]$, then since f is inflationary, we have $a \le f(a)$, which is impossible since, then $f(a) \notin W[a]$. Also, if $a < b$, then $W[a]$ is an initial segment of the well-ordered set $W[b]$ and so is not isomorphic to $W[b]$.

For part 3), if $f: W \approx W$ is an automorphism, then f and f^{-1} are inflationary and so $x \le f(x)$ and $x \le f^{-1}(x)$. Applying f to the latter gives $f(x) \le x$ and so $f(x) = x$, that is, f is the identity map. Hence, if $f, g: W_1 \approx W_2$, then $g^{-1}f$ is an automorphism of W_1 and so $g^{-1}f$ is the identity, whence $f = g$.□

The fact that a strictly monotone function $f: W \to W$ is inflationary has further important consequences. In particular, the following result about pasting together isomorphisms is very useful. For instance, it shows immediately that if W_1 and W_2 are nonisomorphic well-ordered sets, then one is isomorphic to an initial segment of the other.

Theorem 2.3 *Let* W_1 *and* W_2 *be well-ordered sets.*

1) *The family*

$$P = \{f: D \approx E \mid D \in \mathcal{O}(W_1), E \in \mathcal{O}(W_2)\}$$

is nonempty. Also:

a) *If* $f: D \to E$ *and* $g: D' \to E'$ *are in* P *and* $D \subseteq D'$ *then* $g|_D = f$.
b) *For each* $D \in \mathcal{O}(W_1)$*, there is at most one* $f \in P$ *with domain* D *and so we may totally order* P *by*

$$(f: D \approx E) \le (g: D' \approx E') \quad \textit{iff} \quad D \subseteq D'$$

2) *If*

$$A := \bigcup_{\substack{D=\mathrm{dom}(f)\\ \text{where } f\in P}} D \quad \text{and} \quad B := \bigcup_{\substack{E=\mathrm{im}(f)\\ \text{where } f\in P}} E$$

then the function $F: A \approx B$ *defined by*

$$F(x) = f(x)$$

for any f *with* $x \in \mathrm{dom}(f)$ *is the largest element of* P. *Moreover, either* $A = W_1$ *or* $B = W_2$.

3) **(Cantor, 1897)** *Exactly one of the following holds:*
   a) $W_1 \approx W_2$
   b) $W_1 \approx W_2[a]$ *for some* $a \in W_2$
   c) $W_2 \approx W_1[b]$ *for some* $b \in W_1$

   Moreover, the isomorphism in question is unique.

Proof. For part 1), the family P is nonempty since the function that takes the smallest element of W_1 to the smallest element of W_2 is in P. Next, we observe that there is at most one function in P for each $D \in \mathcal{O}(W_1)$, for if

$$f: D \approx E \quad \text{and} \quad g: D \approx E'$$

then $E \approx E'$ are isomorphic down-sets in W_2 and so Theorem 2.2 implies that $E = E'$. Then $f, g: D \approx E$ and so $f = g$. Finally, if

$$f: D \approx E \quad \text{and} \quad g: D' \approx E'$$

are in P and $D \subseteq D'$, then $g|_D: D \approx g(D)$ is in P and so $g|_D = f$.

For part 2), the function F is well defined, since $x \in \mathrm{dom}(f) \cap \mathrm{dom}(g)$ implies that $f(x) = g(x)$ and so the definition of $F(x)$ is independent of the choice of f. Since

$$x \le y \text{ in } A \Leftrightarrow f(x) \le f(y) \text{ for some } f \in P \Leftrightarrow F(x) \le F(y)$$

and since F is clearly surjective, it is an order isomorphism. Since A and B are down-sets, it follows that $F \in P$. Clearly, F is the largest element of P.

Finally, if $A = W_1[a]$ and $B = W_2[b]$, then we can extend $F: W_1[a] \approx W_2[b]$ to an isomorphism $\overline{F}: {\downarrow}a \approx {\downarrow}b$ of down-sets by setting $\overline{F}(a) = b$. This contradicts the fact that F is the largest element of P. Hence, either $A = W_1$ or $B = W_2$.

For part 3), Theorem 2.2 shows that at most one of 3a)–3c) can hold and that the isomorphism must be unique. But the isomorphism F establishes that one of 3a)–3c) must hold.□

Ordinal Numbers

The ordered set

$$X = (\emptyset, \{\emptyset\}, \{\emptyset, \{\emptyset\}\}, \{\emptyset, \{\emptyset\}, \{\emptyset, \{\emptyset\}\}\})$$

has the property that each element is the set of all elements that came before it. Thus, every element of X is also a subset of X. There is a name for such sets.

Definition *A set X is* **transitive** *if every element of X is a subset of X, that is, if*

$$x \in a \in X \quad \Rightarrow \quad x \in X$$ □

Note that to say that a set is transitive is not the same as saying that membership is a transitive relation on X, that is, for any $x, y, z \in X$,

$$x \in y \in z \quad \Rightarrow \quad x \in z$$

This is two different uses of the term transitive: one as it relates to a set and one as it relates to a binary relation. However, both senses are used in the following definition of the principle object of study of this chapter.

Definition *A set α is an* **ordinal number** *or* **ordinal** *if the following hold:*
1) α is transitive, that is,

$$a \in b \in \alpha \quad \Rightarrow \quad a \in \alpha$$

2) Membership is a strict (asymmetric and transitive) well-ordering of α.
We will reserve lowercase Greek letters for ordinal numbers and denote the fact that α is an ordinal by $\alpha \in$ ord. □

Somewhat as an aside, the family ord of all ordinals is too large to be considered a set, since doing so leads to certain paradoxes of set theory. To avoid these paradoxes, set theorists have defined *classes*. Thus, ord is a class. Much, but not all, of set theory can be applied to classes. For more on this, we refer the reader to a book on set theory.

Thus, for elements of an ordinal number α, the phrase "less than" means "element of." Let us make a few observations that follow directly from the definition. Let $\alpha \in$ ord.

1) Since membership is a *strict* order on α, it follows that

$$\alpha \notin \alpha$$

which implies that the elements of α are *proper* subsets of α.
2) Consider the chain

$$a \in b \in c \in \alpha$$

where $\alpha \in$ ord. The transitivity of the *set* α implies that $b \in \alpha$ and so

$a \in b \in \alpha$, whence $a \in \alpha$. More generally,

$$a_1 \in a_2 \in \cdots \in a_n \in \alpha \quad \Rightarrow \quad a_i \in \alpha \text{ for all } i = 1, \ldots, n$$

Also, the transitivity of the membership relation gives

$$a_1 \in a_2 \in \cdots \in a_n \in \alpha \quad \Rightarrow \quad a_i \in a_j \text{ for all } i < j$$

3) **(The elements of α are the initial segments of α)** The initial segment $\alpha[a]$ of α is *by definition* the set of all elements of α that are *elements* of a, that is,

$$\alpha[a] = a$$

In words, each element $a \in \alpha$ is its own initial segment.
4) **(The elements of α are ordinals)** Any element $a \in \alpha$ is also an ordinal. In fact, the transitivity of membership in α shows that a is a transitive set, for if $x \in y \in a$, then $x \in a$ and so $y \subseteq a$. Also, since $a \subseteq \alpha$, it follows that membership is a strict well-ordering on a.
5) **(No other proper subsets of α are ordinals)** A proper subset $S \subset \alpha$ is an ordinal if and only if S is an initial segment (element of) α. For if $S \in$ ord, then S is a proper down-set in α, since if $s \in S$ and $a \in s$ implies that $a \in S$. Hence, Theorem 2.1 implies that S is an initial segment of α.
6) **(In ord, proper inclusion is membership)** If $\alpha, \beta \in$ ord, then

$$\alpha \subset \beta \quad \Leftrightarrow \quad \alpha = \beta[a] \text{ for some } a \in \beta \quad \Leftrightarrow \quad \alpha \in \beta$$

Thus, for ordinals, membership is equivalent to proper inclusion.

A word on notation: It is customary to use the symbol $<$ as an alternative notation for membership. Thus, if $\alpha, \beta \in$ ord, then

$$\alpha < \beta \quad \text{means} \quad \alpha \in \beta$$

The symbol $<$ reminds us more of a strict order, but the symbol $\in$ is more meaningful in this context. However, there is no commonly accepted notation involving $\in$ to express the fact that $\alpha \in \beta$ or $\alpha = \beta$ and so we use $\alpha \leq \beta$.

The Natural Numbers

The prototypical examples of ordinals are the natural numbers. Indeed, the natural number 0 is *defined* to be the empty set, which is an ordinal. The other natural numbers are defined, in their natural order, by specifying that the natural number n is the set of all natural numbers that have already been defined. Thus,

$$\begin{aligned}
&0 = \emptyset \\
&1 = \{0\} = \{\emptyset\} \\
&2 = \{0, 1\} = \{\emptyset, \{\emptyset\}\} \\
&3 = \{0, 1, 2\} = \{\emptyset, \{\emptyset\}, \{\emptyset, \{\emptyset\}\}\} \\
&4 = \{0, 1, 2, 3\} = \{\emptyset, \{\emptyset\}, \{\emptyset, \{\emptyset\}\}, \{\emptyset, \{\emptyset\}, \{\emptyset, \{\emptyset\}\}\}\} \\
&\vdots \\
&n = \{0, 1., \ldots, n-1\} = n - 1 \cup \{n-1\} \\
&\vdots
\end{aligned}$$

By definition, an ordinal is **finite** if it is a natural number; otherwise, it is **infinite**. The set $\mathbb{N}$, under the usual order

$$0 < 1 < 2 < \cdots$$

is an ordinal number, customarily denoted by ω when thought of as an ordinal. (We leave proof to the reader.) Also, ω is the smallest infinite ordinal, in the sense that any proper subset of ω that is also an ordinal must be a natural number. For if α is an ordinal and $\alpha \subset \omega$, then $\alpha = \omega[n] = n$ for some $n \in \mathbb{N}$.

Characterizing Ordinals by Initial Segments

We have seen that the initial segments of $\alpha \in \text{ord}$ coincide with the elements of α, that is,

$$\alpha[a] = a$$

for all $a \in \alpha$. Conversely, suppose that $(W, <)$ is a well-ordered set for which

$$W[a] = a$$

for all $a \in W$. Then W is clearly transitive and the strict well-ordering $<$ must be the membership relation, since

$$a \in b \quad \Leftrightarrow \quad a \in W[b] \quad \Leftrightarrow \quad a < b$$

Hence, W is an ordinal. This gives a very useful characterization of ordinals which is sometimes taken as the definition of ordinal number.

Theorem 2.4 *A well-ordered set $(W, <)$, where $<$ is a strict well-ordering, is an ordinal if and only if*

$$a = W[a]$$

for all $a \in W$. In this case, the strict order relation $<$ is membership, that is,

$$a < b \text{ in } W \quad \Leftrightarrow \quad a \in b$$

□

The Successor of an Ordinal

The **successor** of a set X is the set

$$\operatorname{succ}(X) = X \cup \{X\}$$

Note that the natural number $n+1$ is defined to be the successor of the natural number n.

Theorem 2.5 *If α is an ordinal, then so is its successor.*
Proof. The set $\operatorname{succ}(\alpha) = \alpha \cup \{\alpha\}$ is transitive, since if $x \in a \in \alpha \cup \{\alpha\}$, then either $a = \alpha$, in which case $x \in \alpha$ or else $x \in a \in \alpha$, in which case $x \in \alpha$. Also, α is the largest element in $\operatorname{succ}(\alpha)$ and so membership is a strict well-ordering of $\operatorname{succ}(\alpha)$.□

Sets of Ordinals

We have seen that membership, which is the same as strict inclusion, is a strict order on the class ord. It follows, of course, that membership is a strict order on any nonempty *set* S of ordinals. But we can say more, namely, the union $\bigcup S$ and intersection $\bigcap S$ of S are also ordinals and the intersection λ is actually the least element of S and the union is the least upper bound for S. It follows that any nonempty *set* S of ordinals is *well-ordered* by membership (strict inclusion).

To see this, note first that the union or intersection of transitive sets is transitive. Also, since $\bigcap S$ and $\bigcup S$ are both sets of ordinals, membership is a strict order on these sets. Since the intersection $\lambda = \bigcap S$ is a subset of any ordinal $\alpha \in S$, it follows that λ is well-ordered by membership and so λ is an ordinal. Moreover, since $\lambda \subseteq \alpha$ for all $\alpha \in S$, it follows that $\lambda = \alpha$ or $\lambda \in \alpha$ for every $\alpha \in S$. Thus, λ is a lower bound for S. Also, if $\lambda \in \alpha$ for all $\alpha \in S$, then $\lambda \in \bigcap S = \lambda$, which is false. Hence, $\lambda = \alpha$ for some $\alpha \in S$ and so $\lambda \in S$ is the least element of S.

Finally, since every nonempty set of ordinals has a least element, it follows that the union $\bigcup S$ is well-ordered by membership and is therefore an ordinal.

Theorem 2.6 *Let S be a nonempty set of ordinals, strictly ordered by membership.*

1) $\bigcap S$ *is an ordinal and is the least element of S.*
   a) *Therefore, every nonempty set of ordinals is well-ordered by membership. In particular, the trichotomy law holds, that is, if* $\alpha, \beta \in$ ord, *then exactly one of the following holds:*

$$\alpha \in \beta, \quad \alpha = \beta \quad \textit{or} \quad \beta \in \alpha$$

   b) *A set of ordinals is an ordinal if and only if it is transitive.*
2) $\bigcup S$ *is an ordinal and is the least upper bound (supremum) of S.*

a) *Hence, for every set S of ordinals, there is an ordinal that is greater than every ordinal in S, namely,* $\mathrm{succ}(\bigcup S)$.

3) *The set of finite ordinals is the set of natural numbers.*

Proof. For part 3), the definition of natural number implies that all natural numbers are finite ordinals. Conversely, if $\alpha \in \mathrm{ord}$ is finite, then the trichotomy law implies that $\alpha \in \omega$ and so $\alpha = \omega[n] = n$, for some $n \in \mathbb{N}$.□

Ordinals Are Order Types

We can now show that every well-ordered set is isomorphic to exactly one ordinal.

Theorem 2.7 *Every well-ordered set W is isomorphic to a unique ordinal number, which is called the* **order type** *of W.*

Proof. Since if $\alpha, \beta \in \mathrm{ord}$, then one of α or β is an initial segment of the other, it follows that no two distinct ordinals are order isomorphic and this proves the uniqueness part of the theorem. For the existence part, if there is an ordinal α for which $W \approx \alpha$ or $W \approx \alpha[a]$ for some $a \in \alpha$, then we are done, since $\alpha[a]$ is also an ordinal. So let us assume that this is not the case.

Then Theorem 2.3 implies that for every $\alpha \in \mathrm{ord}$, there is a necessarily unique $a \in W$ for which $\alpha \approx W[a]$. It follows from an axiom schema of set theory that ord is a *set*, which is false, since for any set S of ordinals, there is an ordinal that is strictly greater than all elements in S. (The axiom schema to which we refer is the Axiom Schema of Replacement. We refer the interested reader to a book on set theory for more details.)□

Successor and Limit Ordinals

The successor of an ordinal α is denoted by $\alpha + 1$, that is,

$$\alpha + 1 := \mathrm{succ}(\alpha) = \alpha \cup \{\alpha\}$$

An ordinal β of the form $\beta = \alpha + 1$ is called a **successor ordinal**. A nonsuccessor ordinal is called a **limit ordinal**. Note that

$$\alpha \le \beta \quad \Leftrightarrow \quad \alpha < \beta + 1$$

We leave proof of the following as an exercise.

Theorem 2.8 *Let $\alpha, \beta \in \mathrm{ord}$.*

1) *If $\alpha \in \beta$, then either $\alpha + 1 \in \beta$ or $\alpha + 1 = \beta$.*
2) *$\alpha + 1$ is the unique cover of α.*
3) *The following are equivalent for an ordinal λ:*
   a) *λ is a limit ordinal*
   b) $\alpha \in \mathrm{ord}, \alpha \in \lambda \Rightarrow \alpha + 1 \in \lambda$

c) *λ is the union of all ordinals contained in λ, that is,*

$$\lambda = \bigcup_{\alpha \in \lambda} \alpha$$ □

Transfinite Sequences

The notion of sequence can be extended into the transfinite.

Definition

1) *Let A be a set and let $\delta \in$ ord. A function $f: \delta \to A$ is called a* **transfinite sequence** *of length δ in A and is often denoted by*

$$\mathcal{S} = \langle a_\alpha \mid \alpha < \delta \rangle$$

where $\alpha_\alpha = f(\alpha)$. The set $\operatorname{im}(\mathcal{S}) = \{a_\alpha \mid \alpha < \delta\}$ is called the **underlying set**, **range** *or* **image** *of $\mathcal{S}$ and δ is called the* **index ordinal** *of $\mathcal{S}$.*

2) *By analogy, a "function" f: ord $\to A$ is called a* **sequence** *on* ord *and is denoted by*

$$\langle a_\alpha \mid \alpha \in \text{ord} \rangle$$ □

We will say that an element a is *in* a sequence or *belongs to* a sequence $\mathcal{S}$ if a is in the underlying set of $\mathcal{S}$. The image $\{a_\alpha \mid \alpha < \delta\}$ of an *injective* transfinite sequence $\langle a_\alpha \mid \alpha < \delta \rangle$ is well-ordered by the index ordinal δ:

$$a_\alpha \le a_\beta \quad \text{if} \quad \alpha \in \beta$$

Conversely, if X is a well-ordered set, then it is isomorphic to a unique ordinal δ and the unique isomorphism $f: \delta \approx X$ is a transfinite sequence. Thus, an injective transfinite sequence with range X is essentially just a way to well-order the set X.

Subsequences

Suppose that $f: \delta \to A$ is a transfinite sequence. If $\beta \le \delta$ and if $g: \beta \to \delta$ is a strictly increasing function, then the composition $f \circ g: \beta \to A$ is called a **subsequence** of f. A subsequence of a sequence f is just a way of selecting a portion of the image of f, while at the same time preserving the original order that f imparts to its image.

We can take a slightly different viewpoint of subsequences as follows. Suppose that $f: \delta \to A$ is a transfinite sequence and $W \subseteq \delta$ is a nonempty subset of δ, with the well-order inherited from δ. Then there is a unique order isomorphism $g: \alpha \approx W$, where $\alpha \in$ ord. The map $(f \circ g): \alpha \to A$ is a subsequence of f, but it will sometimes be convenient to think of the map $f|_W: W \to A$ as a subsequence of f, which does no harm since W uniquely determines α and g.

As mentioned, the notation $\mathcal{S} = \langle a_\alpha \mid \alpha < \delta \rangle$ is often used to denote a transfinite sequence $f: \delta \to A$, where $f(\alpha) = a_\alpha$. Then a subsequence

$(f \circ g): \beta \to A$ of f is denoted by $\mathcal{T} = \langle a_{g(\alpha)} \mid \alpha < \beta \rangle$. Thus, to say that an element a_λ in $\mathcal{S}$ is in $\mathcal{T}$ is to say that $a_\lambda = a_{g(\alpha)}$ for some $\alpha \in \beta$.

Transfinite Induction

The well-known principle of induction for the natural numbers can be generalized as follows.

Theorem 2.9 (Transfinite Induction: Version 1) *Let $P(x)$ be a property of ordinals, that is, for each $\alpha \in$ ord, the statement $P(\alpha)$ is either true or false. If the following* **inductive hypothesis** *holds:*

$$P(\beta) \text{ holds for all } \beta < \alpha \quad \Rightarrow \quad P(\alpha) \text{ holds}$$

then $P(\alpha)$ holds for all ordinals α.

Proof. If $P(\alpha)$ does not hold for some $\alpha \in$ ord, then let X be the nonempty set of all ordinals $\beta \in \alpha$ for which $P(\beta)$ fails and let λ be the least element of X. Thus, $P(\lambda)$ fails, but $P(\beta)$ holds for every $\beta \in \lambda$, contradicting the inductive hypothesis.□

There is another common version of transfinite induction.

Theorem 2.10 (Transfinite Induction: Version 2) *Let $P(x)$ be a property of ordinals, that is, for each $\alpha \in$ ord, the statement $P(\alpha)$ is either true or false. Assume that*

1) $P(0)$ holds

2) If α is any ordinal, then

$$P(\alpha) \text{ holds} \quad \Rightarrow \quad P(\alpha + 1) \text{ holds}$$

3) If α is a limit ordinal, then

$$P(\beta) \text{ holds for all } \beta < \alpha \quad \Rightarrow \quad P(\alpha) \text{ holds}$$

Then $P(\alpha)$ holds for all ordinals.

Proof. If $P(\alpha)$ does not hold for some $\alpha \in$ ord, then let X be the nonempty set of all ordinals $\beta < \alpha$ for which $P(\beta)$ fails and let λ be the least element of X. Thus, $P(\lambda)$ fails but $P(\beta)$ holds for every $\beta < \lambda$. If $\lambda = 0$, then we are in violation of 1); if $\lambda = \alpha + 1$ is a successor ordinal, then we are in violation of 2) and if λ is a limit ordinal, then we are in violation of 3).□

Cardinal Numbers

Cardinal numbers are special types of ordinal numbers.

Definition *Two sets X and Y are* **equipotent** *(also called* **equipollent** *or* **equinumerous***) if there is a bijection between X and Y.*□

We use the notation $X \hookrightarrow Y$ to denote the fact that there is an injection from X to Y and the notation $X \leftrightarrow Y$ to denote the fact that there is a bijection from X onto Y, that is, that X and Y are equipotent.

Definition *An ordinal* α *is an* **initial ordinal** *or a* **cardinal** *if* α *is not equipotent to any smaller ordinal* $\beta \in \alpha$. *We denote the class of all cardinal numbers by* Cn.□

The natural numbers are initial ordinals, as is the ordinal ω. Also, it is not hard to show that any infinite cardinal number β is a limit ordinal.

The Cardinality of a Set

In general, a well-orderable set X may have many well-orderings. For each well-ordering $\leq$ on X, the well-ordered set $(X, \leq)$ has a unique order type $\beta_{\leq}$. The family of well-orderings on X is a set and therefore so is the set $\mathrm{OrdTypes}(X)$ of all order types corresponding to the well-orderings of X. Since an order isomorphism is also a bijection, each order type is equipotent to X and so

$$\mathrm{OrdTypes}(X) \subseteq \{\beta \in \mathrm{ord} \mid \beta \text{ is equipotent to } X\}$$

On the other hand, if $\alpha \in \mathrm{ord}$ is equipotent to X, then any bijection $f\colon \alpha \leftrightarrow X$ can be used to transfer the well-ordering from α to X, making $(X, \leq)$ a well-ordered set. Then f is an order isomorphism and so $\alpha \in \mathrm{OrdTypes}(X)$. Thus,

$$\mathrm{OrdTypes}(X) = \{\beta \in \mathrm{ord} \mid \beta \text{ is equipotent to } X\}$$

Therefore, the least member α of $\mathrm{OrdTypes}(X)$ is an initial ordinal, since $\gamma \in \alpha$ implies that γ is not equipotent to X and therefore is not equipotent to α.

Theorem 2.11 *Let* X *be a well-orderable set. Then*

$$\mathrm{OrdTypes}(X) = \{\beta \in \mathrm{ord} \mid \beta \textit{ is equipotent to } X\}$$

and the smallest order type for X *is an initial ordinal, called the* **cardinality** *of* X *and denoted by* $|X|$ *or* $\mathrm{card}(X)$.□

Example 2.12 Consider the set $\mathbb{N}$ of natural numbers. When $\mathbb{N}$ is given the usual order

$$0 < 1 < 2 < \cdots$$

then the resulting well-ordered set is denoted by ω. We may also well-order $\mathbb{N}$ by giving the positive integers the usual order and declaring that 0 is the largest element, that is,

$$1 < 2 < \cdots \quad \text{and} \quad n < 0 \text{ for all positive integers } n$$

Let us denote this well-ordered set by W. Then the map

$$f: W \to \omega + 1 = \omega \cup \{\omega\}$$

defined by

$$f(k) = \begin{cases} k-1 & \text{for } k > 0 \\ \omega & \text{for } k = 0 \end{cases}$$

is an order isomorphism and so $W \approx \omega + 1$. In a similar way, we can well-order $\mathbb{N}$ so that $W \approx \omega + k$ for $k \geq 1$. However, ω is the smallest infinite ordinal and so it is the least ordinal that is equipotent to $\mathbb{N}$, that is, $|\mathbb{N}| = \omega$.□

In this book, we will assume the axiom of choice and its equivalents (unless otherwise noted). In particular, we assume that any set is well-orderable. Hence, any set has a cardinal number, as we have defined the concept. (If one does not assume that every set is well-orderable, then one can define a more general notion of cardinal number or simply restrict the definition to well-orderable sets.)

The Cantor–Schröder–Bernstein Theorem

The well known Cantor–Schröder–Bernstein theorem says that

$$X \hookrightarrow Y \quad \text{and} \quad Y \hookrightarrow X \quad \Rightarrow \quad X \leftrightarrow Y$$

This theorem is usually proved by actually constructing a bijection between $X$and Y. However, with our background in ordinal numbers, we can also prove the theorem by proving that

$$X \hookrightarrow Y \quad \Leftrightarrow \quad |X| \leq |Y|$$

since the theorem, then follows from the trichotomy law

$$|X| \leq |Y| \quad \text{and} \quad |Y| \leq |X| \quad \Rightarrow \quad |X| = |Y|$$

It is easy to see that

$$X \leftrightarrow Y \quad \Leftrightarrow \quad |X| = |Y|$$

since $X \leftrightarrow |X|$ for all sets X and no two distinct cardinal numbers are equipotent. Now suppose that $f: X \hookrightarrow Y$. If we well-order Y to have order type $\beta = |Y|$, then fX inherits this well-ordering and so has an order type α. Moreover,

$$\alpha \approx fX \subseteq Y \approx \beta$$

and so α is order isomorphic to a subset of β. This implies that β cannot be an initial segment of α and so $\alpha \leq \beta$. Since α is an order type of X, we have

$$|X| \leq \alpha \leq \beta = |Y|$$

For the converse, if $X \leq Y$, then $|X| \subseteq |Y|$ and there are maps

$$X \leftrightarrow |X| \hookrightarrow |Y| \leftrightarrow Y$$

where the center map is inclusion. The composition is an injection from X into Y. Thus,

$$X \hookrightarrow Y \quad \Leftrightarrow \quad |X| \leq |Y|$$

Incidentally, this justifies the introduction of the expression $|X| \leq |Y|$ to mean that $X \hookrightarrow Y$, which one often finds in books that introduce the concept of the "cardinality" of a set before introducing cardinal numbers.

Theorem 2.13 *If Y and X are sets then*
1) $|X| \leq |Y|$ *if and only if* $X \hookrightarrow Y$
2) $|X| = |Y|$ *if and only if* $X \leftrightarrow Y$
3) **(Cantor–Schröder–Bernstein theorem)**

$$X \hookrightarrow Y \quad \text{and} \quad Y \hookrightarrow X \quad \Rightarrow \quad X \leftrightarrow Y$$ □

Cantor's Theorem

Another famous theorem associated with Cantor shows that there are arbitrarily large cardinal numbers.

Theorem 2.14 (Cantor's theorem) *If X is a set then $|X| < |\wp(X)|$.*
Proof. Let $f\colon X \to \wp(X)$ be defined by $f(x) = \{x\}$. This is an injection, showing that $|X| \leq |\wp(X)|$. But if $|X| = |\wp(X)|$, then there is a bijection $g\colon X \to \wp(X)$. Let

$$S = \{x \in X \mid x \notin g(x)\}$$

The surjectivity of g implies that $S = g(s)$ for some $s \in X$. But $s \in S$ implies that $s \notin g(s) = S$ and $s \notin S$ implies that $s \in g(s) = S$, both a contradiction. Hence g cannot exist and so $|X| = |\wp(X)|$ cannot hold.□

We mention without proof that the cardinality of the set $\wp_0(X)$ of all *finite* subsets of an infinite set X is the same as the cardinality of X.

The Alephs

Cantor's theorem implies that the set

$$\{\Pi \in \mathrm{Cn} \mid \Sigma < \Pi \leq |\wp(\Sigma)|\}$$

is nonempty and so has a least element. This least element is the smallest cardinal greater than Σ.

Definition *If Σ is a cardinal, then the smallest cardinal $h(\Sigma)$ greater than Σ is called the* **Hartogs number** *of Σ.*□

The Hartogs number and transfinite recursion (which we will not discuss formally in this brief introduction) can be used to define a strictly increasing transfinite sequence of cardinal numbers, indexed by ord.

Definition *Define a strictly increasing transfinite sequence* $\langle\aleph_\alpha \mid \alpha \in \text{ord}\rangle$ *as follows:*

1) $\aleph_0 = \omega$

2) If $\alpha+1$ *is a successor ordinal, then*

$$\aleph_{\alpha+1} = h(\aleph_\alpha)$$

3) If λ *is a limit ordinal, then*

$$\aleph_\lambda = \sup\{\aleph_\beta \mid \beta < \lambda\} = \bigcup\{\aleph_\beta \mid \beta < \lambda\}$$

The numbers $\aleph_\alpha$ *are called* **alephs.** □

Theorem 2.15 *The alephs are precisely the infinite cardinal numbers.*
Proof. Since $\aleph_\alpha < h(\aleph_\alpha)$ and since, for any limit ordinal λ,

$$\aleph_\beta < \aleph_{\beta+1} \le \aleph_\lambda$$

for any $\beta < \lambda$, it follows that the sequence of alephs is strictly increasing. Moreover, since $\aleph_0$ is an infinite set, all alephs are infinite cardinals.

For the converse, it is easy to see by transfinite induction on α that all cardinals Σ less than $\aleph_\alpha$ are alephs. If $\alpha = 0$, then the result holds vacuously. If the result holds for α and $\Sigma < \aleph_{\alpha+1}$, then $\Sigma \le \aleph_\alpha$. But if $\Sigma = \aleph_\alpha$, then Σ is an aleph and if $\Sigma < \aleph_\alpha$, then the inductive hypothesis implies that Σ is an aleph. Finally, if λ is a limit ordinal, then $\Sigma < \aleph_\lambda$ implies that $\Sigma < \aleph_\beta$ for some $\beta < \lambda$ and the inductive hypothesis implies that Σ is an aleph.

Thus, it is sufficient to show that any cardinal Σ is less than some aleph. For this, we show by transfinite induction that any ordinal α satisfies $\alpha \le \aleph_\alpha$, since then $\Sigma \le \aleph_\Sigma < \aleph_{\Sigma+1}$. This is clear for $\alpha = 0$. Also, $\aleph_\alpha < h(\aleph_\alpha)$ gives

$$\alpha \le \aleph_\alpha \quad \Rightarrow \quad \alpha + 1 \le \aleph_\alpha + 1 \le h(\aleph_\alpha) = \aleph_{\alpha+1}$$

Finally, if λ is a limit ordinal and $\alpha < \aleph_\alpha$ for all $\alpha < \lambda$, then

$$\lambda = \bigcup_{\alpha\in\lambda} \alpha \le \bigcup_{\alpha\in\lambda} \aleph_\alpha = \aleph_\lambda$$ □

The question of how big a "jump" occurs in going from the cardinality of a set X and the cardinality of its power set $\wp(X)$ is a very famous one. The **continuum hypothesis** is the statement that the cardinality of the power set $\wp(\mathbb{N})$, which is the same as the cardinality of the real numbers $\mathbb{R}$, is equal to $\aleph_1$. Kurt Gödel showed in 1940 that the continuum hypothesis is *consistent* with, that is, cannot be disproven from, the usual Zermelo–Fraenkel axioms of set theory, with or without the axiom of choice. In 1963, Paul Cohen showed that the continuum hypothesis is *independent* of, that is, cannot be proven from, the Zermelo–Fraenkel axioms, with the axiom of choice. Hence, the continuum hypothesis is independent of the Zermel–Fraenkel axioms and the axiom of choice. The **generalized continuum hypothesis** is the statement that $2^{\aleph_\alpha} = \aleph_{\alpha+1}$ for all ordinals α.

Ordinal and Cardinal Arithmetic

The arithmetic of ordinal numbers is defined as follows.

Definition (Ordinal arithmetic)
Addition:
1) $\alpha + 0 = \alpha$
2) $\alpha + 1 = \text{succ}(\alpha)$
3) $\alpha + (\beta + 1) = (\alpha + \beta) + 1$ *for all* $\beta \in \text{ord}$
4) $\alpha + \lambda = \bigcup_{\beta\in\lambda}\{\alpha + \beta\}$ *for all limit ordinals* λ.
Multiplication:
5) $\alpha \cdot 0 = 0$
6) $\alpha \cdot (\beta + 1) = \alpha \cdot \beta + \alpha$ *for all* $\beta \in \text{ord}$
7) $\alpha \cdot \lambda = \bigcup_{\beta\in\lambda}\{\alpha \cdot \beta\}$ *for all limit ordinals* λ.
Exponentiation:
8) $\alpha^0 = 1$
9) $\alpha^{\beta+1} = \alpha^\beta \cdot \alpha$ *for all* $\beta \in \text{ord}$
10) $\alpha^\lambda = \bigcup_{\beta\in\lambda}\{\alpha^\beta\}$ *for all limit ordinals* λ.□

Here is a short list of the first few infinite ordinals:

$$
\begin{aligned}
&\omega, \omega + 1, \omega + 2, \ldots, \omega + \omega = \omega \cdot 2, \\
&\omega \cdot 2 + 1, \omega \cdot 2 + 2, \ldots, \omega \cdot 2 + \omega = \omega \cdot 3, \\
&\vdots \\
&\omega \cdot k + 1, \ldots, \omega \cdot (k + 1) \\
&\vdots
\end{aligned}
$$

The next limit ordinal is

$$\omega^2 = \omega \cdot \omega = \bigcup_k \omega \cdot k$$

Now we begin again, replacing each ω above by ω^2 and continuing:

$$\omega^2, \ldots, \omega^2 \cdot \omega = \omega^3, \omega^3 + 1, \ldots, \omega^k, \ldots, \omega^\omega = \bigcup_k \omega^k$$

This process continues:

$\omega^\omega + 1, \ldots, \omega^{(\omega^\omega)}, \ldots$

Note that

$$\omega \cdot 2 \neq 2 \cdot \omega$$

since the latter is

$$2 \cdot \omega = \bigcup_{k \in \mathbb{N}} (2 \cdot k) = \omega$$

Thus, ordinal multiplication is *not* commutative.

Since cardinal numbers are ordinal numbers, we already have a definition of sum, product and exponential for cardinal numbers. However, as measures of the *size* of sets regardless of any order structure, there is a more appropriate set of operations, called *cardinal sum*, *cardinal product* and *cardinal exponentiation.*

These operations are defined in terms of set operations.

Definition (Cardinal arithmetic)
1) $\aleph_\alpha + \aleph_\beta = |\aleph_a \sqcup \aleph_\beta|$, *where* $\sqcup$ *denotes the disjoint union.*
2) $\aleph_\alpha \cdot \aleph_\beta = |\aleph_a \times \aleph_\beta|$, *where* $\times$ *denotes the cartesian product.*
3) $\aleph_\alpha^{\aleph_\beta} = |\{f : \aleph_\beta \to \aleph_\alpha\}|$.□

We omit the proof of the following theorem.

Theorem 2.16 *For infinite cardinal numbers* Σ *and* Π, *we have*

$$\Sigma + \Pi = \Sigma \cdot \Pi = \max\{\Sigma, \Pi\}$$

Also, if $\wp_0(X)$ *is the set of all finite subsets of an infinite set* X *then*

$$|\wp_0(X)| = |X|$$ □

Complete Posets

It is well known that the union of an arbitrary family of subgroups of a group G need not be a subgroup of G. However, the union of a *chain* of subgroups of G is a subgroup of G. More generally, the union of a *directed family* of subgroups of G is a subgroup of G. Similar statements can be made in the context of other algebraic structures, such as vector spaces or modules.

Actually, the statement about directed sets is *not* more general than the statement about chains. We will prove that if every chain in a poset P has a join, then every directed set also has a join.

Definition *Let P be a poset.*

1) *P is* **chain-complete** *or* **inductive** *if every chain in P, including the empty chain, has a join. Thus, P has a smallest element.*
2) *P is* **complete** *or a* **CPO** *if P has a smallest element and if every directed subset of P has a join.*□

We should mention that the adjective *complete* has a different meaning when applied to posets than when applied to lattices. (We will define complete lattices in a later chapter.)

The proof that chain-completeness implies completeness uses the fact that any infinite directed set D is the union of a strictly increasing transfinite sequence of directed subsets of D of cardinality less than that of D. Our proof of this result follows the lines of Markowsky [43], which is a sharpening of Iwamura [35]; see also Maeda [41], Skornyakov [57] and Mayer-Kalkschmidt [44].

Theorem 2.17 *An infinite directed poset D is the union of a strictly increasing transfinite sequence*

$$\mathcal{S} = \langle D_\alpha \mid \alpha < |D| \rangle$$

of directed subsets D_α for which $|D_\alpha| < |D|$. In particular, if α is finite, then $|D_\alpha|$ is finite and if α is infinite, then $|D_\alpha| \le |\alpha|$.

Proof. Well-order D using the cardinal $\Sigma = |D|$ as index set, that is,

$$D = \langle d_\alpha \mid \alpha < \Sigma \rangle$$

We will define the sequence $\mathcal{S}$ in such a way that for all $\alpha < \Sigma$,

$$\{d_\epsilon \mid \epsilon < \alpha\} \subseteq D_\alpha \tag{2.18}$$

Then for any $\beta < \Sigma$, we have $d_\beta \in D_{\beta+1} \subseteq \bigcup D_\alpha$ and so $\bigcup D_\alpha = D$.

The finite part of the sequence is easy to define: At each step, just include the smallest element of D (under the well-ordering) not yet in the sequence, along with an upper bound (under the partial order) for what has come before. Specifically, let $D_0 = \{d_0\}$ and let

$$D_{k+1} = D_k \cup \{e_{k+1}, u_{k+1}\}$$

where e_{k+1} is the least element not in D_k and u_{k+1} is any upper bound for the finite set $D_k \cup \{e_{k+1}\}$. It is clear that $\langle D_k \mid k \in \mathbb{N} \rangle$ is a strictly increasing sequence of finite directed (in fact, bounded above) sets. Moreover,

$$\{d_0, \ldots, d_k\} \subseteq D_k$$

It follows that if $\Sigma = \omega$, then $\bigcup D_k = D$ and the proof is complete. So assume that $\Sigma > \omega$.

Assume that for $\beta < \Sigma$, we have constructed a strictly increasing sequence $\langle D_\alpha \mid \alpha < \beta \rangle$ of directed sets D_α for which $|D_\alpha| \le |\alpha|$ and (2.18) holds for all $\alpha < \beta$. This has been done for $\beta = \omega$. Construction of the next element D_β in the sequence depends on the type of the index β.

For a limit ordinal β, we accumulate all that came before:

$$D_\beta = \bigcup_{\alpha<\beta} D_\alpha$$

Then D_β is directed and $\alpha < \beta$ implies that $D_\alpha \subset D_{\alpha+1} \subseteq D_\beta$. Also,

$$|D_\beta| \le \sum_{\alpha<\beta} |D_\alpha| \le |\beta| \cdot |\beta| = |\beta|$$

Moreover, (2.18) holds since if $\epsilon < \beta$, then $\epsilon + 1 < \beta$ and so

$$d_\epsilon \in D_{\epsilon+1} \subseteq D_\beta$$

If $\beta = \delta + 1$ is an infinite successor ordinal, we define $D_{\delta+1}$ in a manner that is similar in spirit to the finite case. First, since $|D_\delta| < |D|$, the set $D \setminus D_\delta$ is nonempty and so we may adjoin a new element by setting

$$D_{\delta+1,0} = D_\delta \cup \{e_\delta\}$$

where e_δ is the least element (under the well-ordering) not in D_δ. Although $D_{\delta+1,0}$ may not have an upper bound, we can include upper bounds (under the partial order) for all *finite* subsets of $D_{\delta+1,0}$. But this process must be repeated, since the inclusion of these upper bounds requires that additional upper bounds be included as well. Specifically, for $k \in \mathbb{N}$, let

$$D_{\delta+1,k+1} = D_{\delta+1,k} \cup \{\text{an upper bound } u_F \text{ for each finite subset } F \subseteq D_{\delta+1,k}\}$$

and let

$$D_{\delta+1} = \bigcup_{k\in\mathbb{N}} D_{\delta+1,k}$$

Now, $D_{\delta+1}$ is directed since any $x, y \in D_{\delta+1}$ are also in some set $D_{\delta+1,k}$ and so have an upper bound in $D_{\delta+1,k+1}$. Also, $|D_{\delta+1,k}| = |D_{\delta+1,0}| = |D_\delta|$ and so

$$|D_{\delta+1}| \le \aleph_0 \cdot |D_\delta| = |D_\delta| \le |\delta| = |\delta + 1|$$

Moreover, (2.18) holds since if $\epsilon < \delta + 1$ then $\epsilon < \delta$ or $\epsilon = \delta$. If $\epsilon < \delta$, the induction hypothesis implies that $d_\epsilon \in D_\delta \subseteq D_{\delta+1}$. If $\epsilon = \delta$, then since D_δ contains all d_ϵ with $\epsilon < \delta$, it follows that if $d_\delta \notin D_\delta$, then $e_\delta = d_\delta$ and so $d_\delta \in D_{\delta+1,0} \subseteq D_{\delta+1}$. Thus, the sequence $\langle D_\alpha \mid \alpha < \Sigma \rangle$ has the desired properties.□

Theorem 2.19 *A poset P is complete if and only if it is chain-complete.*

Proof. If P is complete, then it is chain-complete, since every nonempty chain is a directed set and since a complete poset has a smallest element by definition. For the converse, we use transfinite induction on the cardinality Σ of the directed set D. If Σ is finite, then D has a join, so we may assume that $\Sigma = \aleph_\gamma$ is infinite and that all finite directed sets and all directed sets of cardinality $\aleph_\alpha$ for $\alpha < \gamma$ have a join.

Theorem 2.17 implies that there is a strictly increasing transfinite sequence

$$\mathcal{S} = \langle D_\alpha \mid \alpha < \Sigma \rangle$$

of directed subsets with $|D_\alpha| < |D|$ and

$$\bigcup D_\alpha = D$$

The induction hypothesis implies that each set D_α has a join a_α.

Since $\alpha < \beta$ implies that $D_\alpha \subset D_\beta$, the sequence $\langle a_\alpha \mid \alpha < \Sigma \rangle$ is a chain and so it has a join a. Since $D_\alpha \leq a_\alpha \leq a$, it is clear that a is an upper bound for D. However, if $D \leq b$, then $D_\alpha \leq b$ and so $a_\alpha \leq b$, whence $a \leq b$. Thus, a is the join of D.□

Cofinality

The following concept will prove useful in our study of the completeness (or rather the incompleteness) of lattices.

Definition *Let P be a poset.*

1) *A set $A \subseteq P$ is* **cofinal** *in P if for any $x \in P$, there is an $a \in A$ such that $x \leq a$, that is, if*

   $$P = \downarrow A$$

   A transfinite sequence

   $$\langle a_\alpha \mid \alpha < \beta \rangle$$

   is **cofinal** *in P if the underlying set $\{a_\alpha \mid \alpha < \beta\}$ is cofinal in P.*
2) *Dually, a set $A \subseteq P$ is* **coinitial** *if for any $x \in P$ there is an $a \in A$ such that $a \leq x$, that is, if*

   $$P = \uparrow A$$

   A transfinite sequence $\langle a_\alpha \mid \alpha < \beta \rangle$ is **coinitial** *in P if the underlying set $\{a_\alpha \mid \alpha < \beta\}$ is coinitial in P.*□

If A is cofinal in P, then any upper bound u for A must be in A and must be the largest element of P. In particular, if P has no largest element, then a cofinal set in P is unbounded.

Of course, a poset $(P, \leq)$ is cofinal in itself. However, we are more interested in cofinal subsets that have some additional order properties. In particular, given a well-ordering of P via an injective transfinite sequence

$$\mathcal{P} = \langle a_\alpha \mid \alpha < \delta \rangle$$

we can always find a subsequence $\mathcal{C}$ of $\mathcal{P}$ with the following properties:

1) The underlying set C of $\mathcal{C}$ is cofinal in P under the partial order, that is, for any $p \in P$, there is an $a_\alpha \in C$ for which $p \leq a_\alpha$.
2) The well-order given by $\mathcal{C}$ is compatible with the partial order on P, that is,

$$a_\alpha, a_\beta \in \mathcal{C} \quad \text{and} \quad \alpha < \beta \quad \Rightarrow \quad a_\alpha < a_\beta \quad \text{or} \quad a_\alpha \parallel a_\beta$$

In words, if a_α comes before a_β in the sequence $\mathcal{C}$, then a_α is less than or parallel to a_β in the original partial order.

The subsequence $\mathcal{C}$ is constructed by selecting the elements of $\mathcal{P}$ as follows:

$$a_\beta \in \mathcal{C} \quad \Leftrightarrow \quad a_\beta \text{ is } \leq\text{-maximal in } \{a_\alpha \mid \alpha \leq \beta\}$$

To see that the order of $\mathcal{C}$ is compatible with the partial order on P, if $a_\alpha, a_\beta \in \mathcal{C}$ where $\alpha < \beta$, then the inclusion of a_β implies that a_β is maximal in a set that includes a_α and so $a_\alpha < a_\beta$ or $a_\alpha \parallel a_\beta$.

To see that C is cofinal in $\mathcal{P}$, if $a_\beta \notin C$, then a_β is not maximal in $\langle a_\alpha \mid \alpha \leq \beta \rangle$, which therefore contains an element greater than a_β. But then the *first* such element (in sequential order) a_δ is $\leq$-maximal in $\langle a_\alpha \mid \alpha \leq \delta \rangle$, since otherwise it would not be the first element greater than a_β. Hence, $a_\delta \in C$ and $a_\delta > a_\beta$.

Theorem 2.20 *Let $(P, \leq)$ be a poset, well-ordered by an injective transfinite sequence*

$$\mathcal{P} = \langle a_\alpha \mid \alpha < \delta \rangle$$

Then $\mathcal{P}$ has a cofinal subsequence $\mathcal{C}$ with the following properties:
1) The well-order of $\mathcal{C}$ is compatible with the order in $(P, \leq)$, that is,

$$a_\alpha, a_\beta \in \mathcal{C} \quad \textit{and} \quad \alpha < \beta \quad \Rightarrow \quad a_\alpha < a_\beta \quad \textit{or} \quad a_\alpha \parallel a_\beta$$

2) The underlying set C of $\mathcal{C}$ is cofinal in $(P, \leq)$, that is, if $p \in P$ then there is an $a_\alpha \in C$ for which $p \leq a_\alpha$.□

Exercises

1. Give an example of a linearly ordered set P for which there is a proper down-set that is not an initial segment of P.
2. Let W be a well-ordered set for which $\infty \notin W$. Let $W' = W \cup \{\infty\}$ and extend the order on W so that $a < \infty$ for all $a \in W$. Show that W' is a well-ordered set.

3. Find an example of a strictly monotone function that is not inflationary.
4. Prove that a set X is transitive if and only if $\bigcup X \subseteq X$.
5. Prove that the intersection of transitive sets is transitive.
6. Prove that the union of transitive sets is transitive.
7. Prove that $\alpha + 1$ covers α for any $\alpha \in \mathrm{ord}$.
8. If $\alpha \in \mathrm{ord}$ show that $\mathcal{O}(\alpha) = \alpha + 1$.
9. Prove that if $\alpha \in \beta \in \mathrm{ord}$ then either $\alpha + 1 \in \beta$ or $\alpha + 1 = \beta$.
10. Prove that the following are equivalent for an ordinal α:
    a) α is a limit ordinal
    b) $\beta \in \mathrm{ord}, \beta \in \alpha \Rightarrow \beta + 1 \in \alpha$
    c) α is the union of all ordinals contained in α, that is,

$$\alpha = \bigcup_{\beta < \alpha} \beta$$

11. Prove that if a set S of ordinals has no greatest element, then $\bigcup S$ is a limit ordinal.
12. Let $f \colon W_1 \approx W_2$ be an isomorphism between well-ordered sets.
    a) Show that the restriction of f to $W_1[a]$ is an isomorphism $f \colon W_1[a] \approx W_2[f(a)]$.
    b) Show that the induced map $\overline{f} \colon \wp(W_1) \to \wp(W_2)$ gives rise to a bijection between the families of initial segments, given by

$$\overline{f}(W_1[a]) = W_2[f(a)]$$

    for all $a \in W_1$.
13. Let $(W, \leq)$ be a well-ordered set and let $\infty \notin W$. Let $W' = W \cup \{\infty\}$ and order W' by extending the order on W and specifying that $\infty > a$ for all $a \in W$. Show that the order type of W' is greater than that of W.
14. Prove that any natural number is an initial ordinal.
15. Prove that an infinite cardinal number β is a limit ordinal.
16. Prove that if λ is a limit ordinal then $\alpha + \lambda$ is also a limit ordinal.
17. Prove that the union of limit ordinals is a limit ordinal.
18. Prove that any ordinal α has the form $\alpha = \lambda + n$ where λ is a limit ordinal and n is a finite ordinal.

Chapter 3
Lattices

Closure and Inheritance

As we have remarked, a nonempty subset S of a poset P inherits the order relation from P. However, S need not inherit existing meets and joins from P, in particular, a subset $A \subseteq S$ may have a meet or join as a subset of S that is different from its meet or join as a subset of P.

Closure

We will generally avoid the somewhat ambiguous term *closure* and instead use the following terminology.

Definition *Let P be a poset.*
1) P **has finite meets** *if any nonempty finite subset of P has a meet.*
2) P **has arbitrary meets** *if any subset of P has a meet.*
3) P **has arbitrary nonempty meets** *if any nonempty subset of P has a meet.*
Dual definitions can be made with the word join *in place of* meet.□

If every pair of elements of P has a meet, then P has finite meets. A similar statement holds for joins.

Inheritance

As mentioned, if P is a poset, then a subset $A \subseteq S \subseteq P$ may have a meet as a subset of S and a *different* meet as a subset of P. We refer to the meet of A as a subset of S as the S-meet of A and write $\bigwedge_S A$, and similarly for P. Similar statements hold for joins as well.

Definition *A nonempty subset S of a poset P* **inherits** *finite meets from P if whenever the P-meet of a finite nonempty subset $A \subseteq S$ exists, this P-meet is also in S (and is therefore the S-meet of A). Similar definitions can be made for the inheritance of arbitrary and arbitrary nonempty meets, as well as finite, arbitrary and arbitrary nonempty joins.*□

S. Roman (ed.), *Lattices and Ordered Sets*, doi: 10.1007/978-0-387-78901-9_3,

Note that it is possible that a subset A of S has an S-meet but no P-meet. For example, consider the poset $P = \mathbb{N} \cup \{a, b\}$, where $\mathbb{N}$ has the usual order and the incomparable elements a and b are greater than all members of $\mathbb{N}$. Then $\{a, b\}$ has no meet in P but has a meet in $S = \{0, a, b\}$.

Primeness

A concept related to inheritance is *primeness*.

Definition *Let P be a poset and let S be a nonempty subset of P.*

1) *S is* **join-prime** *if*

$$a \vee b \in S \quad \Rightarrow \quad a \in S \quad \textit{or} \quad b \in S$$

2) *Dually, S is* **meet-prime** *if*

$$a \wedge b \in S \quad \Rightarrow \quad a \in S \quad \textit{or} \quad b \in S$$

□

At the element level, we have the following definition.

Definition

1) *Let P be a poset with finite joins. An element $x \in P$ is* **join-prime** *if*

$$x \le a \vee b \quad \Rightarrow \quad x \le a \quad \textit{or} \quad x \le b$$

for all $a, b \in P$.

2) *Dually, let P be a poset with finite meets. An element $x \in P$ is* **meet-prime** *if*

$$a \wedge b \le x \quad \Rightarrow \quad a \le x \quad \textit{or} \quad b \le x$$

for all $a, b \in P$.□

There is a simple connection between the two types of primeness.

Theorem 3.1 *Let P be a poset.*

1) *An element $a \in P$ is join-prime if and only if the principal filter $\uparrow x$ is join-prime.*
2) *Dually, an element $a \in P$ is meet-prime if and only if the principal ideal $\downarrow x$ is meet-prime.*□

Complementary Properties

The connection between inheritance and primeness is also a simple one: In a word, they are *complementary*. Let us say that two properties p and q of subsets of a poset P are **complementary** if

$$S \text{ has property } p \quad \Leftrightarrow \quad P \setminus S \text{ has property } q$$

Theorem 3.2 *Let P be a poset and let $S \subseteq P$. The following properties are complementary:*

1) *Up-set and down-set.*
2) *Inheriting finite meets and being meet-prime.*
3) *Inheriting finite joins and being join-prime.* □

Semilattices

Semilattices (and lattices) can be defined in two ways: one based on the existence of an order relation satisfying certain properties and one based on the existence of binary operations satisfying certain algebraic properties.

The order-based definition of semilattice is the following.

Definition
1) *A poset P that has finite meets is called a* **meet-semilattice**.
2) *A poset P that has finite joins is called a* **join-semilattice**. □

Note that a meet-semilattice cannot have two distinct minimal elements, since their meet contradicts the minimality of the elements. Hence, a meet-semilattice either has no minimal elements or else it has a unique minimal element, that is, a smallest element. In particular, a meet-semilattice with the DCC, such as a finite meet-semilattice, has a smallest element. Similar statements can be made for join-semilattices.

Semilattices Defined Algebraically

Semilattices can also be defined algebraically as follows.

Definition *A* **semilattice** *is a nonempty set P with a binary operation $\circ$ that satisfies the following properties:*
1) **(associativity)**

$$(a \circ b) \circ c = a \circ (b \circ c)$$

2) **(commutativity)**

$$a \circ b = b \circ a$$

3) **(idempotency)**

$$a \circ a = a$$

□

Here is the connection between order semilattices and algebraic semilattices.

Theorem 3.3
1) *A meet-semilattice P is a semilattice under the meet operation.*
2) *A join-semilattice P is a semilattice under the join operation.*
3) *a)* *A semilattice P is a meet-semilattice with order defined by*

$$a \le b \quad \text{if} \quad a \circ b = a$$

In this case, meet is

$$a \wedge b = a \circ b$$

b) *A semilattice P is a join-semilattice with order defined by*

$$a \leq b \quad \textit{if} \quad a \circ b = b$$

In this case, join is

$$a \vee b = a \circ b$$

Proof. Part 1) and part 2) are left to the reader. For part 3a), let P be a meet-semilattice. The idempotency law shows that $a \leq a$. If $a \leq b$ and $b \leq a$, then

$$a = a \circ b = b$$

Finally, if $a \leq b$ and $b \leq c$, then

$$a \circ c = (a \circ b) \circ c = a \circ (b \circ c) = a \circ b = a$$

and so $a \leq c$. Thus, P is a poset. As to the meet, $a \circ b$ is a lower bound for a and b since

$$(a \circ b) \circ a = (b \circ a) \circ a = b \circ (a \circ a) = b \circ a = a \circ b$$

and so $a \circ b \leq a$, and

$$(a \circ b) \circ b = a \circ (b \circ b) = a \circ b$$

and so $a \circ b \leq b$.

Also, if $u \leq a$ and $u \leq b$, then

$$u \circ a = u = u \circ b$$

and so

$$u \circ (a \circ b) = (u \circ a) \circ b = u \circ b = u$$

whence $u \leq a \circ b$. Thus, $a \circ b$ is the greatest lower bound of a and b. The proof of part 3b) is similar.□

Arbitrary Meets Equivalent to Arbitrary Joins

It may seem at first a bit surprising that a poset has arbitrary meets if and only if it has arbitrary joins. However, there is a simple explanation for this: The join of a family must be the meet of all elements that contain the members of the family (and dually).

Theorem 3.4 *A poset P has arbitrary meets if and only if it has arbitrary joins. In particular:*

1) *If P has arbitrary meets, then the join is given by*

$$\bigvee A = \bigwedge A^u$$

for any subset $A \subseteq P$.

2) *Dually, if P has arbitrary joins, then the meet is given by*

$$\bigwedge A = \bigvee A^\ell$$

for any subset $A \subseteq P$.

Proof. Recall that for subsets X and Y, the expression $X \leq Y$ means that $x \leq y$ for all $x \in X$ and $y \in Y$. For part 1), let A be a nonempty subset of P. To see that $\bigwedge A^u$ is the least upper bound of A, we first note that

$$A \leq A^u \quad \Rightarrow \quad A \leq \bigwedge A^u$$

so $\bigwedge A^u$ is an upper bound for A. Second, if $x \geq A$, then $x \in A^u$, and so $x \geq \bigwedge A^u$, whence $\bigwedge A^u$ is the *least* upper bound of A, that is, the join of A. □

The previous theorem does not hold for closure under finite meets (or joins): If P is bounded and has finite meets, it does not necessarily have finite joins. A simple example is $P = \{0, a, b\}$ where $a \parallel b$ and 0 is the smallest element of P. Here is a more interesting example, where P has a largest element.

Example 3.5 Let $P = \{0, x, y, (1,2] \subseteq \mathbb{R}\}$ be the poset with $\leq$ defined as follows:

1) 0 is the smallest element
2) x and y are less than any element of the half-open interval $(1,2]$ but not comparable to each other
3) the order in $(1,2]$ is the usual one for real numbers.

Let S be a finite subset of P. If at least one of x or y is not in S, then the meet of S is the smallest element of S. If $x, y \in S$, then the meet of S is 0. On the other hand, 2 is the largest element of P, but the set $\{x, y\}$ has no join. □

Lattices

We now come to the principal object of study of the rest of this book.

Definition *Let P be a poset.*

1) *P is a* **lattice** *if every pair of elements of P has a meet and a join.*
2) *P is a* **complete lattice** *if P has arbitrary meets and arbitrary joins.* □

Note that a complete lattice is bounded. Also, we have seen that a poset P is complete if and only if it has arbitrary meets *or* arbitrary joins, in which case it has both.

We remind the reader that the adjective *complete* has a different meaning when applied to posets than when applied to lattices. (Recall that a poset P is

complete or a **CPO** if P has a smallest element and if every *directed* subset of P, or equivalently by Theorem 2.19, every chain of P has a join.)

Example 3.6

1) A singleton set $L = \{a\}$ is a lattice under the only possible order on L. This is a **trivial lattice**. Any lattice with more than one element is a **nontrivial lattice**.
2) Any totally ordered set is a lattice, but not necessarily a complete lattice. For example, the set $\mathbb{Z}$ of integers under the natural order is a lattice, but not a complete lattice.
3) The set $\mathbb{N}$ of natural numbers is a lattice under division, where
$$a \wedge b = \gcd(a,b) \quad \text{and} \quad a \vee b = \operatorname{lcm}(a,b)$$
4) If S is a nonempty set, then the power set $\wp(S)$ is a complete lattice under union and intersection.
5) If P is a poset, then the collection $\mathcal{O}(P)$ of down-sets of P is a complete lattice, where meet is intersection and join is union.
6) If G is a group, then the family $\mathcal{S}(G)$ of subgroups of G is a complete lattice under set inclusion, where meet is intersection. Similar statements can be made for other algebraic objects, such as the ideals (or subrings) of a ring, the submodules of a module or the subfields of a field.
7) Let S be a nonempty set. The **finite-cofinite algebra** $FC(S)$ is the set of all subsets of S that are either finite or cofinite. (A subset A is **cofinite** if its complement is finite.) The set $FC(S)$ is a lattice under set inclusion and as we will see in a later chapter, it is also a Boolean algebra.□

If a lattice L does not have a 0, then we can adjoin a 0 to the poset L. The set $L_\perp$ is also a lattice since finite meets and finite joins of elements of L are not affected and since $0 \wedge a = 0$ and $0 \vee a = a$. Of course, the adjoining of a smallest element does create additional arbitrary meets. Similarly, we can adjoin a 1 if L does not have 1.

The Direct Product of Lattices

The cartesian product of any nonempty family of lattices is a lattice under componentwise operations.

Definition *Let $\mathcal{F} = \{L_i \mid i \in I\}$ be a nonempty family of lattices. The* **direct product** $\prod L_i$ *is the lattice consisting of the cartesian product of the family $\mathcal{F}$, with componentwise operations, that is, if $f, g \in \prod L_i$, then*
$$(f \wedge g)(i) = f(i) \wedge g(i) \quad \textit{and} \quad (f \vee g)(i) = f(i) \vee g(i)$$
□

Meet-Structures and Closure Operators

Before discussing the properties of lattices, let us take a look at one important source of these structures.

Definition

1) Let X be a nonempty set. A subset S of the power set $\wp(X)$ is called an **intersection-structure** *on X, abbreviated $\cap$-structure, if S inherits arbitrary intersections from $\wp(X)$, that is, if*

$$\{S_i \mid i \in I\} \subseteq S \quad \Rightarrow \quad \bigcap_{i \in I} \{S_i\} \subseteq S$$

2) More generally, a subset S of a complete lattice L is called a **meet-structure**, *abbreviated $\wedge$-structure, if S inherits arbitrary meets from L.*□

Note that some authors define an $\cap$-structure as a subset that inherits arbitrary *nonempty* intersections. Intersection-structures are also called **closure systems**, but we prefer to avoid this rather nonspecific terminology.

Since a meet-structure has arbitrary meets, it is a complete lattice.

The main reason why meet-structures are important is that they characterize the closed sets of a *closure operator*, and closure operators are quite common in mathematics.

Definition *A* **closure operator** *on a poset P is a unary operator* cl *on P with the following properties:*

1) **(extensive)**

$$p \le \mathrm{cl}(p)$$

2) **(monotone)**

$$p \le q \quad \Rightarrow \quad \mathrm{cl}(p) \le \mathrm{cl}(q)$$

3) **(idempotent)**

$$\mathrm{cl}(\mathrm{cl}(p)) = \mathrm{cl}(p)$$

The element $\mathrm{cl}(p)$ is called the **closure** *of p. An element $p \in P$ is* **closed** *if it is equal to its own closure, that is, if $p = \mathrm{cl}(p)$. We denote the closed elements in P by $\mathrm{Cl}(P)$. A closure operator on a power set $\wp(X)$ is often referred to as a closure operator* **on** *X.*□

We leave it to the reader to check that $\mathrm{cl}(p)$ is the smallest closed element of P containing p and that if $1 \in P$, then 1 is closed.

Now let L be a complete lattice. To see that $\mathrm{Cl}(P)$ is a meet-structure, it is clear that the empty meet (being the largest element) is closed and if

$$\mathcal{F} = \{k_i \mid i \in I\}$$

is a nonempty family in $\mathrm{Cl}(L)$, then

$$\bigwedge_L k_i \leq k_j$$

for all j. Taking closures gives

$$\mathrm{cl}\left(\bigwedge_L k_i\right) \leq \mathrm{cl}(k_j) = k_j$$

and so

$$\bigwedge_L k_j \leq \mathrm{cl}\left(\bigwedge_L k_i\right) \leq \bigwedge_L k_j$$

which shows that $\bigwedge_L k_i$ is closed. Thus, $\mathrm{Cl}(L)$ is a meet-structure.

It follows that the closure of an element $a \in L$ is equal to the meet of all closed elements containing a, that is,

$$\mathrm{cl}(a) = \bigwedge \{k \in \mathrm{Cl}(L) \mid a \leq k\}$$

Thus, the family $\mathrm{Cl}(L)$ of closed sets completely determines the closure operator, that is, the map $\Gamma\colon \mathrm{cl} \mapsto \mathrm{Cl}(L)$ is an injection from closure operators on L to meet-structures in L.

To see that Γ is surjective, we take the hint from the previous display.

Theorem 3.7 *If C is a meet-structure in a complete lattice L, the* ***C*-closure** *of $a \in L$, defined by*

$$\overline{a} = \bigwedge \{c \in C \mid a \leq c\}$$

is a closure operator on L.

Proof. It is easy to see that this map is extensive and monotone. As to idempotence, we have

$$\overline{a} \leq \overline{(\overline{a})} = \bigwedge \{c \in C \mid \overline{a} \leq c\} \leq \overline{a}$$

where the final inequality holds because $\overline{a}$ is a term in the meet.□

Thus, the map $\Gamma\colon \mathrm{cl} \mapsto \mathrm{Cl}(L)$ is a bijection from closure operators on L to meet-structures in L and its inverse is given by the C-closure. To see this, if cl is a closure operator on L and $a \mapsto \overline{a}$ is the $\mathrm{Cl}(L)$-closure operator corresponding to

$\text{Cl}(L)$, then

$$\overline{a} = \bigwedge \{c \in \text{Cl}(L) \mid a \leq c\} = \text{cl}(a)$$

Theorem 3.8 *Let L be a complete lattice.*

1) The maps

$$\Gamma\colon \text{cl} \to \text{Cl}(L) \quad \textit{and} \quad \Pi\colon C \to C\textit{-closure}$$

are inverse bijections between the closure operators on L and the meet-structures on L.

2) If cl is a closure operator on L, then the $\text{Cl}(L)$-join of $A \subseteq \text{Cl}(L)$ is the closure of the L-join of A, that is,

$$\bigvee\nolimits_{\text{Cl}(L)} A = \text{cl}\left(\bigvee\nolimits_L A\right)$$

Proof. For part 2), the right hand side is an upper bound for A in L that happens to be in $\text{Cl}(L)$ and so

$$\bigvee\nolimits_{\text{Cl}(L)} A \leq \text{cl}\left(\bigvee\nolimits_L A\right) \leq \text{cl}\left(\bigvee\nolimits_{\text{Cl}(L)} A\right) = \bigvee\nolimits_{\text{Cl}(L)} A \qquad \square$$

Example 3.9 Let G be a group and let P be the collection of nonempty subsets of G. Set $\text{cl}(S) = \langle S \rangle$, the subgroup generated by S. Then cl is a closure operator on P, with corresponding $\cap$-structure $\mathcal{S}(G)$, the complete lattice of subgroups of G. Put another way, the closed subsets of G are the subgroups of G. Note that if H and K are subgroups of G, then the union $H \cup K$ is not closed in general.□

Example 3.10 Let $n = p_1^{e_1} \cdots p_m^{e_m}$ be the prime decomposition of a positive integer. Let $L = (\mathbb{N}_n, \mid)$ be the lattice of all positive integers less than or equal to n, under division. Fix an integer $1 \leq k \leq m$. For $a \in L$, let $\text{cl}_k(a)$ be the integer formed by replacing the exponent of p_k in a by e_k. Then cl_k is a closure operator on L. In loose terms, cl_k is "max out the exponent of p_k." Thus, the closed integers a are those for which $p_k^{e_k} \mid a$. Note that if a and b are closed, then $a \wedge b = \gcd(a, b)$ and $a \vee b = \text{lcm}(a, b)$ are also closed.□

Example 3.11 Let (X, τ) be a topological space. The topological closure operator $A \mapsto \overline{A}$ is a closure operator in the sense defined above and so the closed subsets of X form an $\cap$-structure, as is well known.□

Closure operators on a power set $\wp(X)$ can be divided into two classes.

Definition *Let X be a nonempty set. A closure operator* cl *on $\wp(X)$ is* **finitary** *if for any $S \subseteq X$,*

$$\mathrm{cl}(S) = \bigcup\{\mathrm{cl}(F) \mid F \subseteq S, F \textit{ finite}\}$$

A closure operator that is not finitary is **infinitary.**□

The closure operators generally encountered in algebra are finitary. For example, the subgroup generated by a nonempty subset S of a group G is the union of the finitely-generated (even cyclic) subgroups generated by elements of S. However, if (X,τ) is a T_1 topological space (points are closed), then any finite set is closed and so the closure operator is finitary if and only if for any $S \subseteq X$,

$$\overline{S} = \bigcup\{F \mid F \subseteq S, F \text{ finite}\} = S$$

Of course, this holds if and only if all sets are closed, that is, if and only if τ is the discrete topology. In summary, for a T_1 space, topological closure is finitary if and only if the topology is discrete.

We will have much more to say about the types of closure operators, and corresponding meet-structures, related to the substructures of an algebraic structure in a later chapter.

Galois Connections

We next describe an important construction that leads to closure operators.

Definition *Let P and Q be posets. A* **Galois connection** *on the pair (P,Q) is a pair (Π,Ω) of maps $\Pi\colon P \to Q$ and $\Omega\colon Q \to P$, where we write $\Pi(p) = p^*$ and $\Omega(q) = q'$, with the following properties:*

1) **(Order-reversing)** *For all $p, q \in P$ and $r, s \in Q$,*

$$p \le q \Rightarrow p^* \ge q^* \quad \textit{and} \quad r \le s \Rightarrow r' \ge s'$$

2) **(Extensive)** *For all $p \in P$, $q \in Q$,*

$$p \le p^{*\prime} \quad \textit{and} \quad q \le q^{\prime *}$$ □

Example 3.12 If P is a poset, then the maps $u, \ell\colon \wp(P) \to \wp(P)$ defined by

$$u(A) = A^u \quad \text{and} \quad \ell(A) = A^\ell$$

for any $A \subseteq P$ form a Galois connection on the power set $\wp(P)$, that is, the pair (u,ℓ) is a Galois connection for $(\wp(P),\wp(P))$.□

Example 3.13 Let X and Y be nonempty sets and let $R \subseteq X \times Y$ be a relation on $X \times Y$. Then the maps

$$S \subseteq X \mapsto S^* = \{y \in Y \mid xRy \text{ for all } x \in S\}$$

and

$$T \subseteq Y \mapsto T' = \{x \in X \mid xRy \text{ for all } y \in T\}$$

form a Galois connection on $(\wp(X), \wp(Y))$.□

Example 3.14 Let F and E be fields with $F \subseteq E$. Let $G = G(E/F)$ be the group of all automorphisms of E that fix each element of F. The group G is called the **Galois group** of the field extension $F \leq E$. The most famous example of a Galois connection is the **Galois correspondence** between the intermediate fields $F \subseteq K \subseteq E$ and the subgroups H of the Galois group G. This correspondence is given by

$$K \mapsto G(E/K) := \{\sigma \in G \mid \sigma x = x \text{ for all } x \in K\}$$

and

$$H \mapsto \text{fix}(H) := \{x \in E \mid \sigma x = x \text{ for all } \sigma \in H\}$$ □

Example 3.15 Let $n > 1$ and let F be a field. Let $F[x_1, \ldots, x_n]$ denote the ring of polynomials in n variables over F and let F^n denote the ring of all n-tuples over F. Let

$$P = \wp(F[x_1, \ldots, x_n]) \quad \text{and} \quad Q = \wp(F^n)$$

Let $\Pi: F[x_1, \ldots, x_n] \to \mathcal{P}(F^n)$ be defined by

$$\begin{aligned} \Pi(S) &= \text{Set of all common roots of the polynomials in } S \\ &= \{x \in F^n \mid p(x) = 0 \text{ for all } p \in S\} \end{aligned}$$

and let $\Omega: \mathcal{P}(F^n) \to F[x_1, \ldots, x_n]$ be defined by

$$\begin{aligned} \Omega(T) &= \text{Set of all polynomials whose root set includes } T \\ &= \{p \in F[x_1, \ldots, x_n] \mid p(t) = 0 \text{ for all } t \in T\} \end{aligned}$$

The pair (Π, Ω) is a Galois connection on (P, Q).□

There are two closure operators associated with a Galois connection.

Theorem 3.16 *Let (Π, Ω) be a Galois connection on (P, Q). Then*

$$p^{*\prime *} = p^* \quad \textit{and} \quad q^{\prime * \prime} = q'$$

for all $p \in P$ and $q \in Q$. It follows that the composite maps

$$p \to p^{*\prime} \quad \text{and} \quad q \to q^{\prime *}$$

are closure operators on P and Q, respectively. Furthermore,

1) *p^* is closed for all $p \in P$.*
2) *q' is closed for all $q \in Q$.*
3) *The maps $\Pi: P \to \text{Cl}(Q)$ and $\Omega: Q \to \text{Cl}(P)$ are surjective and the restricted maps $\Pi: \text{Cl}(P) \to \text{Cl}(Q)$ and $\Omega: \text{Cl}(Q) \to \text{Cl}(P)$ are inverse*

bijections. Thus,

$$\mathrm{Cl}(P) = \{q' \mid q \in Q\} \quad \textit{and} \quad \mathrm{Cl}(Q) = \{p^* \mid p \in P\}$$

Proof. Since $p \leq p^{*\prime}$, the order-reversing property of * gives

$$p^{*\prime *} \leq p^* \leq (p^*)^{\prime *}$$

and so $p^{*\prime *} = p^*$. We leave the rest of the proof to the reader.□

Properties of Lattices

We now resume our discussion of the properties of lattices.

Lattices as Algebraic Structures

Since a lattice L is a meet-semilattice and a join-semilattice, Theorem 3.3 implies that meet and join are associative, commutative and idempotent operations. The link between the two operations is provided by the **absorption laws**

$$a \wedge (a \vee b) = a$$

and

$$a \vee (a \wedge b) = a$$

In fact, these properties characterize lattices.

Theorem 3.17 *The following properties hold in a lattice L:*
L1) **(associativity)**

$$(a \vee b) \vee c = a \vee (b \vee c)$$
$$(a \wedge b) \wedge c = a \wedge (b \wedge c)$$

L2) **(commutativity)**

$$a \vee b = b \vee a$$
$$a \wedge b = b \wedge a$$

L3) **(idempotency)**

$$a \vee a = a$$
$$a \wedge a = a$$

L4) **(absorption)**

$$a \vee (a \wedge b) = a$$
$$a \wedge (a \vee b) = a$$

Conversely, if L is a nonempty set with two binary operations $\wedge$ and $\vee$ satisfying L1)–L4), then L is a lattice where meet is $\wedge$, join is $\vee$ and the order relation is given by

$$a \leq b \quad \textit{if} \quad a \vee b = b$$

or equivalently,

$$a \leq b \quad \textit{if} \quad a \wedge b = a$$

Moreover, since the set of axioms L1)–L4) is self-dual, it follows that if a statement holds in every lattice, then any dual statement holds in every lattice.
Proof. As to the equivalence of the statements $a \vee b = b$ and $a \wedge b = a$, the absorption laws imply that

$$a \wedge b = a \quad \Rightarrow \quad a \vee b = (a \wedge b) \vee b = b$$

and

$$a \vee b = b \quad \Rightarrow \quad a \wedge b = a \wedge (a \vee b) = a$$

Next we show that $\leq$ is a partial order. The idempotency law implies that $a \leq a$. If $a \leq b$ and $b \leq a$, then

$$b = a \vee b = a$$

For transitivity, if $a \leq b$ and $b \leq c$, then

$$a \vee c = a \vee (b \vee c) = (a \vee b) \vee c = b \vee c = c$$

and so $a \leq c$. Finally, Theorem 3.3 implies that $\wedge$ and $\vee$ are the meet and join in L.□

Varieties of Lattices

Theorem 3.17 shows that lattices can be defined *equationally*, that is, by identities. More generally, a **variety** or **equational class** of lattices is a family of lattices defined by a set of lattice identities. Theorem 3.17 says that the family of all lattices is an 8-based variety of lattices, since there are 8 identities in Theorem 3.17.

As another example, *distributive lattices* are those lattices L that satisfy the two distributive identities

$$a \wedge (b \vee c) = (a \wedge b) \vee (a \wedge c)$$
$$a \vee (b \wedge c) = (a \vee b) \wedge (a \vee c)$$

for all $a, b, c \in L$. Thus, the distributive lattices form a 10-based variety of lattices.

There has been work done on identifying minimal sets of identities for varieties of lattices. Padmanabhan [49] has shown that if $\mathcal{V}$ is a variety of lattices defined by a finite number of identities, then $\mathcal{V}$ can also be defined by just two identities. Also, McKenzie [45] has shown that lattice theory itself, that is, the family of all lattices, is actually 1-based. In other words, there is a single identity that characterizes lattices among all equationally-based "algebras" with two binary operations. However, as McKenzie writes "... although the resulting equation to characterize lattices (equation λ below) is so long, containing over three hundred thousand symbols, that we can only represent it in abbreviated form. This leaves open the problem of finding a really elegant equation to characterize lattices." McKenzie has also shown that no other variety of lattices, except the variety at the other extreme, in which *all* equations are identities, is 1-based.

Tops and Unbounded Sequences

If a lattice L has no largest element, then L must be infinite and it seems reasonable that L should have some form of strictly increasing infinite sequence with no upper bound. This is easy to see when L is countably infinite. For if $L = (a_1, a_2, \ldots)$ has no largest element, the sequence of finite joins

$$\mathcal{S} = \langle s_k \rangle = \langle a_1 \vee \cdots \vee a_k \rangle$$

has no upper bound, since $\mathcal{S} \leq u$ implies that $L \leq u$. Removing all duplicates from $\mathcal{S}$ gives a strictly increasing infinite sequence with no upper bound.

For an arbitrary infinite lattice L with no largest element, we can well-order L and then Theorem 2.20 implies that L contains a cofinal transfinite subsequence

$$\mathcal{S} = \langle b_\alpha \mid \alpha < \delta \rangle$$

compatible with the order of L, that is,

$$b_\alpha < b_\beta \quad \Rightarrow \quad b_\alpha \leq b_\beta \quad \text{or} \quad b_\alpha \parallel b_\beta$$

Let us assume that all duplicates have been removed from this subsequence and so

$$b_\alpha < b_\beta \quad \Rightarrow \quad b_\alpha < b_\beta \quad \text{or} \quad b_\alpha \parallel b_\beta$$

To eliminate the problem of parallel elements, let $\mathcal{K}$ be a maximal chain in $\mathcal{S}$ (with respect to the order of L), which must exist by the Hausdorff maximality principle. Then $\mathcal{K}$ also has no upper bound in L. For if $\mathcal{K} \leq u$ for some $u \in L$, then u must be the largest element of L. Otherwise, there is an $a \in L$ for which $u < a$ or $u \parallel a$. In either case, $\mathcal{K} \leq u < u \vee a$ and the cofinality of $\mathcal{S}$ in L implies that there is an $s \in \mathcal{S}$ for which $\mathcal{K} < u \vee a \leq s$. Hence, $\mathcal{K} \cup \{s\}$ is a chain in $\mathcal{S}$, which contradicts the maximality of $\mathcal{K}$. Thus, $\mathcal{K}$ is a strictly increasing transfinite sequence in L with no upper bound.

Theorem 3.18

1) *A countably infinite lattice L has no largest element if and only if L has a nonempty strictly increasing sequence with no upper bound in L.*
2) *An infinite lattice L has no largest element if and only if it has a nonempty strictly increasing transfinite sequence*

$$\mathcal{B} = \langle b_\alpha \mid \alpha < \kappa \rangle$$

for some ordinal κ, with no upper bound in L.$\square$

Of course, the previous theorem has a dual, which we leave to the reader to formulate.

Bounds for Subsets

On a related matter, given a nonempty subset of a lattice L, can we find a strictly increasing sequence of elements of L that has the same set of upper bounds as A? This is trivial if A is a finite set: Just take the one-element sequence consisting of the join $\bigvee A$.

For a countably infinite set A, we can sequentially order A to get $A = (a_1, a_2, \ldots)$, then take the sequence of finite joins as before

$$\mathcal{S} = \langle s_k \rangle = \langle a_1 \vee \cdots \vee a_k \rangle$$

Removing all duplicate entries gives a sequence that has the same upper bounds as A.

If A is an infinite set of arbitrary cardinality $|A| = \Sigma$, then we can well-order A to get a transfinite sequence

$$\mathcal{A} = \langle a_\alpha \mid \alpha < \Sigma \rangle$$

of length Σ. Since Σ is an initial ordinal, if $\epsilon < \Sigma$, then $|\epsilon| < |\Sigma|$. Thus, to mimic the proof for the countably infinite case, we assume that every nonempty subset of A of cardinality less than Σ has a join. Then for any $\epsilon < \Sigma$, the transfinite subsequence

$$\mathcal{A}_\epsilon = \langle a_\alpha \mid \alpha < \epsilon \rangle$$

of $\mathcal{A}$ has a join

$$u_\epsilon = \bigvee \langle a_\alpha \mid \alpha < \epsilon \rangle$$

Moreover, the transfinite sequence of joins

$$\mathcal{B} = \langle u_\epsilon \mid \epsilon < \Sigma \rangle$$

is nondecreasing and $\mathcal{B}^u = A^u$. As before, we can cast out any duplicate entries in $\mathcal{B}$ to get a strictly increasing sequence.

Theorem 3.19 *Let L be a lattice.*
1) If

$$A = \{a_1, a_2, \ldots\}$$

is a countably infinite subset of L, then there is a strictly increasing sequence $\mathcal{S}$ in L for which $\mathcal{S}^u = A^u$.

2) *If A is an infinite subset of L of cardinality $|A| = \Sigma$ and if every nonempty subset of A of cardinality less than Σ has a join in L, then there is a strictly increasing transfinite sequence*

$$\mathcal{S} = \langle s_\alpha \mid \alpha < \beta \rangle$$

in L for which $\mathcal{S}^u = A^u$.□

Join-Irreducible and Meet-Irreducible Elements

The following types of elements play a special role in the representation theory of lattices.

Definition *Let L be a lattice.*

1) *A nonzero element $j \in L$ is* **join-irreducible** *if j is not the join of two smaller elements, that is, if*

$$j = a \vee b \quad \Rightarrow \quad j = a \quad or \quad j = b$$

The set of join-irreducible elements of L is denoted by $\mathcal{J}(L)$.

2) *Dually, a nonunit element $m \in L$ is* **meet-irreducible** *if m is not the meet of two larger elements, that is, if*

$$m = a \wedge b \quad \Rightarrow \quad m = a \quad or \quad m = b$$

The set of meet-irreducible elements of L is denoted by $\mathcal{M}(L)$.□

Example 3.20

1) The atoms of any lattice are join-irreducible.
2) In a chain, all nonzero elements are join-irreducible.
3) In the power set lattice $\wp(X)$, the join-irreducible elements are the singleton subsets.
4) In $(\mathbb{N}, \mid)$ an element is join-irreducible if and only if it is a prime power. However, $(\mathbb{N}, \mid)$ has no meet-irreducible elements, since for example $n = 2n \wedge 3n$.
5) In the lattice of all open subsets of the plane (under set inclusion), no elements are join-irreducible.
6) If P is a finite poset, the join-irreducible elements of $\mathcal{O}(P)$ are precisely the principal ideals in $\mathcal{O}(P)$. It is clear that $\downarrow p$ is join-irreducible, since if $\downarrow p = D_1 \cup D_2$, then $p \in D_i$ for some i and so $\downarrow p = D_i$. Also, if $D \in \mathcal{O}(P)$ is join-irreducible, then since

$$D = \bigcup_{d \in D} \{\downarrow d\}$$

is a finite join, it follows that $D = \downarrow d$ for some $d \in D$ and so D is principal.□

Completeness

Let us take a closer look at completeness.

Chain Conditions and Completeness

Chain conditions are a form of finiteness condition. One aspect of that finiteness is that the chain conditions imply that an arbitrary join (or meet) is equal to a finite join (or meet), as described in the following theorem.

Theorem 3.21 *Let L be a lattice.*

1) *If L has the ACC, then L has arbitrary nonempty joins. In fact, for every nonempty subset $A \subseteq L$ there is a finite subset $A_0 \subseteq A$ such that*

$$\bigvee A = \bigvee A_0$$

Thus, a lattice with 0 *and with ACC is complete.*

2) *Dually, if L has the DCC then L has arbitrary nonempty meets. In fact, for every nonempty subset $A \subseteq L$ there is a finite subset $A_0 \subseteq A$ such that*

$$\bigwedge A = \bigwedge A_0$$

Thus, a lattice with 1 *and with DCC is complete.*

Proof. For part 1), suppose that L has the ACC and let A be a nonempty subset of L. Consider the family $\mathcal{F}$ of all joins of finite subsets of A. Since $\mathcal{F}$ is nonempty, it has a maximal element, say

$$m = \bigvee A_0$$

Now if $\bigvee A_0$ is not an upper bound for A, there is an $a \in A$ for which $\bigvee A_0 < a$ or $\bigvee A_0 \parallel a$. In either case,

$$\bigvee (A_0 \cup \{a\}) > \bigvee A_0$$

which contradicts the maximality of $\bigvee A_0$. Hence, $\bigvee A_0$ is an upper bound for A. Since it is clearly a least upper bound for A, we have $\bigvee A = \bigvee A_0$.□

Note that the converse does not hold: A complete lattice need not have the ACC. Indeed, the bounded lattice $\mathbb{Z}^*$ consisting of the integers under the usual order with a smallest and a largest element adjoined is complete but has neither chain condition.

Incompleteness and Coalescing Pairs of Sequences

It is not hard to find conditions that characterize the incompleteness of an infinite lattice L. Again, we consider the countable case first, since this does not involve the use of ordinals or transfinite sequences.

The Countable Case

Let L be a countably infinite lattice. Of course, if L is missing either a largest element or a smallest element, then L is not complete. We have seen that the lack of a largest element is equivalent to the existence of a strictly increasing infinite sequence in L with no upper bound. A similar statement holds for smallest elements.

Intuitively speaking, the presence of a strictly increasing sequence with no upper bound indicates that L is missing a "limit point" for the sequence. But a lattice can be missing other limit points as well. This happens when there is a strictly increasing sequence and a strictly decreasing sequence that are headed toward each other on a collision course and, while they never actually meet, they leave no space in between. This happens, for example, in the lattice of all nonzero rational numbers (under the usual order) for the sequences $(-1/n)$ and $(1/n)$.

For convenience, we make the following nonstandard definition.

Definition *Let L be a lattice. A strictly increasing (possibly empty) sequence*

$$B = (b_1 < b_2 < \cdots)$$

and a strictly decreasing (possibly empty) sequence

$$C = (\cdots < c_2 < c_1)$$

coalesce *if the following hold:*

1) $B < C$*, that is,* $b_i < c_j$ *for all* i, j*.*

2) There is no $a \in L$ *for which* $B \leq a \leq C$*, or equivalently,*

$$B^u \cap C^\ell = \emptyset$$

In this case, we say that (B, C) *is a* **coalescing pair** *in L.*□

Note that if (B, C) is a coalescing pair, then B cannot have a join and C cannot have a meet. In particular, B and C are either infinite or else empty. Moreover, B is empty if and only if C is a strictly decreasing sequence with no lower bound, that is, if and only if L has no smallest element. Dually, C is empty if and only if L has no largest element.

Now we can characterize incompleteness in a countably infinite lattice.

Theorem 3.22 *A countably infinite lattice L is incomplete if and only if it has a coalescing pair* (B, C) *of sequences.*

Proof. If L is bounded but incomplete, then there is an infinite subset

$$S = \{s_1, s_2, \ldots\}$$

of L that has no join in L. Since L has a largest element, the set S^u is nonempty and is, in fact, inherits meets and joins from L and so is a lattice. Moreover, S^u has no smallest element, since such an element would be the join of S in L. Now, we may apply Theorem 3.19 to S to get a strictly increasing sequence

$$B = (b_1 < b_2 < \cdots)$$

for which $B^u = S^u$. Then we may apply the dual of Theorem 3.18 to B^u to get a strictly decreasing infinite sequence

$$C = (c_1 > c_2 > \cdots)$$

with no lower bound in B^u. Thus, $B^u \cap C^\ell = \emptyset$. It is also clear that $B < C$ and so the pair (B, C) is a coalescing pair.□

The General Case

The general case is entirely analogous.

Definition *Let L be a lattice. A strictly increasing (possibly empty) transfinite sequence*

$$\mathcal{B} = \langle b_\alpha \mid \alpha < \kappa \rangle$$

and a strictly decreasing (possibly empty) transfinite sequence

$$\mathcal{C} = \langle c_\alpha \mid \alpha < \delta \rangle$$

coalesce *if the following hold:*

1) $\mathcal{B} < \mathcal{C}$, that is, $b_\alpha < c_\beta$ for all $\alpha < \kappa$ and $\beta < \delta$.

2) There is no $a \in L$ for which $\mathcal{B} \le a \le \mathcal{C}$, or equivalently,

$$\mathcal{B}^u \cap \mathcal{C}^\ell = \emptyset$$

We refer to the pair $(\mathcal{B}, \mathcal{C})$ as a **coalescing pair** *in L.*□

As in the countable case, if $(\mathcal{B}, \mathcal{C})$ is a coalescing pair, then $\mathcal{B}$ cannot have a join and $\mathcal{C}$ cannot have a meet. Hence, $\mathcal{B}$ and $\mathcal{C}$ are either infinite or else empty. Moreover, $\mathcal{B}$ is empty if and only if L has no smallest element and $\mathcal{C}$ is empty if and only if L has no largest element.

Theorem 3.23 (Davis, 1955 [13]) *A lattice L is incomplete if and only if L has a coalescing pair $(\mathcal{B}, \mathcal{C})$.*

Proof. If L is bounded but not complete, then there is a nonempty subset of L that has no join. Pick such a subset of least cardinality, say

$$S = \{s_i \mid i \in I\}$$

Thus, any subset of S of smaller cardinality has a join and Theorem 3.19 implies that there is a strictly increasing transfinite sequence

$$\mathcal{B} = \langle b_\alpha \mid \alpha < \delta \rangle$$

in S that has no join in L.

Since L is bounded, the set $\mathcal{B}^u$ is nonempty and inherits meets and joins from L and so is a lattice. Moreover, since $\mathcal{B}$ has no join, $\mathcal{B}^u$ has no smallest element. Hence, Theorem 3.18 implies that $\mathcal{B}^u$ has a strictly decreasing transfinite sequence

$$\mathcal{C} = \langle c_\alpha \mid \alpha < \kappa \rangle$$

with no lower bound in $\mathcal{B}^u$ and so $\mathcal{B}^u \cap \mathcal{C}^\ell = \emptyset$. It is clear that $\mathcal{B} < \mathcal{C}$ and so $(\mathcal{B}, \mathcal{C})$ is a coalescing pair.□

Chain-Completeness

A lattice L is **chain-complete** if every chain in L, including the empty chain, has a join. The previous characterization of incompleteness makes it easy to show that a lattice is complete if and only if it is chain-complete.

Theorem 3.24 *A lattice is complete if and only if it is chain-complete.*
Proof. Assume that L is chain-complete but not complete. Since L is chain-complete, it has a smallest element. Hence, Theorem 3.23 implies that L has a coalescing pair $(\mathcal{B}, \mathcal{C})$, where $\mathcal{B}$ is nonempty. However, since $\mathcal{B}$ is a chain in L, it has a join, which is not possible. Hence, L must be complete.□

Sublattices

A nonempty subset M of a lattice L inherits the order of L, but not necessarily the meets and joins. When it does, it is called a sublattice of L.

Definition *Let L be a lattice and let $M \subseteq L$ be a nonempty subset of L.*
1) *M is a* **sublattice** *of L if M inherits finite meets and finite joins from L.*
2) *If L is complete, then M is a* **complete sublattice** *of L if M inherits arbitrary meets and arbitrary joins from L.*□

Example 3.25 Let $L = \{1, 2, 3, 6, 12\}$ under division (see Figure 3.1).

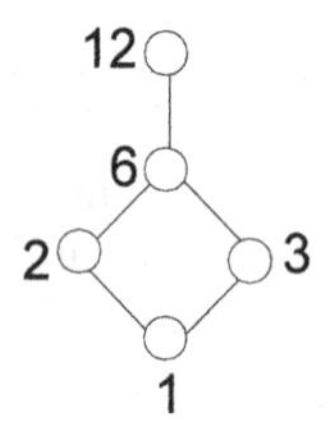

Figure 3.1

The subset $S = \{1, 2, 3, 12\}$ is a lattice under division but not a sublattice of L. The subset $T = \{1, 2, 3, 6\}$ is a sublattice of L.□

Example 3.26 If L is a lattice and $a < b$, then the closed interval $[a, b]$ is a sublattice of L.□

Definition *Let S be a nonempty set. A sublattice of a power set $\wp(S)$ is called a* **power set sublattice**, *or a* **ring of sets.**□

Thus, in a power set sublattice, meet is intersection and join is union. For example, the down-sets $\mathcal{O}(P)$ of a poset form a power set sublattice. On the other hand, the family $\mathcal{S}(V)$ of subspaces of a vector space V is a lattice, where meet is intersection but join is in general not union. Hence, $\mathcal{S}(V)$ is not a power set sublattice.

If L is a lattice, then the intersection of any family of sublattices of L is either empty or it is a sublattice. Thus, for any nonempty subset $A \subseteq L$, there is a smallest sublattice of L containing A: It is just the intersection of all sublattices containing A.

Definition *If L is a lattice and $A \subseteq L$ is nonempty, then the* **sublattice generated** *by A, denoted by $[A]$, is the smallest sublattice of L containing A.*□

Lattice Terms

We can also describe the sublattice generated by a nonempty subset in more constructive terms using lattice polynomials and evaluation maps. Intuitively speaking, $[A]$ is simply the set of all "lattice legal" expressions that can be formed using the elements of A and the meet and join operators (and parentheses). But we wish to be a bit more formal.

Definition *Let X be a nonempty set. A* **lattice term** (*or just* **term**), *also called a* **lattice polynomial**, *over X is any expression defined as follows. Let $\sqcup$ and $\sqcap$ be formal symbols.*

1) *The elements of X are terms of* **weight** 1.
2) *If p and q are terms, then so are*

$$(p \sqcap q) \quad \text{and} \quad (p \sqcup q)$$

whose weights are the sum of the weights of p and q.

3) *An expression in the symbols $X \cup \{\sqcap, \sqcup, (,)\}$ is a term if it can be formed by a finite number of applications of 1) and 2).*

We denote the set of all lattice terms over X by $\mathcal{T}_X$. The elements of X are called **variables.**□

Two terms are equal if and only if they are identical. Thus, for example $x_1 \sqcup x_2$ is not equal to $x_2 \sqcup x_1$.

It is customary to omit the final pair of parentheses when writing lattice terms. A lattice term involving some or all of the variables $x_{i_1}, \ldots, x_{i_n} \in X$ but no others is denoted by $p(x_{i_1}, \ldots, x_{i_n})$.

We now define two binary operations $\wedge$ and $\vee$, called meet and join respectively, on $\mathcal{T}_X$. For $p, q \in \mathcal{T}_X$, let

$$p \wedge q = p \sqcap q \quad \text{and} \quad p \vee q = p \sqcup q$$

Then since $x_i \sqcap x_j = x_i \wedge x_j$ for all $x_i, x_j \in X$ and similarly for join, we can write any lattice term $p \in \mathcal{T}_X$ using only the variables and the meet and join symbols (and parentheses).

Note that we do *not* claim that $\mathcal{T}_X$ is a lattice under meet and join. Indeed, $\mathcal{T}_X$ consists of formal expressions and so, for example, $x_1 \vee x_2 \neq x_2 \vee x_1$.

Now, given a lattice L and a function $f: X \to A$, where $A \subseteq L$, the **evaluation map** $\epsilon_f: \mathcal{T}_X \to L$ is defined, for each $p(x_{i_1}, \ldots, x_{i_n}) \in \mathcal{T}_X$ by

$$\epsilon_f p(x_{i_1}, \ldots, x_{i_n}) = p(f x_{i_1}, \ldots, f x_{i_n})$$

Note that for lattice terms p and q,

$$\epsilon_f(p \wedge q) = \epsilon_f p \wedge \epsilon_f q \quad \text{and} \quad \epsilon_f(p \vee q) = \epsilon_f p \vee \epsilon_f q$$

We can now describe the sublattice $[A]$ as the set of all elements of L that are images of all possible lattice terms under all possible evaluation maps into A.

Theorem 3.27 *Let L be a lattice and let $A \subseteq L$ be nonempty. Let $X = \{x_1, x_2, \ldots\}$. Then*

$$[A] = \{\epsilon_f p(x_1, \ldots, x_n) \mid p \in \mathcal{T}_X, f: X \to A\}$$

Proof. Let S be the set on the right above. It is clear that any element of S is in $[A]$ and since S is a sublattice of L, it follows that $S = [A]$.□

Denseness

Definition *Let D be a nonempty subset of a poset P.*

1) *D is* **join-dense** *in P if for every $p \in P$, there is a subset $S \subseteq D$ for which*

$$p = \bigvee_P S$$

2) *D is* **finite-join-dense** *in P if for every $p \in P$, there is a finite nonempty subset $S \subseteq D$ for which*

$$p = \bigvee_P S$$

Meet-dense *and* **finite-meet-dense** *are defined dually.*□

Note that if D is join-dense in P, then any $p \in P$ is the join of *all* elements of D that are contained in p, that is,

$$p = \bigvee D_{\downarrow}(p)$$

Example 3.28 It is clear that the singleton sets are join-dense in the power set $\wp(X)$. However, if X is infinite, then the singleton sets are not finite-join-dense.□

In a lattice with the DCC, the join-irreducible elements are finite-join-dense.

Theorem 3.29 *Let L be a lattice.*

1) *If L has the DCC, then $\mathcal{J}(L)$ is finite-join-dense in L.*
2) *Dually, if L has the ACC, then $\mathcal{M}(L)$ is finite-meet-dense in L.*

Proof. If some element of L is not the join of a finite number of join-irreducible elements, then let a be a minimal such element. Thus, a is not join-irreducible and so it has the form $a = b \vee c$ where $b < a$ and $c < a$. The minimality of a implies that b and c are joins of a finite number of join-irreducible elements and therefore so is a, which is a contradiction.□

Lattice Homomorphisms

A monotone map $f: L \to M$ between lattices need not, in general, preserve meets and joins. Even an order *embedding* need not preserve meets and joins.

Example 3.30 Consider the lattices of integers in Figure 3.2, where the order is division.

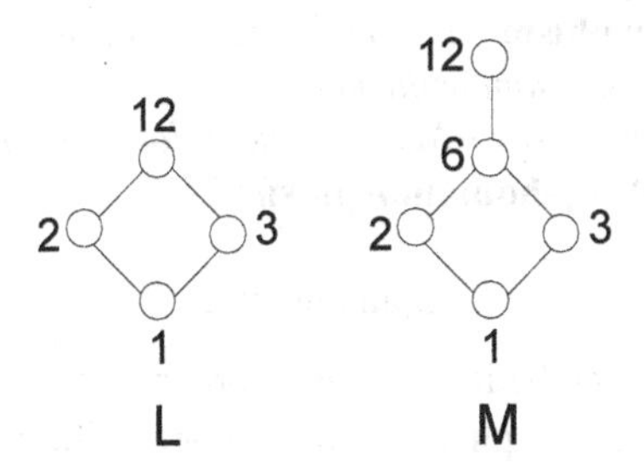

Figure 3.2

The inclusion map $f: L \to M$ defined by $f(n) = n$ is an order embedding, but

$$f(2 \vee 3) = f(12) = 12$$

and

$$f(2) \vee f(3) = 2 \vee 3 = 6$$

Hence, f does not preserve joins.□

A monotone map $f: L \to M$ does send a lower bound for a subset $S \subseteq L$ to a lower bound of $f(S)$ and so

$$f\left(\bigwedge S\right) \leq \bigwedge f(S)$$

and similarly

$$f\left(\bigvee S\right) \geq \bigvee f(S)$$

assuming that the relevant meets and joins exist. In fact, each of these conditions is equivalent to f being monotone. For example, if $a \leq b$, then $a \wedge b = a$ and so

$$f(a) = f(a \wedge b) \leq f(a) \wedge f(b) \leq f(b)$$

On the other hand, it is easy to see that an order *isomorphism* does preserve meets and joins.

Definition *Let L and M be lattices. A function $f: L \to M$ that preserves finite meets and joins, that is, for which*

$$f(a \wedge b) = f(a) \wedge f(b)$$
$$f(a \vee b) = f(a) \vee f(b)$$

is called a **lattice homomorphism**.

1) *A* **lattice monomorphism** *or* **lattice embedding** *is an injective lattice homomorphism.*
2) *A* **lattice epimorphism** *is a surjective lattice homomorphism.*
3) *A* **lattice endomorphism** *of L is a lattice homomorphism from L to itself.*
4) *A* **lattice isomorphism** *is a bijective lattice homomorphism, or equivalently, an order isomorphism.*

A lattice homomorphism $f: L \to M$ of bounded lattices for which $f(0) = 0$ and $f(1) = 1$ is called a **$\{0, 1\}$-homomorphism**. □

To reiterate, we have the following (and no more):

lattice embedding	$\Rightarrow$	order embedding
lattice isomorphism	$\Leftrightarrow$	order isomorphism

A key property of lattice embeddings not shared by order embeddings is the following.

Theorem 3.31 *Let $f: L \to M$ be a lattice embedding.*

1) *$f(L)$ is a sublattice of M.*
2) *If L is complete, then so is $f(L)$, but $f(L)$ need not be a complete sublattice of M.*

Proof. For part 2), the map $f: L \to f(L)$ is a lattice isomorphism. We leave it to the reader to show that $f(L)$ is also complete. On the other hand, let

$L = \mathbb{N} \cup \{\infty\}$ have the usual order on $\mathbb{N}$ and let $n < \infty$ for all $n \in \mathbb{N}$. Let $M = \mathbb{N} \cup \{a, \infty\}$ with the usual order on $\mathbb{N}$ and $n < a < \infty$ for all $n \in \mathbb{N}$. Let $f: L \to M$ be the inclusion map, which is a lattice embedding. However,

$$f\left(\bigvee_L \mathbb{N}\right) = f(\infty) = \infty$$

but

$$\bigvee_M f(\mathbb{N}) = \bigvee_M \mathbb{N} = a \neq \infty$$

□

Meet and Join Homomorphisms

Definition *Let L and M be lattices.*

1) *A function $f: L \to M$ that preserves finite meets, that is,*

$$f(a \wedge b) = f(a) \wedge f(b)$$

is called a **meet-homomorphism***.*

2) *Dually, a function $f: L \to M$ that preserves finite joins, that is,*

$$f(a \vee b) = f(a) \vee f(b)$$

is called a **join-homomorphism***.* □

A meet-homomorphism is order preserving, since

$$a \leq b \quad \Leftrightarrow \quad a \wedge b = a \quad \Rightarrow \quad f(a) \wedge f(b) = f(a) \quad \Leftrightarrow \quad f(a) \leq f(b)$$

Dually, a join-homomorphism is order preserving as well.

The B-Down Map

The B-down map $\phi_{B,\downarrow}: L \to \wp(B)$, defined by

$$\phi_{B,\downarrow}(a) = B_\downarrow(a) := \{b \in B \mid b \leq a\}$$

is a meet-homomorphism, since

$$B_\downarrow(a \wedge b) = B_\downarrow(a) \cap B_\downarrow(b)$$

but it is not a join-homomorphism, since in general

$$B_\downarrow(a \vee b) \neq B_\downarrow(a) \cup B_\downarrow(b)$$

Of course, the B-down map is order preserving, that is,

$$x \leq y \quad \Rightarrow \quad B_\downarrow(x) \subseteq B_\downarrow(y)$$

but it need not be an order embedding.

Theorem 3.32 *Let B be a nonempty subset of a lattice L and let $\phi_{B,\downarrow}: L \to \wp(B)$ be the B-down map. The following are equivalent:*

1) $\phi_{B,\downarrow}$ *is an order embedding, that is, for all* $x, y \in L$,

$$x \leq y \quad \Leftrightarrow \quad B_\downarrow(x) \subseteq B_\downarrow(y)$$

2) $\phi_{B,\downarrow}$ *satisfies the following for all* $x, y \in L$,

$$x < y \quad \Leftrightarrow \quad B_\downarrow(x) \subset B_\downarrow(y)$$

3) $\phi_{B,\downarrow}$ *is injective, equivalently, for all* $x, y \in L$,

$$x = y \quad \Leftrightarrow \quad B_\downarrow(x) = B_\downarrow(y)$$

4) $\phi_{B,\downarrow}$ *preserves strict order, that is, for all* $x, y \in L$,

$$x < y \quad \Rightarrow \quad B_\downarrow(x) \subset B_\downarrow(y)$$

5) B *is join-dense in* L.
If these conditions hold, then B *is said to be a* **base** *for* L.
Proof. We first show that 1) and 5) are equivalent. If 5) holds and $B_\downarrow(x) \subseteq B_\downarrow(y)$, then

$$x = \bigvee B_\downarrow(x) \leq \bigvee B_\downarrow(y) = y$$

and since the B-down map is always order-preserving, it follows that 1) holds. Conversely, if 1) holds, then $B_\downarrow(x) \leq x$ and if $B_\downarrow(x) \leq u$, then $B_\downarrow(x) \subseteq B_\downarrow(u)$ and so 1) implies that $x \leq u$. Thus, $x = \bigvee B_\downarrow(x)$.

To complete the proof, we show that 2) $\Rightarrow$ 4) $\Rightarrow$ 1) $\Rightarrow$ 3) $\Rightarrow$ 2). Note first that

$$B_\downarrow(x) \subseteq B_\downarrow(y) \quad \Leftrightarrow \quad B_\downarrow(x) = B_\downarrow(x \wedge y) \tag{3.33}$$

It is clear that 2) implies 4). If 4) holds and $B_\downarrow(x) \subseteq B_\downarrow(y)$, then (3.33) implies that $B_\downarrow(x) = B_\downarrow(x \wedge y)$. Hence, 4) shows that $x \wedge y < x$ cannot hold and so $x \wedge y = x$, that is, $x \leq y$. Thus, 4) implies 1). It is easy to see that 1) implies 3). Finally, (3.33) shows that 3) implies 1) and together they imply 2).□

One often sees the definition of a base B given in the form

$$x \not\leq y \quad \Leftrightarrow \quad \exists b \in B \text{ with } b \leq x \text{ and } b \not\leq y$$

for all $x, y \in L$. Of course, this is equivalent to

$$x \leq y \quad \Leftrightarrow \quad B_\downarrow(x) \subseteq B_\downarrow(y)$$

Ideals and Filters

The following are important substructures in lattices.

Definition *Let* L *be a lattice.*
1) *An* **ideal** *of* L *is a nonempty down-set in* L *that inherits finite joins, that is,*
 a) $a \in I,\ x \leq a \Rightarrow x \in I$

b) $a, b \in I \Rightarrow a \vee b \in I$
In this case, we write $I \trianglelefteq L$. *A* **proper ideal**, *that is, an ideal* $I \neq L$, *is denoted by* $I \triangleleft L$. *The set of all ideals of L is denoted by* $\mathcal{I}(L)$.

2) *Dually, a* **filter** *in L is a nonempty up-set in L that inherits finite meets, that is,*
a) $a \in F, x \geq a \Rightarrow x \in F$
b) $a, b \in F \Rightarrow a \wedge b \in F$
In this case, we write $F \trianglerighteq L$. *A* **proper filter**, *that is, a filter* $F \neq L$, *is denoted by* $F \triangleright L$. *The set of all filters of L is denoted by* $\mathcal{F}(L)$.□

Note that the properties of being an ideal and being a filter are *not* complementary.

Example 3.34 If L is a lattice and $a \in L$, then the principal ideal

$$\downarrow a = \{x \in L \mid x \leq a\}$$

is an ideal of L and the principal filter $\uparrow a$ is a filter.□

The intersection $I \cap J$ of two ideals I and J of a lattice L is not empty, since if $i \in I$ and $j \in J$, then $i \wedge j \in I \cap J$. Hence, $I \cap J$ is an ideal of L. On the other hand, in the lattice $\mathbb{Z}$ of all integers under the usual order, the intersection of the family $\{\downarrow n \mid n \in \mathbb{Z}\}$ of ideals is empty. This leaves us with two alternatives. We can deal only with lattices with 0 for which $\mathcal{I}(L)$ is an $\cap$-structure or we can define

$$\mathcal{I}_{\emptyset}(L) = \mathcal{I}(L) \cup \{\emptyset\}$$

which is an intersection-structure for any lattice L.

Theorem 3.35 *Let L be a lattice.*

1) *The set* $\mathcal{I}(L)$ *is a lattice, where meet is intersection and join is given by*

$$I \vee J = \{x \in L \mid x \leq i \vee j \textit{ for some } i \in I \textit{ and } j \in J\}$$

2) *The set* $\mathcal{I}_{\emptyset}(L)$ *is an* $\cap$*-structure and therefore a complete lattice, where the join of a family* $\mathcal{F}$ *of ideals is given by*

$$\bigvee \mathcal{F} = \{x \in L \mid x \leq a_1 \vee \cdots \vee a_n, \textit{ where } a_i \in \bigcup \mathcal{F} \textit{ and } n \geq 1\}$$

3) *The down map* $\phi_{\downarrow}: L \to \mathcal{I}_{\emptyset}(L)$ *is a lattice embedding, that is, for all* $a, b \in L$,

$$\downarrow a \cap \downarrow b = \downarrow(a \wedge b) \quad \textit{and} \quad \downarrow a \vee \downarrow b = \downarrow(a \vee b)$$

The image of $\phi_{\downarrow}$ *is the sublattice* $\mathcal{P}(L)$ *of* $\mathcal{I}_{\emptyset}(L)$ *consisting of all principal ideals of L.*

4) *The union of any* directed *family of ideals in L is an ideal in L.*

5) *If L has the ACC, then all ideals are principal. Dually, if L has the DCC then all filters are principal.*

Proof. It is clear that the set

$$I = \{x \in L \mid x \le a_1 \vee \cdots \vee a_n, \text{ where } a_i \in \bigcup \mathcal{F}\}$$

is an ideal containing I_k and so $\bigvee \mathcal{F} \subseteq I$. Also, since $a_1 \vee \cdots \vee a_n \in \bigvee \mathcal{F}$, it follows that $I \subseteq \bigvee \mathcal{F}$ and so $I = \bigvee \mathcal{F}$.

For part 3), if it is easy to see that

$$\downarrow a \cap \downarrow b = \downarrow (a \wedge b)$$

and

$$\begin{aligned} \downarrow a \vee \downarrow b &= \{x \in L \mid x \le i \vee j \text{ for } i \in \downarrow a \text{ and } j \in \downarrow b\} \\ &= \{x \in L \mid x \le a \vee b\} \\ &= \downarrow (a \vee b) \end{aligned}$$

For part 4), let $\mathcal{C}$ be a directed family of ideals in $\mathcal{I}(L)$ and let $U = \bigcup \mathcal{C}$. If $a \in U$ and $b \le a$, then $a \in C \in \mathcal{C}$ and so $b \in C \subseteq U$. Also, if $a, b \in U$, then a and b both lie in one of the ideals $C \in \mathcal{C}$ and so $a \vee b \in C \subseteq U$.

For part 5), Theorem 3.21 implies that if I is an ideal, then there is a finite subset I_0 of I for which $\bigvee I_0 = \bigvee I$. Hence, $I = \downarrow (\bigvee I_0)$ is principal.□

The Ideal Generated by a Subset

Every nonempty subset A of a lattice L is contained in a smallest ideal, namely, the intersection of all ideals of L containing A.

Definition *Let A be a nonempty subset of a lattice L.*

1) *The* **ideal generated** *by A, denoted by $(A]$, is the smallest ideal of L containing A.*
2) *The* **filter generated** *by A, denoted by $[A)$, is the smallest filter of L containing A.*□

We leave proof of the following to the reader.

Theorem 3.36 *Let L be a lattice and let $A \subseteq L$ be a nonempty subset of L.*

1)

$$(A] = \downarrow [A] = \downarrow \mathcal{T}_A$$

2)

$$(A] = \bigvee \{\downarrow a \mid a \in A\} = \{x \in L \mid x \le a_1 \vee \cdots \vee a_n \text{ for } a_i \in A\} \quad \square$$

Prime and Maximal Ideals

The following special types of ideals play a key role in lattice theory.

Definition *Let* L *be a lattice.*

1) *A proper ideal* I *of* L *is* **maximal** *if for any ideal* I,

$$J \subseteq I \subseteq L \quad \Rightarrow \quad I = J \quad \text{or} \quad I = L$$

A proper filter F *of* L *is* **maximal** *if for any filter* X,

$$F \subseteq X \subseteq L \quad \Rightarrow \quad X = F \quad \text{or} \quad X = L$$

A maximal filter is also called an **ultrafilter**.

2) *A proper ideal* J *is* **prime** *if*

$$a \wedge b \in J \quad \Rightarrow \quad a \in J \quad \text{or} \quad b \in J$$

The set of prime ideals of L *is called the* **spectrum** *of* L *and is denoted by* $\operatorname{Spec}(L)$. *Dually, a proper filter* F *is* **prime** *if*

$$a \vee b \in F \quad \Rightarrow \quad a \in F \quad \text{or} \quad b \in F$$

□

Although the properties of being an ideal and being a filter are *not* complementary, Theorem 3.2 does imply the following.

Theorem 3.37 *The properties of being a prime ideal and being a prime filter are complementary, that is,* I *is a prime ideal if and only if* I^c *is a prime filter.*□

We leave proof of the following to the reader.

Theorem 3.38 *Let* $f: L \to M$ *be a lattice homomorphism.*

1) *If* $I \trianglelefteq M$, *then* $f^{-1}(I) \trianglelefteq L$.
2) *If* $P \in \operatorname{Spec}(M)$, *then* $f^{-1}(P) \in \operatorname{Spec}(L)$.□

In ring theory, maximal ideals are prime. This is not true in general for lattices.

Example 3.39 In the lattice M_3 of Figure 3.3,

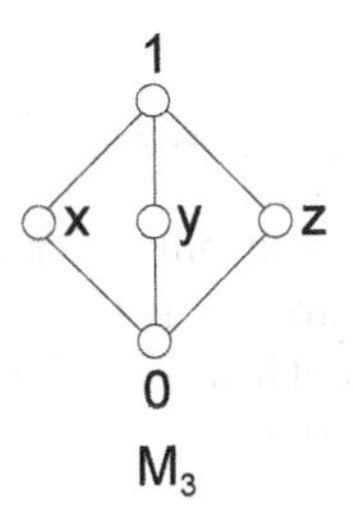

Figure 3.3

the ideal $I = \{0, x\}$ is maximal. However, $y \wedge z \in I$ but $y \notin I$ and $z \notin I$ and so I is not prime.□

Ideal Kernels and Homomorphisms

It is usual in discussions of homomorphisms of the common algebraic structures, such as groups, rings and modules, to discuss the kernel of a homomorphism. We will postpone a careful discussion of kernels of lattice homomorphisms to a later chapter, but make a few remarks here in connection with prime ideals.

If a lattice M has a smallest element, then the **ideal kernel** of a lattice homomorphism $f: L \to M$ is the set

$$\ker(f) = f^{-1}(0) = \{a \in L \mid f(a) = 0\}$$

This is easily seen to be an ideal of L. There is another type of kernel, related to congruence relations, that will be discussed in a later chapter. This accounts for the adjective *ideal* in the term *ideal kernel.*

On the other hand, not all ideals are ideal kernels.

Example 3.40 Consider again the lattice M_3 in Figure 3.3, of which we will have much to say throughout the book. The subset $I = \{0, z\}$ is an ideal of M_3. Suppose that $f: M_3 \to M$ is a lattice homomorphism for which $f(0) = f(z) = 0$. Then

$$f(1) = f(y \vee z) = f(y) \vee f(z) = f(y)$$

Hence,

$$0 = f(0) = f(x \wedge y) = f(x) \wedge f(y) = f(x) \wedge f(1) = f(x \wedge 1) = f(x)$$

and so $x \in \ker(f)$. This shows that I is not the kernel of any lattice homomorphism.□

We will discuss in more detail the question of when an ideal of a lattice is a kernel of some homomorphism in a later chapter. Nevertheless, we can address the issue now for prime ideals. If L is a lattice and $S \subseteq L$, then the **indicator function** of S is the function

$$f_S: L \to \{0, 1\}$$

defined by setting $f_S(x) = 0$ if and only if $x \in S$. Here we consider the set $\{0, 1\}$ as a two-element lattice with $0 < 1$. In general, an indicator function need not be a lattice homomorphism. However, the indicator function of a *prime ideal* P is a lattice epimorphism, since

$$\begin{aligned} f_P(a \wedge b) = 0 &\Leftrightarrow a \wedge b \in P \\ &\Leftrightarrow a \in P \text{ or } b \in P \\ &\Leftrightarrow f_P(a) = 0 \text{ or } f_P(b) = 0 \\ &\Leftrightarrow f_P(a) \wedge f_P(b) = 0 \end{aligned}$$

and since P^c is a prime filter,

$$\begin{aligned} f_P(a \vee b) = 1 &\Leftrightarrow a \vee b \in P^c \\ &\Leftrightarrow a \in P^c \text{ or } b \in P^c \\ &\Leftrightarrow f_P(a) = 1 \text{ or } f_P(b) = 1 \\ &\Leftrightarrow f_P(a) \vee f_P(b) = 1 \end{aligned}$$

Of course, the ideal kernel of f_P is P. Conversely, suppose that the indicator function $f_S: L \to \{0,1\}$ is an epimorphism. Then

$$S = \ker(f_S) = f^{-1}\{0\}$$

is a prime ideal in L, since $\{0\}$ is a prime ideal in $\{0,1\}$. Thus, prime ideals can be characterized as the ideal kernels of indicator epimorphisms.

Theorem 3.41 *In a lattice L, the prime ideals in L are precisely the ideal kernels of the indicator epimorphisms.* □

Lattice Representations

A lattice homomorphism $f: L \to M$ is often referred to as a **representation map**, or simply as the **representation** of L in M. The most useful representations are the embeddings, since in this case, L is isomorphic to a sublattice of M. Embeddings are also called **faithful** representations.

The advantages of a faithful representation appear when M provides a more familiar setting than L. For example, intersections and unions of sets are more familiar than general meets and joins. Thus, it is often desirable to represent the elements of a lattice L by subsets of a set X, with meet and join represented by intersection and union, *if possible*. When representing elements of a lattice by subsets of a set X, the set X is called the **carrier set** for the representation.

As we will see, various versions of the B-down map $\phi_{B,\downarrow}: a \mapsto \downarrow a$, obtained by varying the set B and the range of the map, are valuable tools for finding lattice representations. For instance, Theorem 3.35 implies the following.

Theorem 3.42 (Representation theorem for lattices) *For any lattice L, the down map*

$$\phi_\downarrow: L \hookrightarrow \mathcal{I}_\emptyset(L)$$

is a lattice embedding. If L is complete, then $\phi_\downarrow(L)$ is an intersection-structure and so every complete lattice is isomorphic to an intersection-structure.
Proof. The second statement follows from Theorem 3.31.□

Note that in the range $\mathcal{I}_\emptyset(L)$ of this version of the down map, meet is intersection but join is not in general union. We can change the range of the map so that meet is intersection and join is union by replacing $\mathcal{I}_\emptyset(L)$ with $\mathcal{O}(L)$:

$$\phi_\downarrow : L \to \mathcal{O}(L)$$

However, this map is a meet-homomorphism but not, in general, a join-homomorphism. Indeed, an arbitrary lattice cannot be *faithfully* represented as a power set sublattice, since power set sublattices are distributive (defined below). On the other hand, we will see in a later chapter that an arbitrary *distributive* lattice can be faithfully represented as a power set sublattice, via yet another variation of the down map.

Special Types of Lattices

There are many special types of lattices. We introduce the most important types here and go into more details in later chapters.

Atomic and Atomistic Lattices

Recall that an **atom** in a lattice L with 0 is a element $a \in L$ that covers 0.

Definition
1) *A lattice L is* **atomic** *if every nonzero element of L contains an atom.*
2) *A lattice L is* **atomistic** *if every nonzero element of L is the join of atoms.*□

It is clear that atomistic lattices are atomic. However, a finite chain of length at least 3 is atomic but not atomistic.

For the record, we note some of the more common variations on the concept of atomicity:

Definition
1) *A lattice L is* **strongly atomic** *if every nontrivial interval $[a,b]$ in L is atomic, that is, if for any $a < b$ in L, there exist $u \in L$ for which*

$$a \sqsubset u \leq b$$

2) *A lattice L is* **weakly atomic** *if for any $a < b$ in L, there exist $u, v \in L$ for which*

$$a \leq u \sqsubset v \leq b$$

□

Distributive Lattices

It is not hard to show that if a lattice L satisfies either one of the following **distributive laws**:

$$a \wedge (b \vee c) = (a \wedge b) \vee (a \wedge c)$$
$$a \vee (b \wedge c) = (a \vee b) \wedge (a \vee c)$$

for all elements in L, then it satisfies the other distributive law.

We also note that for each distributive law, one side is greater than or equal to the other in *all* lattices. For instance, we always have

$$a \wedge (b \vee c) \geq (a \wedge b) \vee (a \wedge c)$$

since each term on the left is greater than or equal to each term on the right.

Definition *A lattice L for which the distributive laws hold is called a* **distributive lattice**. □

It is well known that any ring of sets is a distributive lattice, since

$$A \cap (B \cup C) = (A \cap B) \cup (A \cap C)$$

If we adjoin a new largest element or smallest element to a distributive lattice L, the resulting lattice is still distributive. Also, we leave it as an exercise to show that if L is distributive, then so is the lattice $\mathcal{I}(L)$.

As we will see, the distributive property is quite a strong property, making distributive lattices much better behaved than their nondistributive cousins. One example is that in a distributive lattice, all maximal ideals are prime. In fact, in some sense, distributive lattices are closer in behavior to Boolean algebras than to nondistributive lattices. In particular, in 1933, Garrett Birkhoff [6] published his famous representation theorem for distributive lattices, which says that a lattice is distributive if and only if it is isomorphic to a power set sublattice (ring of sets). Thus, for distributive lattices, meet is essentially intersection and join is essentially union.

Modular Lattices

The modular property is a special case of the distributive property.

Definition *A lattice L is* **modular** *if it satisfies the* **modular law**: *For all $a, b, c \in L$,*

$$a \geq c \quad \Rightarrow \quad a \wedge (b \vee c) = (a \wedge b) \vee c \qquad \square$$

Note that any distributive lattice is modular. Also, we leave it as an exercise to show that if L is modular, then so is the lattice $\mathcal{I}(L)$.

One reason for the importance of modular lattices is that some very important lattices are modular but not distributive. For example, the lattice of normal subgroups $\mathcal{N}(G)$ of a group G is modular. (This will be easy to prove in the next chapter.) On the other hand, it is not hard to see that the lattice of (normal) subgroups of the Klein four-group is isomorphic to M_3, which is not distributive. We also note here that the lattice $\mathcal{S}(G)$ of all subgroups of a group need not be modular; the alternating group A_4 is an example.

Note that the modular law is equivalent to the identity

$$a \wedge (b \vee (a \wedge c)) = (a \wedge b) \vee (a \wedge c)$$

for all $a, b, c \in L$. This identity is obtained from the consequent of the modular law by replacing c by $a \wedge d$, which represents an arbitrary element contained in a (and then replacing d by c). Thus, modularity, like distributivity defined below, is a variety of lattices.

Complemented Lattices

Lattice elements may have complements.

Definition *Let L be a bounded lattice. A* **complement** *of $a \in L$ is an element $b \in L$ for which*

$$a \wedge b = 0 \quad \textit{and} \quad a \vee b = 1$$

In this case, we say that a and b are **complementary**.□

Complements need not exist and they need not be unique when they do exist. For instance, in a bounded chain, no elements other than 0 and 1 have complements. Also, in both of the lattices in Figure 3.4, the element y has two complements.

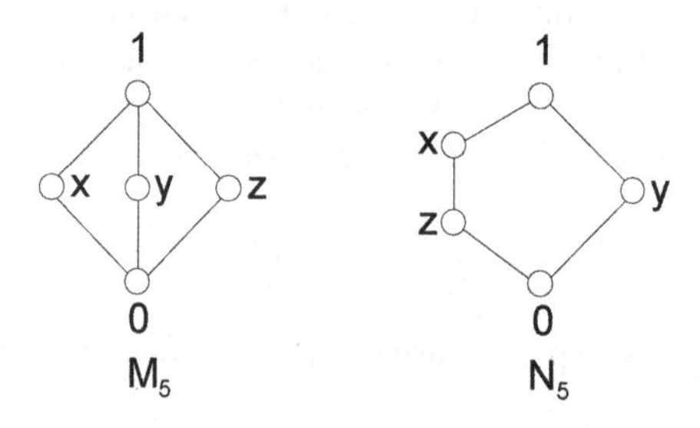

Figure 3.4

However, we will see that if a lattice is distributive, then complements, when they exist, are unique.

Definition

1) *A bounded lattice* L *for which each element has a complement is called a* **complemented lattice**.
2) *A bounded lattice* L *for which each element has a unique complement is called a* **uniquely complemented lattice**. □

It is also easy to see that in a complemented lattice any prime ideal is maximal. In particular, in a complemented distributive lattice, otherwise known as a *Boolean algebra*, the prime ideals are the same as the maximal ideals.

Here are a few simple properties of complements that hold in uniquely complemented lattices that do not hold in arbitrary complemented lattices.

Theorem 3.43 *Let* L *be a uniquely complemented lattice.*
1) *If* $a \in L$ *then* $a'' = a$.
2) *If* a *is an atom in* L, *then* a' *is a coatom.*
3) *If* a *and* b *are distinct atoms in* L, *then* $a \leq b'$.

These statements do not hold in arbitrary complemented lattices.

Proof. The lattices M_3 and N_5 show that these statements do not hold in arbitrary complemented lattices (where a' is interpreted as any complement of a). Part 1) follows from the definition of complement. For part 2), if $a' = 1$, then $a = a'' = 1' = 0$, which is false. Also, if $a' \leq x < 1$, then $a \vee x \geq a \vee a' = 1$ and so $a \vee x = 1$. Also, since a is an atom, $a \wedge x = 0$ or $a \wedge x = a$. But $a \wedge x = a$ implies that $a \leq x$, which gives

$$1 = a \vee a' \leq x$$

which is false. Hence, $a \wedge x = 0$ and so $x = a'$. Thus, 1 covers a', that is, a' is a coatom.

For part 3), since b' is a coatom, $a \vee b' = b'$ or $a \vee b' = 1$. In the former case, $a \leq b'$, as desired. In the case $a \vee b' = 1$, we cannot have $a \wedge b' = 0$, since then $a' = b'$, which implies that $a = b$. Thus, $a \wedge b' = a$, that is, $a \leq b'$.□

Relatively Complemented Lattices

Relative complements are a generalization of complements.

Definition *Let* $u < v$ *be elements of a lattice* L *and let* $a \in [u, v]$. *A* **relative complement** *of* a *with respect to* $[u, v]$ *is a complement of* a *in the sublattice* $[u, v]$, *that is, an element* $x \in [u, v]$ *for which*

$$x \wedge a = u \quad \text{and} \quad x \vee a = v$$

The set of all relative complements of a *with respect to* $[u, v]$ *is denoted by* $a_{[u,v]}$. *If the relative complement is unique, we will simply refer to it as* $a_{[u,v]}$. *Use of the notation* $a_{[u,v]}$ *tacitly implies that* $a \in [u, v]$.□

Theorem 3.44 *Let* L *be a lattice. The following are equivalent:*

1) Relative complements, when they exist, are unique.
2) An element of L is uniquely determined by its meet and join with any other element, that is, for all $x, a, b \in L$

$$\begin{Bmatrix} x \wedge a = x \wedge b \\ x \vee a = x \vee b \end{Bmatrix} \quad \Rightarrow \quad a = b \qquad \square$$

Definition
1) *A lattice L is* **relatively complemented** *if every closed interval $[u, v]$ in L is complemented.*
2) *A lattice L is* **uniquely relatively complemented** *if every closed interval $[u, v]$ in L is uniquely complemented.*$\square$

A complemented lattice need not be relatively complemented. For instance, in the lattice N_5 of Figure 3.4, the element z has no relative complement with respect to the interval $[0, x]$. On the other hand, this cannot happen in a modular lattice.

Theorem 3.45 *Let L be a bounded modular lattice. If $x \in L$ has a complement x', then x also has a relative complement x_r with respect to any interval $[a, b]$ containing x. In fact, x_r is formed by projecting x' into the interval $[a, b]$ as follows:*

$$x_r = (x' \vee a) \wedge b = (x' \wedge b) \vee a$$

Proof. Since $b \geq a$, the modular law shows that the two expressions above are equal. Also,

$$x \wedge [(x' \wedge b) \vee a] = [x \wedge (x' \wedge b)] \vee a = a$$

and

$$x \vee [(x' \wedge b) \vee a] = x \vee (x' \wedge b) = (x \vee x') \wedge b = b$$

and so $(x' \wedge b) \vee a \in x_{[a,b]}$.$\square$

Sectionally Complemented Lattices

In some cases, all we need in order to prove a result is that the lattice in question has relative complements with respect to intervals of the form $[0, v]$.

Definition *A lattice L is* **sectionally complemented** *if it has a smallest element and if every interval of the form $[0, v]$ in L is complemented.*$\square$

Recall that any atom in a lattice is join-irreducible. In sectionally complemented lattices, the converse is also true. Also, a sectionally complemented lattice is atomistic if and only if it is atomic.

Theorem 3.46 *Let L be a sectionally complemented lattice.*

1) An element $a \in L$ is join-irreducible if and only if it is an atom, that is,

$$\mathcal{J}(L) = \mathcal{A}(L)$$

2) L is atomistic if and only if it is atomic.

Proof. For part 1), we have $\mathcal{A}(L) \subseteq \mathcal{J}(L)$ in any lattice. If a is not an atom, then there is a $b \in L$ for which $0 < b < a$. If $c \in b_{[0,a]}$, then $c \vee b = a$ and since $c \wedge b = 0$, it follows that $c \neq a$. Hence, a is not join-irreducible. Thus, $\mathcal{J}(L) \subseteq \mathcal{A}(L)$.

For part 2), assume that L is atomic. For $x \neq 0$, it is clear that $\mathcal{A}_{\downarrow}(x) \leq x$. If $\mathcal{A}_{\downarrow}(x) \leq u$, then

$$\mathcal{A}_{\downarrow}(x) \leq m := x \wedge u \leq x$$

and $m > 0$. But if $0 < m < x$, then m has a relative complement m' in $[0, x]$ and $0 < m' < x$. Since L is atomic, there is an atom $a < m' < x$ and so $a \in \mathcal{A}_{\downarrow}(x)$. Hence, $a \leq m$ and so $a \leq m \wedge m' = 0$, which is false. Thus, $m = x$, which is equivalent to $x \leq u$. This shows that $x = \bigvee \mathcal{A}_{\downarrow}(x)$ and so L is atomistic.□

The Dedekind–MacNeille Completion

Any poset P can be order embedded in a complete lattice L. When this happens, it is customary to say that L is a **completion** of P. For example, the down map $\phi_{\downarrow}: P \to \mathcal{O}(P)$ is an order embedding of P into $\mathcal{O}(P)$. However, although the image $\phi_{\downarrow}(P)$, which is the family of principal ideals, is join-dense in $\mathcal{O}(P)$, it is not generally meet-dense.

On the other hand, by restricting the range $\mathcal{O}(P)$ of the down map, we can get an order embedding of P into a complete lattice in such a way that the image is both join-dense and meet-dense in the range. To this end, note that for any $a \in P$,

$$\downarrow a = a^{u\ell}$$

where $u, \ell: \wp(P) \to \wp(P)$ are the maps defined by

$$u(A) = A^u \quad \text{and} \quad \ell(A) = A^\ell$$

for any $A \subseteq P$. For readability, we also write $\{a\}^u$ as a^u and similarly for ℓ. Thus, the image of the down map belongs to the family of all subsets of P of the form $A^{u\ell}$ for $A \subseteq P$.

Now, it is easy to see that the pair (u, ℓ) is a Galois connection on $\wp(P)$, that is, for the pair $(\wp(P), \wp(P))$. Therefore, the composition $u\ell$ is a closure operator on P and so the family of subsets of the form $A^{u\ell}$ is the family of closed subsets of this closure operator and is therefore an $\cap$-structure.

Theorem 3.47 *Let P be a poset. The maps $u, \ell: \wp(P) \to \wp(P)$ defined by*

$$u(A) = A^u \quad \text{and} \quad \ell(A) = A^\ell$$

for any $A \subseteq P$ form a Galois connection on $\wp(P)$, that is, the following hold:
1) For any $A, B \subseteq P$,

$$A \subseteq A^{\ell u} \quad \text{and} \quad B \subseteq B^{u\ell}$$

2) For any $A, B \subseteq P$,

$$A \subseteq B \quad \Rightarrow \quad B^\ell \subseteq A^\ell \quad \text{and} \quad B^u \subseteq A^u$$

Hence:
3) For all $A \subseteq P$,

$$A^{\ell u \ell} = A^\ell \quad \text{and} \quad A^{u \ell u} = A^u$$

4) The compositions ℓu and $u\ell$ are closure operators on $\wp(P)$ and so the families of closed sets

$$C_{u\ell} := \{S \subseteq P \mid S = S^{u\ell}\} \quad \text{and} \quad C_{\ell u} := \{S \subseteq P \mid S = S^{\ell u}\}$$

are $\cap$-structures and hence complete lattices.□

The complete lattice of closed sets under the closure operator $u\ell$ has a name.

Definition *Let P be a poset. The $\cap$-structure consisting of the closed sets of the operator $u\ell$:*

$$\mathrm{DM}(P) = \{A \subseteq P \mid A^{u\ell} = A\} = \{A^{u\ell} \mid A \subseteq P\}$$

is called the **Dedekind–MacNeille (DM) completion**, *the* **MacNeille completion**, *the* **completion by cuts** *or the* **normal completion** *of P.*□

The next theorem justifies the term completion for $\mathrm{DM}(P)$.

Theorem 3.48 *Let P be a poset and let $\mathrm{DM}(P)$ be its Dedekind–MacNeille completion. Let*

$$\phi_\downarrow : P \to \mathrm{DM}(P)$$

be the down map with range $\mathrm{DM}(P)$, that is,

$$\phi_\downarrow(a) = \downarrow a = a^{u\ell}$$

for all $a \in P$.
1) The image $\phi_\downarrow(P)$ is both join-dense and meet-dense in $\mathrm{DM}(P)$. In particular, for any $A \subseteq P$,

$$A^{u\ell} = \bigcap\{\phi_\downarrow(x) \mid x \in A^u\}$$

and

$$A^{u\ell} = \bigvee_{\mathrm{DM}(P)}\{\phi_\downarrow(a) \mid a \in A\}$$

2) $\phi_\downarrow$ *preserves existing meets and joins in* P*, that is, if* $A \subseteq P$ *has a meet, then*

$$\phi_\downarrow\left(\bigwedge\nolimits_P A\right) = \bigcap\{\phi_\downarrow(a) \mid a \in A\}$$

and if A *has a join, then*

$$\phi_\downarrow\left(\bigvee\nolimits_P A\right) = A^{u\ell} = \bigvee_{\mathrm{DM}(P)}\{\phi_\downarrow(a) \mid a \in A\}$$

In particular:

a) If P *is a lattice, then* $\phi_\downarrow\colon P \hookrightarrow \mathrm{DM}(P)$ *is a lattice embedding.*

b) If P *is a complete lattice, then* $\phi_\downarrow\colon P \approx \mathrm{DM}(P)$ *is an isomorphism.*

Proof. For the first equation in part 1), we have

$$\begin{aligned} \alpha \in A^{u\ell} &\Leftrightarrow \alpha \le A^u \\ &\Leftrightarrow \alpha \le x \text{ for all } x \in A^u \\ &\Leftrightarrow \alpha \in x^{u\ell} \text{ for all } x \in A^u \\ &\Leftrightarrow \alpha \in \bigcap\{x^{u\ell} \mid x \in A^u\} \end{aligned}$$

For the second equation, since $a \in A$ implies that $a^{u\ell} \subseteq A^{u\ell}$, it follows that $A^{u\ell}$ is an upper bound for the family $\mathcal{F} := \{a^{u\ell} \mid a \in A\}$ in $\mathrm{DM}(P)$. But if B is an upper bound for $\mathcal{F}$ in $\mathrm{DM}(P)$, then $A \subseteq B$ and so $A^{u\ell} \subseteq B^{u\ell} = B$. Hence, $A^{u\ell}$ is the least upper bound of $\mathcal{F}$.

The first statement in 2) is clear and the second statement follows from part 1), using the fact that

$$\phi_\downarrow\left(\bigvee\nolimits_P A\right) = \downarrow\left(\bigvee\nolimits_P A\right) = A^{u\ell}$$

Moreover, if P is complete, then for any $A \in \mathrm{DM}(P)$, we have

$$A = A^{u\ell} = \phi_\downarrow\left(\bigvee\nolimits_P A\right)$$

and so $\phi_\downarrow$ is surjective.□

Theorem 3.48 says that P is order isomorphic to a join-dense and meet-dense subset $\phi_\downarrow(P)$ of $\mathrm{DM}(P)$. Moreover, if L is a lattice, then $\phi_\downarrow(L)$ is a sublattice of $\mathrm{DM}(P)$ and if L is complete, then $\phi_\downarrow(L) = \mathrm{DM}(L)$. This is what we expect from a true completion of P.

The Dedekind–MacNeille completion $\mathrm{DM}(P)$ is characterized, up to isomorphism, by the fact that it is a complete lattice L in which P is both join-dense and meet-dense.

Theorem 3.49 *Let L be a complete lattice. If a subset $P \subseteq L$ is both join-dense and meet-dense in L, then $L \approx \mathrm{DM}(P)$, via the P-down map $\phi_{P,\downarrow}: L \approx \mathrm{DM}(P)$ defined by*

$$\phi_{P,\downarrow}(a) = P_{\downarrow}(a)$$

Proof. We must first show that $P_{\downarrow}(a) \in \mathrm{DM}(P)$, that is, that

$$P_{\downarrow}(a)^{u_P \ell_P} = P_{\downarrow}(a)$$

However, since P is join-dense in L, we have

$$a = \bigvee P_{\downarrow}(a)$$

and so

$$x \in P_{\downarrow}(a)^{u_P} \quad \Leftrightarrow \quad x \in P \text{ and } x \geq a \quad \Leftrightarrow \quad x \in P_{\uparrow}(a)$$

that is,

$$P_{\downarrow}(a)^{u_P} = P_{\uparrow}(a)$$

Similarly, the meet-denseness of P in L implies that

$$\begin{aligned} P_{\downarrow}(a)^{u_P \ell_P} &= P_{\uparrow}(a)^{\ell_P} \\ &= P_{\uparrow}(a)^{\ell} \cap P \\ &= \left[\bigwedge P_{\uparrow}(a)\right]^{\ell} \cap P \\ &= P_{\downarrow}(a) \end{aligned}$$

Thus, $\phi_{P,\downarrow}$ does map L into $\mathrm{DM}(P)$. Moreover, $\phi_{P,\downarrow}$ is an order embedding, since

$$a_1 \leq a_2 \quad \Leftrightarrow \quad P_{\downarrow}(a_1) \subseteq P_{\downarrow}(a_2) \quad \Leftrightarrow \quad \phi_{P,\downarrow}(a_1) \subseteq \phi_{P,\downarrow}(a_2)$$

For surjectivity, the elements $A \in \mathrm{DM}(P)$ satisfy $A = A^{u_P \ell_P}$, where $A \subseteq P$. But if $a = \bigvee A$, then

$$A = A^{u_P \ell_P} = a^{u_P \ell_P} = P_{\uparrow}(a)^{\ell_P} = P_{\downarrow}(a) = \phi_{P,\downarrow}(a)$$

and so $\phi_{P,\downarrow}$ is surjective.□

Example 3.50 Consider the rational numbers $\mathbb{Q}$ with the usual order, as a subset of the real numbers $\mathbb{R}$. Adjoin a largest element ∞ and a smallest element $-\infty$ to $\mathbb{R}$ and let $\mathbb{R}' = \{-\infty, \infty\} \cup \mathbb{R}$. Then $\mathbb{R}'$ is a complete lattice.

Moreover, for any $r \in \mathbb{R}$, we have

$$r = \bigvee_{\mathbb{R}} \mathbb{Q}_{\downarrow}(r) \quad \text{and} \quad r = \bigwedge_{\mathbb{R}} \mathbb{Q}_{\uparrow}(r)$$

and

$$\infty = \bigvee_{\mathbb{R}} \mathbb{Q} \quad \text{and} \quad -\infty = \bigwedge_{\mathbb{R}} \mathbb{Q}$$

and so $\mathbb{Q}$ is both join-dense and meet-dense in $\mathbb{R}'$. Hence, $\mathbb{R}'$ is isomorphic to the Dedekind–MacNeille completion of $\mathbb{Q}$.□

Let us remark that if L is a distributive lattice, it does not necessarily follow that its Dedekind–MacNeille completion is distributive. Funayama [21] (see also page 238 of Balbes [1]) has found a rather involved example of a distributive lattice L for which the completion $\mathrm{DM}(L)$ has a sublattice isomorphic to the nondistributive lattice N_5 and is therefore not distributive.

Exercises

1. Prove that if L is a distributive lattice, then any maximal ideal is prime.
2. Prove that if L is a complemented lattice then any prime ideal of L is maximal.
3. Let $f\colon L \to M$ be a lattice embedding. Show that $f(L)$ is a sublattice of M.
4. What are the largest and smallest elements of the lattice $(\mathbb{N}, \mid)$?
5. If $(P, \leq)$ is a meet-semilattice then $(P, \circ)$ is a semilattice where
$$a \circ b = a \wedge b$$
What does the corresponding join-semilattice look like?
6. Prove that the set $\mathbb{Z}^*$ of integers, under the usual order and with a smallest element $-\infty$ and a largest element ∞ adjoined, is complete but has neither chain condition.
7. Show that the following are equivalent for a lattice homomorphism $f\colon L \to M$ between uniquely complemented lattices:
   a) f preserves complements, that is, $f(a') = (fa)'$ for all $a \in L$.
   b) f is a $\{0,1\}$-homomorphism.
8. What are the join-irreducible elements of $(\mathbb{Z}^+, \leq)$? What are the join-irreducible elements of $(\mathbb{Z}^+, \mid)$?
9. Let $A = \{a_1, a_2, \ldots\}$ and $B = \{b_1, b_2, \ldots\}$ be disjoint countably infinite sets. Define a relation R on the union $A \cup B$ by
$$a_{i+1} R a_i, \quad b_{i+1} R b_i \quad \text{and} \quad b_i R a_i$$
for $i = 1, 2, \ldots$. Show that R is a covering relation. Draw a picture of the corresponding poset L. Show that L is a lattice. Find the join-irreducible elements. Find an element of L that is not the join of join-irreducibles.
10. Find a lattice L that has no join-irreducible elements, other than the one in the text.

11. Let L be a lattice and let $a \in L$. Let X be a generating set for L. For any subset S of L, let $P(S)$ be the property that for any finite subset $S_0 \subseteq S$,

$$a \leq \bigvee S_0 \quad \Rightarrow \quad a < s \text{ for some } s \in S_0$$

Show that if $P(X)$ holds then $P(L)$ holds. *Hint*: Show that if $P(F)$ holds for some $X \subseteq F$, then for any $u, v \in F$, it follows that $P(F \cup \{u \vee v\})$ and $P(F \cup \{u \wedge v\})$ hold.

12. Let P be a finite poset. Theorem 1.11 says that for each finite poset P, the down map $\phi_{P,\downarrow}: P \to \mathcal{O}(P)$ defined by $\phi_{P,\downarrow}(a) = \downarrow a$ is an order embedding whose image is the set of principal ideals in $\mathcal{O}(P)$:

$$\operatorname{im}(\phi_{P,\downarrow}) = \{\downarrow a \mid a \in P\}$$

Prove that the principal ideals are precisely the join-irreducibles in $\mathcal{O}(P)$ and so the map

$$\phi_{P,\downarrow}: P \approx \mathcal{J}(\mathcal{O}(P))$$

is an order isomorphism.

13. Let (Π, Ω) be a Galois connection on (P, Q). If P and Q are lattices, show that de Morgan's laws hold in $\mathrm{Cl}(P)$ and $\mathrm{Cl}(Q)$, that is, for $p, q \in \mathrm{Cl}(P)$ and $r, s \in \mathrm{Cl}(Q)$,

$$(p \wedge q)^* = p^* \vee q^*, \quad (p \vee q)^* = p^* \wedge q^*$$

and

$$(r \wedge s)' = r' \vee s', \quad (r \vee s)' = r' \wedge s'$$

14. A lattice L is **atomistic** if every nonzero element of L is the join of atoms. What is the relationship, if any, between atomic and atomistic?
15. Let L be a lattice. Show that if $a \in L$ then the *join map* $j_a: L \to L$ defined by $j_a x = x \vee a$ is monotone. Is it an order embedding?
16. Prove that if we adjoin a 0 (or 1) to a distributive lattice L that does not have a smallest element (or largest element), the resulting lattice is still distributive. Prove similar statement concerning modular lattices.
17. Let P be a poset and let $\{a_{i,j} \mid i \in I, j \in J\}$ be a doubly-indexed subset of P. Show that if for all $u \in I$ and $v \in J$, the joins

$$\bigvee_j a_{u,j}, \quad \bigvee_i a_{i,v} \quad \text{and} \quad \bigvee_i \Big(\bigvee_j a_{i,j}\Big)$$

exist then $\bigvee_{i,j} a_{i,j}$ and $\bigvee_j (\bigvee_i a_{i,j})$ exist and

$$\bigvee_{i,j} a_{i,j} = \bigvee_i \Big(\bigvee_j a_{i,j}\Big) = \bigvee_j \Big(\bigvee_i a_{i,j}\Big)$$

18. Prove that the down map $\phi_\downarrow: L \to \mathcal{O}(L)$ is a lattice homomorphism.
19. Let L be a lattice. Prove that $I \subseteq L$ is an ideal of L if and only if I is a sublattice and that $i \in I$, $a \in L$ imply $a \wedge i \in I$.

20. Prove that a lattice L has the property that every ideal is prime if and only if L is a chain.
21. Let L be a distributive lattice. Prove that if $a \in L$ has a complement a' then for any prime ideal I, exactly one of a or a' is in I.
22. Let L be a lattice.
    a) Show that if L is modular, then so is $\mathcal{I}(L)$.
    b) Show that if L is distributive, then so is $\mathcal{I}(L)$.
23. If $f: L \to M$ is a lattice homomorphism, show that $f^{-1}(I) \trianglelefteq L$ for every $I \trianglelefteq M$.
24. Prove that if $\sigma: P \to Q$ is an order isomorphism, its induced map $\sigma: \mathcal{O}(P) \to \mathcal{O}(Q)$ is a lattice isomorphism.
25. Let I be an ideal of a lattice L and let $a \notin I$. Describe the smallest ideal containing I and a.
26. Let L be a lattice and let $A, B \subseteq L$ be nonempty. Prove that $\downarrow[A] \cap \uparrow[B] = \emptyset$ if and only if for all nonempty finite subsets $A_0 \subseteq A$ and $B_0 \subseteq B$, we have $\bigvee A_0 \not\geq \bigwedge B_0$.
27. Let V be a vector space and let $\sigma: V \to V$ be a linear operator on V. Show that the family of σ-invariant subspaces of V is an $\cap$-structure.
28. Describe the ideals of $(\mathbb{Z}^+, \mid)$.
29. Prove that a nonempty subset S of a lattice L is a convex sublattice of L if and only if $S = I \cap F$, where I is an ideal and F is a filter.
30. Let L be a lattice.
    a) Let S be a nonempty subset of L. Show that S is contained in a directed subset D of L with the same join as S, that is, $\bigvee S$ exists if and only if $\bigvee D$ exists, in which case the two are equal.
    b) Use part a) to prove that if L is chain-complete then L is complete.
31. Let L be a lattice. Prove that the following are equivalent.
    a) Relative complements, when they exist, are unique.
    b) An element of L is uniquely determined by its meet and join with any another element. That is, for all $a, b, c \in L$

$$\left.\begin{array}{l} a \wedge b = a \wedge c \\ a \vee b = a \vee c \end{array}\right\} \quad \Rightarrow \quad b = c$$

32. Let P be a poset and let $\mathcal{O}(P)$ be the poset of down-sets in P, under set inclusion.
    a) Show that any principal ideal $\downarrow a$ is join-irreducible in $\mathcal{O}(P)$.
    b) If P is finite, show that $\mathcal{J}(\mathcal{O}(P))$ is the set of principal ideals.
33. Let L be a relatively complemented lattice. Prove the following:
    a) $\alpha \in a_{[\alpha,\beta]} \Leftrightarrow a = \beta$
    b) $\beta \in a_{[\alpha,\beta]} \Leftrightarrow a = \alpha$
    c) $a \in a_{[\alpha,\beta]} \Leftrightarrow a = \alpha = \beta$
34. Let L be a lattice. Prove that the congruences classes of a congruence relation on L are convex sublattices of L.
35. A lattice L with 0 is **disjunctive** if $a \not\leq b$ implies that there is an $x \in L$ such that

$$x \wedge a > 0 \quad \text{and} \quad x \wedge b = 0$$

Prove that a lattice L with 0 has a base consisting of atoms of L if and only if L is atomic and disjunctive.

36. Prove directly (without using the concept of a base) that if L is a lattice with the DCC then $\mathcal{J}(L)$ is join-dense in L.

The Dedekind–MacNeille Completion

37. Let X be a nonempty set. Find the smallest family $\mathcal{F}$ of subsets of X for which $\mathrm{DM}(\mathcal{F}) \approx \wp(X)$.
38. Find the Dedekind–MacNeille completion of an antichain P.
39. Let X be an infinite set and let L be the finite-cofinite algebra of X. Find the Dedekind–MacNeille completion of L.
40. Describe conditions on P that characterize the statement $\emptyset \in \mathrm{DM}(P)$.
41. Prove that a poset P is a complete lattice if and only if every subset A of P for which $A = A^{u\ell}$ is a principal ideal.
42. Let L be a lattice. Prove that an ideal $I \trianglelefteq L$ is DM-closed if and only if it is the intersection of principal ideals.
43. Let P be a poset.
    a) Suppose that $\sigma: P \to Q$ is an order embedding and that $\sigma(P)$ is both join-dense and meet-dense in Q. Show that there is an order embedding $\tau: Q \to \mathrm{DM}(P)$ such that $\tau \circ \sigma = \phi_{\downarrow}$.
    b) Prove that if Q is a complete lattice then τ is an isomorphism.
44. Prove that if L is a lattice with no infinite chains (ACC and DCC) then L is isomorphic to the DM completion of $\mathcal{J}(L) \cup \mathcal{M}(L)$.

MacNeille Extensions

Let $f: L \to M$ be a monotone function between lattices L and M, thought of as subsets of their DM-completions. Define the **lower MacNeille extension** $f^\circ: \mathrm{DM}(L) \to \mathrm{DM}(M)$ of f by

$$f^\circ(x) = \bigvee\{f(a) \mid a \in L, a \le x\}$$

and the **upper MacNeille extension** $f^\bullet: \mathrm{DM}(L) \to \mathrm{DM}(M)$ of f by

$$f^\bullet(x) = \bigwedge\{f(a) \mid a \in L, x \le a\}$$

The map f is called **smooth** if $f^\circ = f^\bullet$.

45. Show that
    a) f° and $f^\bullet$ are extensions of f
    b) $f^\circ(x) \le f^\bullet(x)$ for all $x \in L$
    c) f° and $f^\bullet$ are monotone.
46. Show that the join map $\bigvee: L \times L \to L$ is smooth and that $\bigvee^\circ = \bigvee^\bullet$ is the join in $\mathrm{DM}(L)$. A dual statement holds for the meet.
47. Let L be the finite-cofinite algebra of $\mathbb{N}$ and let $\mathbf{2} = \{0, 1\}$ be the two-element lattice. Let $f: L \to \mathbf{2}$ be defined by

$$f(A) = \begin{cases} 1 & \text{if } A \text{ is cofinite} \\ 0 & \text{if } A \text{ is finite} \end{cases}$$

Find the Dedekind–MacNeille completion of L. Find the extensions f° and $f^{\bullet}$ of f. Show that f is a lattice homomorphism but that f° does not preserve joins and $f^{\bullet}$ does not preserve meets.

48. Let L be a distributive lattice L containing two elements x and z for which $L' = \mathrm{DM}(L)$ has a sublattice isomorphic to N_5 (see Figure 11.4) and is therefore not distributive. Let $f\colon L \to L$ be defined by

$$f(u) = z \vee u$$

a) Show that f preserves meets.
b) Find f° and show that it does not preserve meets.

Galois Connections

49. Show that a pair of maps (Π, Ω) is a Galois connection for (P, Q) if and only if for all $p \in P$ and $q \in Q$,

$$\Pi p \geq q \quad \Leftrightarrow \quad p \leq \Omega q$$

50. Let P be a poset and let PI be the set of principal ideals and PF be the set of principal filters in P, each of which is ordered by set inclusion. Show that the maps $\Pi(\downarrow x) \mapsto \uparrow x$ and $\omega\colon (\uparrow x) \mapsto \downarrow x$ form a Galois connection.

Chapter 4
Modular and Distributive Lattices

In this chapter, we examine modularity and distributivity in more detail.

Quadrilaterals

At various times throughout the remainder of the book, we will have use for the following concept.

Definition *Let L be a lattice. As shown on the left in Figure 4.1, a* **quadrilateral** *in L is a set consisting of two elements $a, b \in L$, together with their meet and join. We write this in the form of a 4-tuple*

$$Q = (a \wedge b, a, b, a \vee b)$$

A quadrilateral Q is **degenerate** *if Q has fewer than 4 distinct elements (in which case, Q has either 1 or 2 elements).* □

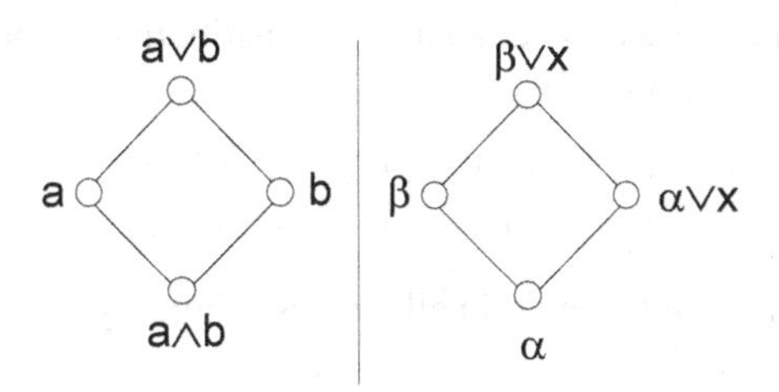

Figure 4.1

Note that the edges in Figure 4.1 are not necessarily meant to indicate the covering relation. Also, a quadrilateral Q is degenerate if and only if a and b are comparable.

The Definitions

For convenience, we repeat the definitions.

Definition

S. Roman (ed.), *Lattices and Ordered Sets*, doi: 10.1007/978-0-387-78901-9_4,

1) *A lattice L is* **modular** *if it satisfies the* **modular law***: For all $a, b, c \in L$,*

$$a \geq c \quad \Rightarrow \quad a \wedge (b \vee c) = (a \wedge b) \vee c$$

2) *A lattice L is* **distributive** *if it satisfies the* **distributive laws***: For all $a, b, c \in L$,*

$$\begin{aligned} a \wedge (b \vee c) &= (a \wedge b) \vee (a \wedge c) \\ a \vee (b \wedge c) &= (a \vee b) \wedge (a \vee c) \end{aligned}$$

□

We leave it to the reader to verify that the modular law is self-dual. It follows that a lattice is modular if and only if its dual is modular.

Theorem 4.1 *If either of the distributive laws holds for all elements of a lattice L, then so does the other.*

Proof. Suppose that the first distributive law holds. Then applying it to the right side of the second distributive law and using absorption gives

$$\begin{aligned} (a \vee b) \wedge (a \vee c) &= [(a \vee b) \wedge a] \vee [(a \vee b) \wedge c] \\ &= a \vee [(a \vee b) \wedge c] \\ &= a \vee [(a \wedge c) \vee (b \wedge c)] \\ &= a \vee (b \wedge c) \end{aligned}$$

which shows that the second law holds.□

Note that the distributive laws are dual to one another and so a lattice is distributive if and only if its dual lattice is distributive. It is clear that a distributive lattice is modular, but the converse is false, as we will see.

The usual definitions of modular and distributive lattices given above are not as "sharp" as possible. For example, a lattice is distributive if and only if either of the following *inequalities* holds:

$$\begin{aligned} a \wedge (b \vee c) &\leq (a \wedge b) \vee (a \wedge c) \\ a \vee (b \wedge c) &\geq (a \vee b) \wedge (a \vee c) \end{aligned}$$

since the reverse inequalities hold in all lattices. Similarly, a lattice is modular if and only if

$$a > c \quad \Rightarrow \quad a \wedge (b \vee c) \leq (a \wedge b) \vee c$$

since the reverse inequality holds in all lattices.

Examples

Let us consider some examples.

Example 4.2 The lattice $\mathcal{S}(G)$ of subgroups of a group need not be modular. The alternating group A_4 provides an example: Let

$$a = \langle(1\,2)(3\,4),(1\,3)(2\,4)\rangle, \quad b = \langle(1\,2\,3)\rangle \quad \text{and} \quad c = \langle(1\,2)(3\,4)\rangle$$

On the other hand, it is easy to see that the sublattice of normal subgroups $\mathcal{N}(G)$ of a group G is modular. In $\mathcal{N}(G)$, meet is intersection and join is the set product, that is, if $A, B \in \mathcal{N}(G)$, then

$$A \wedge B = A \cap B \quad \text{and} \quad A \vee B = AB$$

and the modular law is

$$A \geq C \quad \Rightarrow \quad A \cap BC = (A \cap B)C$$

a fact known in group theory as **Dedekind's law**.

It can be shown that $\mathcal{S}(G)$ is distributive if and only if G is locally cyclic (that is, any finite nonempty subset of G generates a *cyclic* subgroup). The proof is a bit complex (see Hall, *The Theory of Groups* [31]). Thus, if G is finite, then $\mathcal{S}(G)$ is distributive if and only if G is cyclic. The proof that a finite cyclic group has a distributive subgroup lattice is not hard and is left as an exercise.□

Example 4.3 The lattice of subspaces of a vector space of dimension at least 2 is modular but not distributive.□

Example 4.4

1) Any sublattice of a power set lattice $\wp(S)$ is distributive.
2) Any chain is distributive.
3) The lattice $(\mathbb{N}, \mid)$ is distributive.
4) The lattice of subgroups of $\mathbb{Z}$ is distributive.
5) The lattices M_3 and N_5 of Figure 4.2 are not distributive.□

Of course, distributivity extends to finite joins and meets, but not necessarily to infinite ones.

Example 4.5 The complete lattice $(\mathbb{N}, \mid)$ is distributive. However, let A be the set of odd natural numbers. Then $\bigvee A = 0$ and so

$$2 \wedge \left(\bigvee A\right) = 2$$

On the other hand, $2 \wedge a = 1$ for all odd numbers and so

$$\bigvee_{a \in A} (2 \wedge a) = 1 \neq 0$$

This example indicates one way in which arbitrary distributive lattices differ fundamentally from power sets: In a power set, for example, meet distributes over infinite joins, that is,

$$A \cap \left(\bigcup B_i\right) = \bigcup (A \cap B_i)$$ □

Inheritance by $\mathcal{I}(L)$

We have seen (Theorem 3.35) that if L is a lattice, then so is the set $\mathcal{I}(L)$ of all ideals of L. Moreover, the down map $\phi_{\downarrow}: L \to \mathcal{I}(L)$ is a lattice embedding. Therefore, if $\mathcal{I}(L)$ is modular, then so is L and if $\mathcal{I}(L)$ is distributive, then so is L. The converse is also true.

Theorem 4.6 *Let L be a lattice.*
1) L is modular if and only if $\mathcal{I}(L)$ is modular.
2) L is distributive if and only if $\mathcal{I}(L)$ is distributive.
Proof. For part 1), suppose that $A \supseteq C$. We must show that

$$A \cap (B \vee C) \subseteq (A \cap B) \vee C$$

If $x \in A \cap (B \vee C)$, then $x = a \in A$ and $x \le b \vee c$ for $b \in B$ and $c \in C$. Thus, since $a \vee c \in A$, we have

$$x \le a \wedge (b \vee c) \le (a \vee c) \wedge (b \vee c) = [(a \vee c) \wedge b] \vee c \in (A \cap B) \vee C$$

and so $\mathcal{I}(L)$ is modular.

For part 2), if L is distributive, we must show that

$$(A \vee B) \cap C \subseteq (A \cap C) \vee (B \cap C)$$

If $x \in (A \vee B) \cap C$, then $x \le a \vee b$ for some $a \in A$ and $b \in B$ and $x = c \in C$, whence

$$x \le (a \vee b) \wedge c = (a \wedge c) \vee (b \wedge c) \in (A \cap C) \vee (B \cap C)$$

for $a \in A$, $b \in B$ and $c \in C$. Hence, $\mathcal{I}(L)$ is distributive.□

Characterizations

There are several useful ways to characterize modularity and distributivity. Let us look at some of the most important ones.

Characterizations by Forbidden Sublattice

Modularity and distributivity can be characterized by the absence of certain sublattices. As we shall see, these characterizations can be extremely useful in showing that certain properties imply modularity or distributivity. The offending lattices are shown in Figure 4.2. From now on, we will reserve the notations M_3 and N_5 for these lattices. The lattice M_3 is called a **diamond** and the lattice N_5 is called a **pentagon**. Note that some authors use the notation M_5 rather than M_3. (Grätzer [24] uses M_3 but Birkhoff [5] uses M_5.)

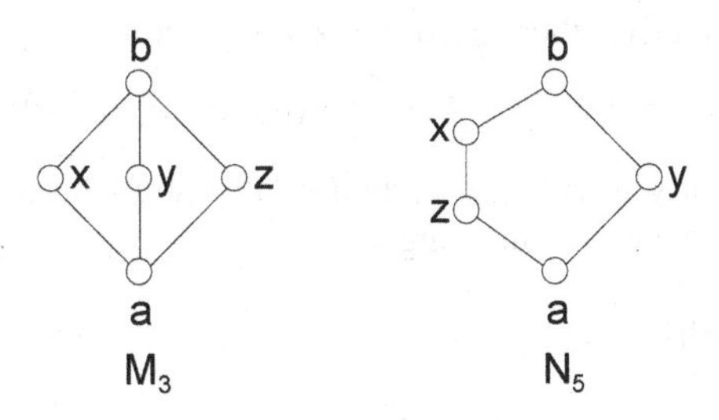

Figure 4.2

Theorem 4.7

1) A lattice L is modular if and only if it does not have a sublattice isomorphic to N_5.

2) A lattice L is distributive if and only if it does not have a sublattice isomorphic to either N_5 or M_3.

Proof. For part 1), it is easy to see that N_5 is not modular and so any lattice with a sublattice isomorphic to N_5 is not modular. Conversely, if L is not modular, then there exist $a, b, c \in L$ such that

$$a \geq c \quad \text{and} \quad x := a \wedge (b \vee c) > z := (a \wedge b) \vee c$$

To see that $S = \{x, b, z, x \wedge b, x \vee b\}$ is an N_5 sublattice of L, we have $z < x$ and

$$z \vee b = (a \wedge b) \vee c \vee b = c \vee b \geq x$$

and so $z \vee b \geq x \vee b$. But the reverse inequality is clear and so $z \vee b = x \vee b$. Also,

$$x \wedge b = a \wedge (b \vee c) \wedge b = a \wedge b \leq z$$

and so $x \wedge b \leq z \wedge b$. But the reverse inequality is clear and so $z \wedge b = x \wedge b$. Finally, to show that $x \parallel b$ and $z \parallel b$, we have the following:

1) If $b \leq x$, then $b \leq a$ and since $c \leq a$, it follows that $x = z$, a contradiction.
2) If $z \leq b$, then $c \leq b$ and since $c \leq a$, it follows that $x = z$, a contradiction.
3) If $x < b$, then $z < x < b$, which is false by 2).
4) If $b < z$ implies $b < z < x$, which is false by 1).

Thus, b is parallel to both x and z and S is N_5.

For part 2), if L contains one of M_3 or N_5, then L is not distributive since these lattices are not distributive. Conversely, suppose that L is not distributive. If L is not modular, then it has an embedded N_5, so we may assume that L is modular.

Since L is not distributive, there exist $a, b, c \in L$ such that

$$x := a \wedge (b \vee c) > z := (a \wedge b) \vee (a \wedge c)$$

We use this inequality to rule out certain relationships between a, b and c. In particular, we must have the following:

1) a, b, c must be distinct
2) $a \not\leq c$ and $a \not\leq b$
3) $a > c$ and $a > b$ cannot both hold
4) $b \parallel c$.

Thus, we are left with two possibilities:

5) $a > c, a \parallel b, b \parallel c$ (or symmetrically, $a \parallel c, a > b, b \parallel c$)
6) $a \parallel c, a \parallel b, b \parallel c$.

In case 5), we have $a > c$ and the modularity of L implies that

$$x = a \wedge (b \vee c) = (a \wedge b) \vee c = z$$

which is false. Thus, 5) can be eliminated as well.

Thus, we are left with the case in which a, b and c are pairwise incomparable. We clain that the elements

$$\begin{aligned} p &:= (a \wedge b) \vee (a \wedge c) \vee (b \wedge c) \\ q &:= (a \vee b) \wedge (a \vee c) \wedge (b \vee c) \\ u &:= (a \wedge q) \vee p \\ v &:= (b \wedge q) \vee p \\ w &:= (c \wedge q) \vee p \end{aligned}$$

form an M_3 sublattice S of L, where u, v and w are the incomparable elements of S, p is the smallest element and q is the largest element. We need to show that these elements are distinct and that

$$p = u \wedge v = u \wedge w = v \wedge w$$

and

$$q = u \vee v = u \vee w = v \vee w$$

It is clear that $p \leq q$. To see that $p \neq q$, note that

$$a \wedge q = a \wedge (b \vee c) = x$$

and, using the modular law,

$$\begin{aligned} a \wedge p &= a \wedge [(a \wedge b) \vee (a \wedge c) \vee (b \wedge c)] \\ &= [a \wedge (b \wedge c)] \vee [(a \wedge b) \vee (a \wedge c)] \\ &= (a \wedge b) \vee (a \wedge c) \\ &= z \end{aligned}$$

and since $x > z$, we must have $p < q$.

To show that $u \wedge v = p$, we have

$$u \wedge v = [(a \wedge q) \vee p] \wedge [(b \wedge q) \vee p]$$

and since $(a \wedge q) \vee p \geq p$ and $q > p$, multiple applications of the modular law give

$$\begin{aligned} u \wedge v &= [[(a \wedge q) \vee p] \wedge (b \wedge q)] \vee p \\ &= [[q \wedge (a \vee p)] \wedge (b \wedge q)] \vee p \\ &= [(b \vee p) \wedge (a \wedge q)] \vee p \end{aligned}$$

Now, $b \vee p = b \vee (a \wedge c)$ and $a \wedge q = a \wedge (b \vee c)$ and so a bit of rearranging gives

$$\begin{aligned} u \wedge v &= \{[b \vee (a \wedge c)] \wedge [a \wedge (b \vee c)]\} \vee p \\ &= \{a \wedge [(b \vee c) \wedge [(a \wedge c) \vee b]]\} \vee p \end{aligned}$$

Another application of the modular law using the fact that $b \vee c \geq b$ and then yet another using the fact that $a \geq a \wedge c$ gives

$$\begin{aligned} u \wedge v &= \{a \wedge [[(b \vee c) \wedge (a \wedge c)] \vee b]\} \vee p \\ &= \{a \wedge [(a \wedge c) \vee b]\} \vee p \\ &= \{(a \wedge b) \vee (a \wedge c)\} \vee p \\ &= p \end{aligned}$$

In a similar way, it can be shown that the other equalities hold.

As to uniqueness, if $u = v$, for example, then $p = u \wedge v = u \vee v = q$, which is false. Thus, we see that u, v and w are distinct. If $u = q$, then $u = u \vee v$ and so $v \leq u$ and similarly, $w \leq u$, whence $v = v \wedge u = w \wedge u = w$, which is false. In this way, we see that all five elements are distinct.□

Characterization by Relative Complements

The characterizations by forbidden sublattice can be rephrased in terms of relative complements. Indeed, saying that N_5 is a sublattice of L is equivalent to saying that there are elements x, y, z, a, b in L for which $x > z$ and $x, z \in y_{[a,b]}$. Thus, N_5 is not a sublattice of L if and only if no set $y_{[a,b]}$ of relative complements has two distinct comparable elements.

Similarly, saying that either one of M_3 or N_5 is a sublattice of L is equivalent to saying that there are elements x, y, z, a, b in L for which $x, z \in y_{[a,b]}$, where either $x > z$ or $x \parallel z$. Hence, neither M_3 nor N_5 is a sublattice of L if and only if the sets $y_{[a,b]}$ are either empty or singletons.

Theorem 4.8

1) *A lattice L is modular if and only if every set $a_{[u,v]}$ of relative complements is either empty or is an antichain.*
2) *A lattice L is distributive if and only if relative complements in L, when they exist, are unique or equivalently, if and only if every element of L is uniquely determined by its meet and join with any other element, that is,*

$$\left\{\begin{array}{l} x \wedge b = x \wedge c \\ x \vee b = x \vee c \end{array}\right\} \quad \Rightarrow \quad b = c$$

In particular, uniquely relatively complemented lattices are distributive. □

Thus, uniquely relatively complemented lattices are distributive. However, the converse fails: Distributive lattices need not be complemented, let alone relatively complemented, as witnessed by any chain with more than two elements, or by the lattice in Figure 4.3.

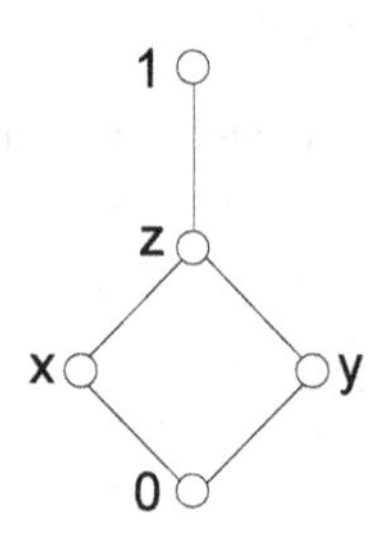

Figure 4.3

Unique Complementation and Distributivity

In 1904, E. V. Huntington [34] conjectured that any uniquely complemented lattice was distributive. For over 40 years this remained one of the main problems of lattice theory. It was not until 1945 that Dilworth [16], in a rather grueling 32 pages, showed that every lattice is a sublattice of a uniquely complemented lattice. Hence, unique complementation cannot imply distributivity, since otherwise every lattice would be distributive.

Much work has been done on the question of what additional properties are required of a uniquely complemented lattice in order to imply distributivity. We will discuss this question in some detail when we discuss Boolean algebras in a later chapter. (A complemented distributive lattice is a Boolean algebra.) As a preview, for a uniquely complemented lattice L, any one of the following conditions implies distributivity:

1) L is modular
2) L satisfies de Morgan's laws
3) L is atomic
4) L has finite width.

In view of these conditions, it is not surprising that it is hard to find a "natural" example of a nondistributive, uniquely complemented lattice.

Characterization Using the Join and Meet Maps

The modular and distributive laws can also be described in terms of the **join maps** $j_a: L \to L$ and the **meet maps** $m_a: L \to L$, defined by

$$j_a(x) = x \vee a \quad \text{and} \quad m_a(x) = x \wedge a$$

Distributivity has a simple characterization in terms of these maps.

Theorem 4.9 *Let L be a lattice.*
1) L is distributive if and only if the join map j_a is a lattice homomorphism for all $a \in L$.
2) Dually, L is distributive if and only if the meet map m_a is a lattice homomorphism for all $a \in L$.

Proof. It is easy to see that j_a preserves joins. As to preservation of meets, the equation

$$j_a(x \wedge y) = j_a(x) \wedge j_a(y)$$

is none other than the distributive law

$$a \vee (x \wedge y) = (a \vee x) \wedge (a \vee y) \qquad \square$$

As to modularity, with reference to the quadrilateral in Figure 4.4, if L is modular, then for any $a, b \in L$ and $x \in [a \wedge b, a]$,

$$m_a \circ j_b(x) = (x \vee b) \wedge a = x \vee (b \wedge a) = x$$

and so $m_a \circ j_b = \iota$ is the identity map on the interval $[a \wedge b, a]$.

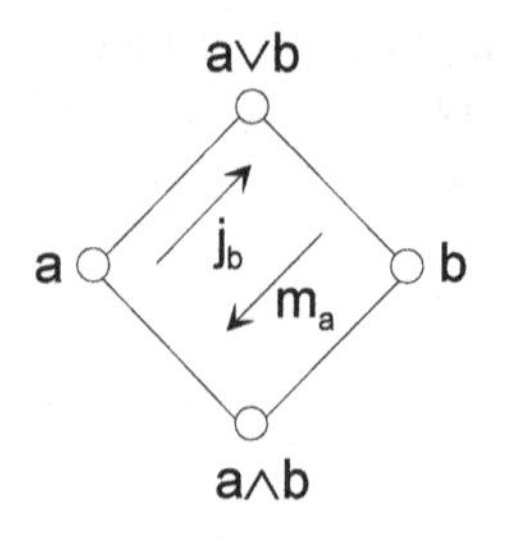

Figure 4.4

Conversely, if L is not modular, then L contains a sublattice N_5, where $z \in [x \wedge y, x]$. But

$$m_x \circ j_y(z) = (z \vee y) \wedge x = x \neq z$$

Hence, $m_x \circ j_y \neq \iota$. Thus, L is modular if and only if $m_a \circ j_b = \iota$ on $[a \wedge b, a]$, for all $a, b \in L$.

Theorem 4.10 *Let L be a lattice and let*

$$j_b\colon [a \wedge b, a] \to [b, a \vee b] \quad \textit{and} \quad m_a\colon [b, a \vee b] \to [a \wedge b, a]$$

be the join and meet maps, restricted to the opposite sides of the quadrilateral in Figure 4.4. The following are equivalent:

1) *L is modular.*
2) *For all $a, b \in L$,*

$$m_a \circ j_b = \iota$$

on $[a \wedge b, a]$.

3) *Dually, for all $a, b \in L$,*

$$j_b \circ m_a = \iota$$

on $[b, a \vee b]$.

4) *For all $a, b \in L$, the restricted maps m_a and j_b are inverse lattice isomorphisms on these intervals.*

Proof. We have seen that 1)–3) are equivalent and since m_a and j_b are order-preserving, 2) and 3) imply that these maps are inverse order (lattice) isomorphisms and so 4) holds. If 4) holds, then clearly 2) holds.□

Modularity and Semimodularity

Recall that $a \sqsubset b$ means that a covers b and $a \sqsubseteq b$ means that $a \sqsubset b$ or $a = b$. It is too much to expect that the join map $j_x\colon L \to L$ preserves covering, that is, that

$$a \sqsubset b \quad \Rightarrow \quad a \vee x \sqsubset b \vee x$$

for if $x \geq b$, then $a \vee x = b \vee x = x$. On the other hand, it is possible that the join map **weakly preserves covering**, that is,

$$a \sqsubset b \quad \Rightarrow \quad a \vee x \sqsubseteq b \vee x$$

Indeed, this happens in any modular lattice L. To see this, suppose that $a \sqsubset b$ and that $a \vee x \leq u \leq b \vee x$. Since $u > a \vee x$, the modular law gives

$$u = u \wedge (b \vee (a \vee x)) = (u \wedge b) \vee (a \vee x)$$

But since $a \sqsubset b$ and $a \leq u \wedge b \leq b$, it follows that $u \wedge b = a$ or $u \wedge b = b$. Hence, $u = a \vee x$ or $u = b \vee x$, which shows that $a \vee x \sqsubseteq b \vee x$.

The weak preservation of covering is equivalent to the condition

$$\alpha \wedge \beta \sqsubset \alpha \quad \Rightarrow \quad \beta \sqsubset \alpha \vee \beta \tag{4.11}$$

for all $\alpha, \beta \in L$. To see that (4.11) is implied by the weak preservation of covering, let $\alpha \wedge \beta \sqsubset \alpha$ and set $a = \alpha \wedge \beta$, $b = \alpha$ and $x = b$. Then $a \vee x \sqsubseteq b \vee x$, where

$$a \vee x = (\alpha \wedge \beta) \vee \beta = \beta \quad \text{and} \quad b \vee x = \alpha \vee \beta$$

and so $\beta \sqsubseteq \alpha \vee \beta$. But $\beta = \alpha \vee \beta$ implies that $\alpha \wedge \beta = \alpha$, which contradicts the fact that $\alpha \wedge \beta \sqsubset \alpha$. Hence, $\beta \sqsubset \alpha \vee \beta$.

Conversely, assume that (4.11) holds and that $a \sqsubset b$. Then

$$a \leq b \wedge (a \vee x) \leq b$$

and so $b \wedge (a \vee x) = a$ or $b \wedge (a \vee x) = b$. But in the latter case, $b \leq a \vee x$ and so $b \vee x = a \vee x$. In the case $b \wedge (a \vee x) = a$, we have $b \wedge (a \vee x) \sqsubset b$ and so (4.11) implies that

$$a \vee x \sqsubset b \vee (a \vee x) = b \vee x$$

as desired. The two equivalent conditions are shown in Figure 4.5 and have a name (see Grätzer [24]).

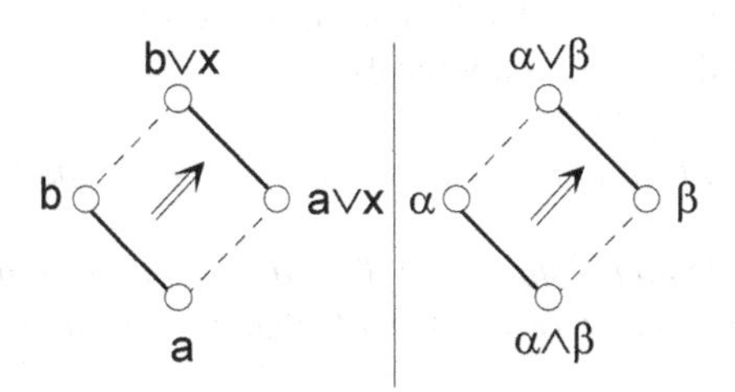

Figure 4.5

Definition *Let L be a lattice.*

1) *As shown in Figure 4.5, the* **upper covering condition** *on L is the condition that the join map $j_x\colon L \to L$ weakly preserves covering, that is,*

$$a \sqsubset b \quad \Rightarrow \quad a \vee x \sqsubseteq b \vee x$$

for all $a, b, x \in L$, *or equivalently*

$$a \wedge b \sqsubset a \quad \Rightarrow \quad b \sqsubset a \vee b$$

for all $a, b \in L$. *A lattice that has the upper covering condition is called an* **upper semimodular** *lattice.*

2) *Dually, the* **lower covering condition** *on* L *is the condition that the meet map* $m_x: L \to L$ *weakly preserves covering, that is,*

$$a \sqsubset b \quad \Rightarrow \quad a \wedge x \sqsubseteq b \wedge x$$

for all $a, b, x \in L$, *or equivalently*

$$b \sqsubset a \vee b \quad \Rightarrow \quad a \wedge b \sqsubset a$$

for all $a, b \in L$. *A lattice that has the lower covering condition is called a* **lower semimodular** *lattice.*□

Theorem 4.12 *Every modular lattice is upper semimodular and lower semimodular.*□

Example 4.13 The lattice in Figure 4.6 is upper semimodular but not modular.□

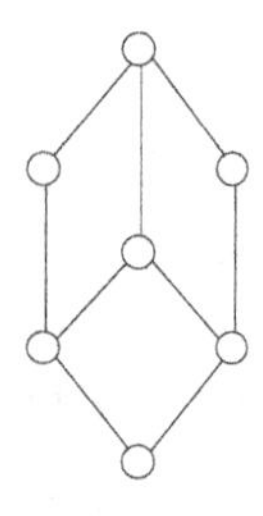

Figure 4.6

For lattices with no infinite chains, the converse of Theorem 4.12 holds. To explore this, we need a preliminary result.

Theorem 4.14 *Let* L *be a lattice with no infinite chains and let* $h: L \to \mathbb{N}$ *be the height function.*

1) *If* L *is upper semimodular, then* L *satisfies the Jordan–Dedekind chain condition, that is, any two maximal chains from* a *to* b *have the same length.*
2) *Dually, if* L *is lower semimodular, then* L *satisfies the Jordan–Dedekind chain condition.*
3) L *is upper semimodular if and only if*

$$h(a) + h(b) \geq h(a \wedge b) + h(a \vee b)$$

4) *Dually, L is lower semimodular if and only if*

$$h(a)+h(b)\le h(a\wedge b)+h(a\vee b)$$

Proof. Note first that since L has no infinite chains, a chain from a to b is maximal if and only if each step in the chain is a covering. We prove part 1) by induction on n that if L has a maximal chain of length n from a to b, then all maximal chains from a to b have length n. If $n=1$, then $a\sqsubset b$. But if $a\sqsubset u_1\sqsubset\cdots\sqsubset u_m=b$ is a maximal chain from a to b, then $a<u_1\le b$ and so $u_1=b$ and this chain also has length 1.

Assume that the result is true for all chains of length less than n. With reference to Figure 4.7,

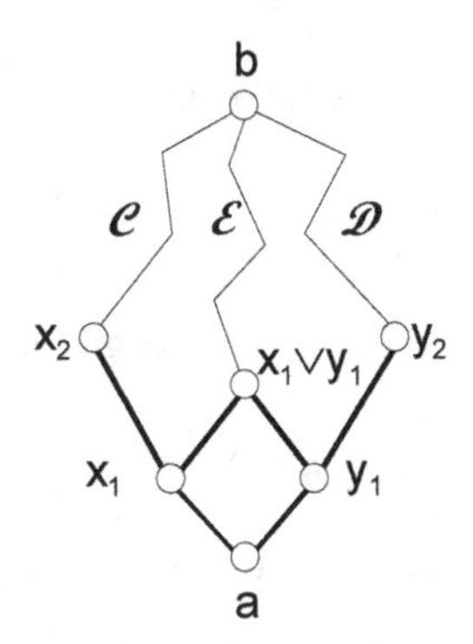

Figure 4.7

let

$$\mathcal{C}=\{a\sqsubset x_1\sqsubset x_2\sqsubset\cdots\sqsubset x_{n-1}\sqsubset b\}$$

and

$$\mathcal{D}=\{a\sqsubset y_1\sqsubset y_2\sqsubset\cdots\sqsubset y_{m-1}\sqsubset b\}$$

be maximal chains from a to b. If $x_1=y_1$, then the induction hypothesis implies that $n-1=m-1$, so assume that $x_1\ne y_1$.

Since $a\sqsubset x_1$ and $a\sqsubset y_1$, the upper covering condition implies that $x_1\sqsubset x_1\vee y_1$ and $y_1\sqsubset x_1\vee y_1$. Let $\mathcal{E}$ be a maximal chain from $x_1\vee y_1$ to b. Prefixing $\mathcal{E}$ with x_1 gives a maximal chain from x_1 to b and so the induction hypothesis implies that $\operatorname{len}(\mathcal{E})=n-1$. Similarly, $\operatorname{len}(\mathcal{E})=m-1$ and so $m=n$.

Part 3) can be rephrased by saying that L is upper semimodular if and only if

$$d(a\wedge b,a)\ge d(b,a\vee b)$$

for all $a, b \in L$, where $d(x, y)$ is the length of *any* maximal chain from x to y. But, as shown in Figure 4.8,

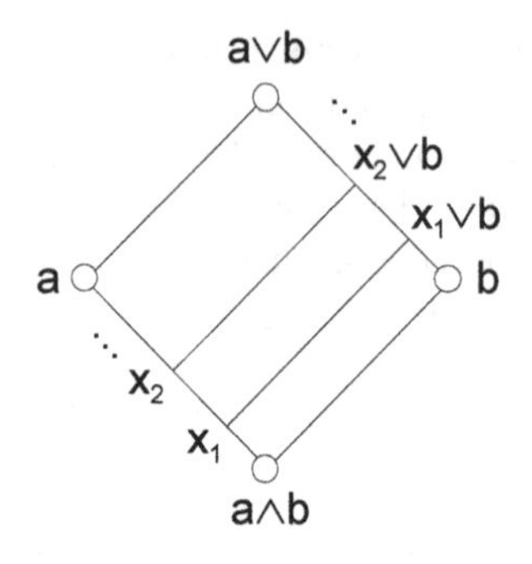

Figure 4.8

if

$$a \wedge b \sqsubset x_1 \sqsubset x_2 \sqsubset \cdots \sqsubset x_{n-1} \sqsubset a$$

is a maximal chain from $a \wedge b$ to a of length n, then the upper covering condition gives

$$b \sqsubseteq x_1 \vee b \sqsubseteq x_2 \vee b \sqsubseteq \cdots \sqsubseteq x_{n-1} \vee b \sqsubseteq a \vee b$$

and so there is a maximal chain from b to $a \vee b$ of length at most n. Hence,

$$d(a \wedge b, a) \geq d(b, a \vee b)$$

Conversely, if 3) holds and $a \wedge b \sqsubset a$, then

$$1 = d(a \wedge b, a) \geq d(b, a \vee b)$$

But $b \neq a \wedge b$ and so $d(b, a \vee b) = 1$, that is, $b \sqsubset a \vee b$ and so L is upper semimodular.□

Theorem 4.15 *Let L be a lattice with no infinite chains. The following are equivalent:*
1) L is modular
2) L is upper and lower semimodular
3) The height function h satisfies

$$h(a) + h(b) = h(a \wedge b) + h(a \vee b)$$

or, equivalently,

$$d(a \wedge b, a) = d(b, a \vee b)$$

for all $a, b \in L$.

Proof. We have seen that 1) implies 2) and that 2) implies 3). Suppose that 3) holds. If L is not modular, then it contains a sublattice N_5, as shown in Figure

4.2. Hence,

$$h(x) + h(y) = h(x \vee y) + h(x \wedge y)$$

and

$$h(z) + h(y) = h(z \vee y) + h(z \wedge y)$$

But the right-hand sides of these two equations are equal and so $h(x) = h(z)$, which is false.□

The definition of upper and lower covering condition is that of Grätzer [24]. Birkhoff [5] employs a somewhat stronger form of covering condition that is equivalent when L has no infinite chains. We will refer to it by the nonstandard name *double upper covering condition.*

Definition *Let L be a lattice. With reference to Figure 4.9, we make the following definitions.*

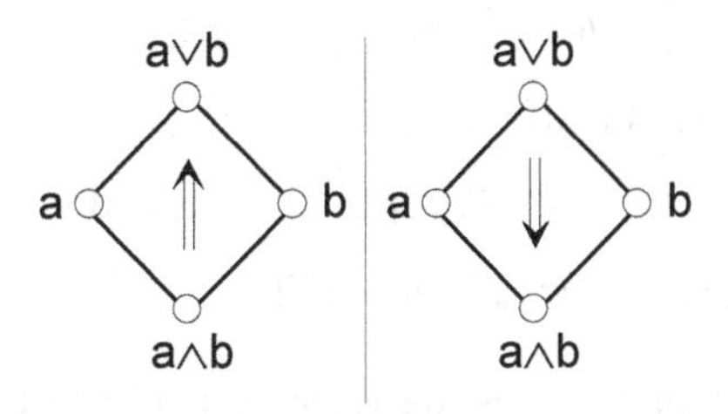

Figure 4.9

1) The **double upper covering condition** *is*

$$a \wedge b \sqsubset a,\ a \wedge b \sqsubset b \quad \Rightarrow \quad a \sqsubset a \vee b,\ b \sqsubset a \vee b$$

2) Dually, the **double lower covering condition** *is*

$$a \sqsubset a \vee b,\ b \sqsubset a \vee b \quad \Rightarrow \quad a \wedge b \sqsubset a,\ a \wedge b \sqsubset b$$ □

It is clear that the upper covering condition implies the double upper covering condition. The converse holds if L has no infinite chains.

Theorem 4.16 *A lattice with no infinite chains has the upper covering condition if and only if it has the double upper covering condition. A similar statement holds for lower in place of upper.*

Proof. Suppose that the double upper covering condition holds and that $a \sqsubset b$. We show that $a \vee x \sqsubseteq b \vee x$ for any $x \in L$. We may assume that $a \vee x \neq b \vee x$ and so $b \notin [a, a \vee x]$. Consider a maximal chain

$$a = u_0 \sqsubset u_1 \sqsubset \cdots \sqsubset u_{n-1} \sqsubset u_n = a \vee x$$

from a to $a \vee x$, as shown in Figure 4.10.

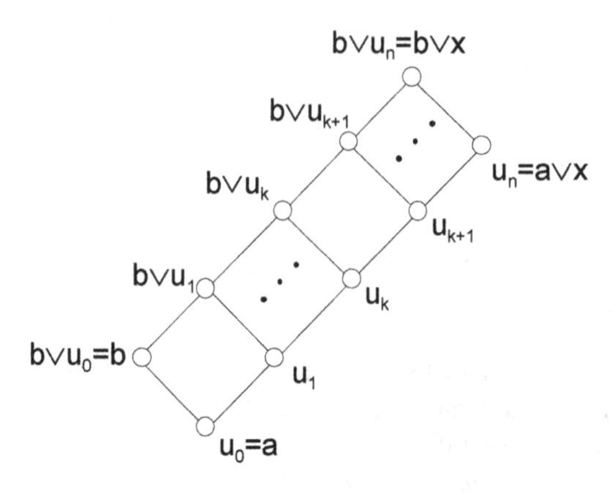

Figure 4.10

Since $u_0 \sqsubset b \vee u_0$ and $u_0 \sqsubset u_1$, the double upper covering condition implies that

$$u_1 \sqsubset b \vee u_1 \quad \text{and} \quad b \vee u_0 \sqsubset b \vee u_1$$

More generally, we have for all $k = 0, \ldots, n-1$,

$$u_k \sqsubset b \vee u_k \text{ and } u_k \sqsubset u_{k+1} \Rightarrow u_{k+1} \sqsubset b \vee u_{k+1} \text{ and } b \vee u_k \sqsubset b \vee u_{k+1}$$

Hence, $u_n \sqsubset b \vee u_n$, that is, $a \vee x \sqsubset b \vee x$.□

For more on semimodular lattices, we refer the reader to Grätzer [24] or Stern [58].

Partition Lattices and Representations

We begin with the definition of a partition.

Definition *A* **partition** *of a nonempty set X is a collection $\{A_i \mid i \in I\}$ of nonempty subsets of S, called the* **blocks** *of the partition, for which*
1) $A_i \cap A_j = \emptyset$ *for all* $i \neq j$
2) $S = \bigcup A_i$.
A block is **trivial** *if it has size* 1.□

It is well known that there is a bijection between the family $\text{Equ}(X)$ of equivalence relations on a nonempty set X and the family $\text{Part}(X)$ of partitions of S. When we write $\text{Equ}(X)$ or $\text{Part}(X)$, it is with the understanding that X is a nonempty set.

Thinking of equivalence relations as subsets of X^2, it is clear that $\text{Equ}(X)$ is a poset under set inclusion. For $\text{Part}(X)$, we have the following order.

Definition *Let P and Q be partitions of a nonempty set X. Then P is a* **refinement** *of Q, written $P \le Q$, if each block of Q is a union of one or more blocks of P. If $P \le Q$, we say that P is* **finer** *than Q and Q is* **coarser** *than P.*□

Theorem 4.17 *Let X be a nonempty set. The bijection $\Pi\colon \mathrm{Equ}(X) \to \mathrm{Part}(X)$ that maps an equivalence relation θ to the corresponding partition P_θ is an order isomorphism, that is,*

$$\theta \subseteq \sigma \quad \Leftrightarrow \quad P_\theta \le P_\sigma$$ □

In view of this theorem, we will describe our results in terms of either partitions or equivalence relations, whichever seems clearer at the time.

Since $\mathrm{Equ}(X)$ is an intersection-structure, it is a complete lattice. To describe the join of equivalence relations, we use the following concept.

Definition *Let $\mathcal{F}$ be a family of binary relations on a nonempty set X. If $a, b \in X$, then a* **witness sequence** *from a to b is a sequence*

$$a = a_1, a_2, \ldots, a_{n-1}, a_n = b$$

in X for which

$$a = a_1 \overset{\theta_1}{\equiv} a_2 \overset{\theta_2}{\equiv} a_3 \equiv \cdots \equiv a_{n-1} \overset{\theta_{n-1}}{\equiv} a_n = b$$

where $\theta_i \in \mathcal{F}$. If there is a witness sequence from a to b, we say that b is **reachable** *from a using $\mathcal{F}$. The* **length** *of a witness sequence is the number of terms in the sequence, not including the first and last. Thus, $a \equiv x \equiv b$ has length 1.*□

Theorem 4.18 *Let X be a nonempty set. The poset $\mathrm{Equ}(X)$ is an intersection-structure and therefore a complete lattice. Let $\mathcal{F} = \{\theta_i \mid i \in I\}$ be a family of equivalence relations on X, with corresponding partitions $\mathcal{P} = \{P_i \mid i \in I\}$, where $P_i = P_{\theta_i}$.*

1) The meet is given by

$$\bigwedge \mathcal{F} = \bigcap \mathcal{F}$$

and so

$$a\Big(\bigwedge \mathcal{F}\Big)b \quad \text{if} \quad a\theta b \text{ for all } \theta \in \mathcal{F}$$

The meet of the family $\mathcal{P}$ can be described as the coarsest common refinement of $\mathcal{P}$. Thus, the blocks of $\bigwedge \mathcal{P}$ are the largest subsets of X that fit inside a single block from each member of $\mathcal{P}$.

2) *The join*

$$\theta = \bigvee \mathcal{F}$$

of $\mathcal{F}$ is the transitive closure of the union $\bigcup \mathcal{F}$. Specifically, $a\theta b$ if and only if there is a witness sequence

$$a = a_1 \overset{\theta_{i_1}}{\equiv} a_2 \overset{\theta_{i_2}}{\equiv} a_3 \equiv \cdots \equiv a_{n-1} \overset{\theta_{i_{n-1}}}{\equiv} a_n = b$$

from a to b using $\mathcal{F}$. The join of $\mathcal{P}$ can be described as follows: Two elements $a, b \in X$ are equivalent under the join if there is a finite sequence of "jumps" starting with a and ending with b, where each jump must lie within a single block of some *member of $\mathcal{P}$.*

Proof. We leave the proof that $\text{Equ}(X)$ is an intersection-structure to the reader. For part 2), let $\theta = \bigvee \mathcal{F}$. The reachable relation R is easily seen to be an equivalence relation. Moreover, since $\theta_i \subseteq \theta$, it follows that aRb implies $a\theta b$, that is, $R \subseteq \theta$. But $\theta_k \subseteq R$ for all k and so $\theta \subseteq R$. Thus, $\theta = R$.□

We leave proof of the following for the exercises.

Theorem 4.19 *Let X be a nonempty set.*

1) *The atoms in* $\text{Part}(X)$ *are the partitions with only one nontrivial block, which has size* 2.
2) *For the covering relation in* $\text{Part}(X)$, *we have $\sigma \sqsubset \theta$ if and only if θ is obtained from σ by taking the union of exactly two distinct blocks of σ.*
3) $\text{Part}(X)$ *is atomistic and strongly atomic.*
4) $\text{Part}(X)$ *is semimodular.*
5) $\text{Part}(X)$ *is relatively complemented.*□

Representations

Cayley's theorem of group theory says that any group can be embedded in (is isomorphic to a subgroup of) a permutation group. Thus, permutation groups are universal, in the sense that they "contain" all groups. Partition lattices play the analogous role for lattices.

Definition *Let L be a lattice.*

1) *A* **representation** *of a lattice L is a lattice embedding*

$$\Lambda: L \hookrightarrow \text{Part}(X)$$

   for some nonempty set X.
2) *A representation Λ has*
   a) **type** 1 *if for all $a, b \in L$ and $x, y \in X$,*

$$x(\Lambda a \vee \Lambda b)y \quad \Rightarrow \quad x(\Lambda a)z(\Lambda b)y$$

   for some $z \in X$.

b) **type** 2 *if for all* $a, b \in L$ *and* $x, y \in X$,

$$x(\Lambda a \vee \Lambda b)y \quad \Rightarrow \quad x(\Lambda a)z_1(\Lambda b)z_2(\Lambda a)y$$

for some $z_1, z_2 \in X$.

c) **type** 3 *if for all* $a, b \in L$ *and* $x, y \in X$,

$$x(\Lambda a \vee \Lambda b)y \quad \Rightarrow \quad x(\Lambda a)z_1(\Lambda b)z_2(\Lambda a)z_3(\Lambda b)y$$

for some $z_1, z_2, z_3 \in X$.□

Note that the definition of type and the commutativity of join imply that there is a witness sequence from x to y that starts with Λb as well.

In 1946, P. M. Whitman [67] proved that every lattice has a representation. In 1953, Bjarni Jónsson [37] proved that every lattice has a type 3 representation. We follow the proof more or less as given in Grätzer [24] and for this proof, we require two definitions.

Definition *A* **meet representation** *of a lattice L is a meet-preserving bijection from L to* Part(X), *for some nonempty set X*.□

Definition *Let L be a lattice with* 0 *and let X be a nonempty set. A function* $\delta: X \times X \to L$ *is a* **distance function** *if*

1) *δ is surjective.*

2) *δ is symmetric, that is,*

$$\delta(x, y) = \delta(y, x)$$

for all $x, y \in X$.

3) *δ is normalized, that is,*

$$\delta(x, x) = 0$$

for all $x \in X$.

4) *δ satisfies the triangle inequality, that is,*

$$\delta(x, z) \leq \delta(x, y) \vee \delta(y, z)$$

for all $x, y, z \in X$.□

For example, the function $\delta: L \times L \to L$ given by

$$\delta(a, b) = \begin{cases} a \vee b & \text{if } a \neq b \\ 0 & \text{if } a = b \end{cases}$$

is a distance function.

Theorem 4.20 *Let L be a lattice. A distance function* $\delta: X \times X \to L$ *defines a meet-representation*

$$\overline{\delta}: L \to \text{Part}(X)$$

of L by (in equivalence relation language)

$$x\overline{\delta}(a)y \quad \textit{if} \quad \delta(x,y) \leq a$$

for all $a \in A$ and $x, y \in X$.

Proof. It is clear that $\overline{\delta}(a)$ is reflexive and symmetric. If $x\overline{\delta}(a)y\overline{\delta}(a)z$, then $\delta(x,y) \leq a$ and $\delta(y,z) \leq a$ and so the triangle inequality shows that $\delta(x,z) \leq a$. Hence, $x\overline{\delta}(a)z$ and so $\overline{\delta}(a)$ is transitive and therefore an equivalence relation.

Now,

$$\begin{aligned} x\overline{\delta}(a \wedge b)y &\Leftrightarrow x \vee y \leq a \wedge b \\ &\Leftrightarrow x \vee y \leq a \text{ and } x \vee y \leq b \\ &\Leftrightarrow x\overline{\delta}(a)y \text{ and } x\overline{\delta}(b)y \\ &\Leftrightarrow x[\overline{\delta}(a) \cap \overline{\delta}(b)]y \end{aligned}$$

and so δ preserves meets. To see that $\overline{\delta}$ is injective, if $\overline{\delta}(a) = \overline{\delta}(b)$, then $\delta(x,y) \leq a$ if and only if $\delta(x,y) \leq b$. However, if $a \neq b$, we may assume that $a \not\leq b$ and since δ is surjective, there are $x, y \in X$ for which $\delta(x,y) = a$, but then $\delta(x,y) \not\leq b$. Hence, $\overline{\delta}$ is a meet representation of L.□

Definition *Let $X \subseteq Y$ and let $\delta: X \times X \to L$ and $\epsilon: Y \times Y \to L$ be distance functions. Then ϵ* **extends** *δ if $\epsilon|_{X \times X} = \delta$.*□

It is clear that if ϵ extends δ, then $\overline{\delta} \leq \overline{\epsilon}$ on X.

Theorem 4.21 (Jónsson [37])

1) Every lattice L has a type 3 representation.

2) A lattice L has a type 2 representation if and only if it is modular.

Proof. We first prove that any lattice L with a type 2 representation is modular. For convenience, let us identify an element $a \in L$ with its image Λa under the representation $\Lambda: L \to \text{Part}(X)$. We must show that for $a, b, c \in L$ with $a \geq c$,

$$a \wedge (b \vee c) \leq (a \wedge b) \vee c$$

that is, for any $x, y \in X$,

$$x[a \wedge (b \vee c)]y \leq x[(a \wedge b) \vee c]y$$

Now, the left side implies that xay and $x(b \vee c)y$ and since the representation has type 2, there are $z_1, z_2 \in X$ for which

$$xcz_1bz_2cy$$

Thus, $z_1cxaycz_2$ and so $c \leq a$ implies that z_1az_2 and since z_1bz_2, it follows that

$z_1(a \wedge b)z_2$, whence $x[(a \wedge b) \vee c]y$, which is the right-hand side of the modular inequality. Hence, L is modular.

For part 1) and the remaining direction of part 2), suppose that $\delta: X \times X \to L$ is a distance function, with corresponding meet representation $\overline{\delta}: L \to \text{Part}(X)$. Suppose further that $u\overline{\delta}(a \vee b)v$ for some $a, b \in L$ and $u, v \in X$, that is,

$$\delta(u, v) \leq a \vee b$$

Then it may fail to hold that $u[\overline{\delta}(a) \vee \overline{\delta}(b)]v$, but we can adjoin some elements to X (3 elements in general and 2 elements when L is modular) and extend δ so that this is the case, with a witness sequence from u to v of length 3 in general and of length 2 when L is modular. This is done as follows.

For any L (modular or not), let z_1, z_2 and z_3 be symbols not in X and let

$$X^* = X \cup \{z_1, z_2, z_3\}$$

Extend δ to δ^* by setting

$$\begin{aligned} \delta^*(z_1, z_2) &= b \\ \delta^*(z_2, z_3) &= a \\ \delta^*(z_1, z_3) &= a \vee b \end{aligned}$$

and for all $x \in X$,

$$\begin{aligned} \delta^*(x, z_1) &= \delta(x, u) \vee a \\ \delta^*(x, z_2) &= \delta(x, u) \vee a \vee b \\ \delta^*(x, z_3) &= \delta(x, v) \vee b \end{aligned}$$

and require also that δ^* be symmetric and normalized.

These extensions may seem arbitrary at first, but they are designed so that the resulting function $\delta^*: X^* \times X^* \to L$ will satisfy the triangle inequality and therefore be a distance function. In particular, by matching a and b as above, the right side of the triangle inequality

$$\delta^*(x, z_j) \leq \delta^*(x, z_i) \vee \delta^*(z_i, z_j) \tag{4.22}$$

has the form

$$\delta(x, u \text{ or } v) \vee a \vee b$$

whereas the left side has the form

$$\delta(x, u \text{ or } v) \vee c$$

where $c \leq a \vee b$. Moreover, since δ satisfies the triangle inequality, we have

$$\delta(x, u) \leq \delta(x, v) \vee \delta(v, u) \leq \delta(x, v) \vee a \vee b$$

and similarly,

$$\delta(x, v) \leq \delta(x, u) \vee a \vee b$$

Hence, it is clear that (4.22) holds. We leave it to the reader to check the other cases of the triangle inequality.

Thus, δ^* is a distance function. Moreover, the extension also satisfies

$$\begin{aligned} \delta^*(u, z_1) &\leq a \\ \delta^*(z_1, z_2) &\leq b \\ \delta^*(z_2, z_3) &\leq a \\ \delta^*(z_3, v) &\leq b \end{aligned}$$

that is,

$$u\overline{\delta^*}(a)z_1\overline{\delta^*}(b)z_2\overline{\delta^*}(a)z_3\overline{\delta^*}(b)v$$

which is a witness sequence from u to v under $\overline{\delta^*}(a) \vee \overline{\delta^*}(b)$ of length 3, as promised. Let us refer to $(X^*, \overline{\delta^*})$ as the 3-**extension** of (X, δ) by (a, b, u, v) and write

$$(X^*, \overline{\delta^*}) = (X, \delta) \uparrow (a, b, u, v)$$

When L is modular, we can accomplish the same goal by adjoining only two new symbols,

$$X^* = X \cup \{z_1, z_2\}$$

and extending δ to δ^* by setting

$$\begin{aligned} \delta^*(z_1, z_2) &= (\delta(u, v) \vee a) \wedge b \\ \delta^*(x, z_1) &= \delta(x, u) \vee a \\ \delta^*(x, z_2) &= \delta(x, v) \vee a \end{aligned}$$

for all $x \in X$ and by requiring that δ^* be symmetric and normalized. This is also a distance function. As for the only nontrivial case of the triangle inequality,

$$\delta^*(x, z_2) \leq \delta^*(x, z_1) \vee \delta^*(z_1, z_2)$$

the right-hand side is

$$\delta^*(x, z_1) \vee \delta^*(z_1, z_2) = \delta(x, u) \vee a \vee [(\delta(u, v) \vee a) \wedge b]$$

Since $a \leq \delta(u, v) \vee a$, the modular law gives

$$= \delta(x, u) \vee [(\delta(u, v) \vee a) \wedge (a \vee b)]$$

But $\delta(u, v) \leq a \vee b$ and so $\delta(u, v) \vee a \leq a \vee b$, which reduces this to

$$= \delta(x, u) \vee \delta(u, v) \vee a$$

Finally, the triangle inequality for d gives

$$= \delta(x, v) \vee a = \delta^*(x, z_2)$$

which is the left-hand side. Thus, δ^* is a distance function. Moreover,

$$\begin{aligned} \delta^*(u, z_1) &\le a \\ \delta^*(z_1, z_2) &\le b \\ \delta^*(z_2, v) &\le a \end{aligned}$$

that is,

$$u\overline{\delta^*}(a)z_1\overline{\delta^*}(b)z_2\overline{\delta^*}(a)v$$

which is a witness sequence from u to v under $\overline{\delta^*}(a) \vee \overline{\delta^*}(b)$ of length 2, as promised. Let us refer to this pair $(X^*, \overline{\delta^*})$ as the 2-**extension** of (X, δ) by (a, b, u, v) and also write

$$(X^*, \overline{\delta^*}) = (X, \delta) \uparrow (a, b, u, v)$$

Now, for either case, we need an extension X^+ of X and δ^+ of δ so that given *any* $a, b \in L$ and $u, v \in X$ for which $u\overline{\delta^+}(a \vee b)v$, that is, for which

$$\delta^+(u, v) \le a \vee b$$

then $u[\overline{\delta^+}(a) \vee \overline{\delta^+}(b)]v$, with witness sequence of length 3 in general and length 2 when L is modular. To this end, we first well-order the set of all quadruples (a, b, u, v) where $a, b \in L$, $u, v \in X$ and

$$\delta^+(u, v) \le a \vee b$$

to form a transfinite sequence

$$\langle (a_\alpha, b_\alpha, u_\alpha, v_\alpha) \mid \alpha < \lambda \rangle$$

Then we accumulate extensions. First, let

$$(X_0, \delta_0) = (X, \delta) \uparrow (a_0, b_0, u_0, v_0)$$

where $\uparrow$ stands for the 3-extension in general or the 2-extension when L is modular.

For each ordinal $\beta < \alpha$, suppose we have defined (X_β, δ_β) in such a way that if $\gamma < \beta < \alpha$, then $X_\gamma \subseteq X_\beta$ and $\delta_\gamma \subseteq \delta_\beta$. Then the union of the δ_β's is a distance function

$$\bigcup_{\beta<\alpha} \delta_\beta \colon \left[\bigcup_{\beta<\alpha} X_\beta\right] \times \left[\bigcup_{\beta<\alpha} X_\beta\right] \to L$$

and we let

$$(X_\alpha, \delta_\alpha) = \left(\bigcup_{\beta<\alpha} X_\beta, \bigcup_{\beta<\alpha} \delta_\beta\right) \uparrow (a_\alpha, b_\alpha, u_\alpha, v_\alpha)$$

and

$$X^+ = \bigcup_{\alpha<\lambda} X_\alpha \quad \text{and} \quad \delta^+ = \bigcup_{\alpha<\lambda} \delta_\alpha$$

Then δ^+ is also a distance function.

Thus, for any set X and any distance function $\delta: X \times X \to L$, there is a set $X^+ \supseteq X$ and $\delta^+ \supseteq \delta$ with the property that if $u[\overline{\delta}(a \vee b)]v$, then $u[\overline{\delta^+}(a) \vee \overline{\delta^+}(b)]v$, with witness sequence of length 3 in general or length 2 when L is modular. It follows that δ^+ preserves join in the restricted sense that

$$\overline{\delta}(a \vee b) = \overline{\delta^+}(a) \vee \overline{\delta^+}(b)$$

To finish the proof, note that if L does not have a smallest element, we can adjoin one to get a lattice L_0. If L_0 is isomorphic to a sublattice of Part(X), then so is L. Therefore, we may as well assume that L has a smallest element. Define a distance function $\delta_0: L \times L \to L$ by

$$\delta_0(a, b) = \begin{cases} a \vee b & \text{if } a \neq b \\ 0 & \text{if } a = b \end{cases}$$

and let $X_0 = L$. Define a sequence of pairs (X_k, δ_k) for $k = 0, 1, \ldots$ by

$$X_{k+1} = X_k^+ \quad \text{and} \quad \delta_{k+1} = \delta_k^+$$

and then set

$$X = \bigcup \{X_k \mid k = 0, 1, \ldots\}$$

and

$$\delta = \bigcup \{\delta_k \mid k = 0, 1, \ldots\}$$

Then $\delta: X \times X \to L$ is a distance function.

Moreover, if $\uparrow$ is the 3-extension and $u\overline{\delta}(a \vee b)v$ for $u, v \in X$, then $u, v \in X_k$ for some k and so $u\overline{\delta_k}(a \vee b)v$, which implies that

$$u\overline{\delta_k^+}(a)z_1\overline{\delta_k^+}(b)z_2\overline{\delta^+}(a)z_3\overline{\delta^+}(b)v$$

for some $z_1, z_2, z_3 \in X$. Since $\delta_k^+ = \delta_{k+1} \subseteq \delta$, we have

$$u\overline{\delta}(a)z_1\overline{\delta}(b)z_2\overline{\delta}(a)z_3\overline{\delta}(b)v$$

and so $u[\overline{\delta}(a) \vee \delta(b)]v$. Hence, the embedding $\overline{\delta}: L \to$ Part(X) is a lattice embedding of type 3. A similar argument holds for the 2-extension when L is modular.□

Theorem 4.21 prompts us to wonder about type 1 representations.

Definition *Let $x_0, x_1, x_2, y_0, y_1, y_2$ be variables and let*

$$w_{i,j} = (x_i \vee x_j) \wedge (y_i \vee y_j)$$
$$w = w_{0,1} \wedge (w_{0,2} \vee w_{1,2})$$

The **Arguesian identity** *is*

$$(x_0 \vee y_0) \wedge (x_1 \vee y_1) \wedge (x_2 \vee y_2) \leq ((w \vee x_1) \wedge x_0) \vee ((w \vee y_1) \wedge y_0)$$

and a lattice that satisfies the Arguesian identity is **Arguesian.**□

It can be shown that the following implications hold, but their converses do not hold:

Distributive ⇒ Arguesian ⇒ Modular

Theorem 4.23 (Jónsson [37]) *Any lattice that has a type 1 representation is Arguesian.*

Proof. We may assume that L is a sublattice of Part(X). Let $a_i, b_i \in L$ for $i = 0, 1, 2$. If $x, y \in X$ are related under

$$(a_0 \vee b_0) \wedge (a_1 \vee b_1) \wedge (a_2 \vee b_2)$$

then $x(a_i \vee b_i)y$ and so there exist u_0, u_1 and u_2 in X for which

$$x a_i u_i b_i y$$

for $i = 0, 1, 2$. Let

$$c_{i,j} = (a_i \vee a_j) \wedge (b_i \vee b_j)$$

for $0 \leq i < j \leq 3$. Then

$$u_i c_{i,j} u_j$$

for all $0 \leq i < j \leq 3$. Set

$$c = c_{0,1} \wedge (c_{0,2} \vee c_{1,2})$$

Then $u_0 c u_1$ and so

$$x[((c \vee a_1) \wedge a_0)]u_0$$
$$u_0[((c \vee b_1) \wedge b_0)]y$$

and so x and y are related under the join

$$((c \vee a_1) \wedge a_0) \vee ((c \vee b_1) \wedge b_0)$$

as required by the Arguesian identity.□

The converse of Theorem 4.23 is not true. M. Haiman [29] and [30] has shown that there are Arguesian lattices that do not have type 1 representations.

Note that the embedding $\Lambda\colon L \hookrightarrow \text{Part}(X)$ of Theorem 4.21 involves an infinite set X and for many years, it remained an open question as to whether or not a finite lattice could be embedded into a finite partition lattice. In 1980, Pudlák and Tůma [50] showed that this is the case, but the proof is quite difficult.

One of the most interesting consequences of Theorem 4.23 is the following.

Theorem 4.24 (Whitman [67]) *Every lattice L is isomorphic to a sublattice of the lattice $\mathcal{S}(G)$ of all subgroups of some group G.*

Proof. We may assume that L is a sublattice of $\text{Part}(X)$. Let S_X be the group of all permutations of X. For each $a \in L$, let S_a be the subgroup of S_X generated by all transpositions $(x\,y)$ for which xay. If $\tau \in S_a$, then τ is a product of transpositions

$$\tau = (x_1\,y_1)\cdots(x_n\,y_n)$$

where x_iay_i. If the transpositions on the right are not disjoint, then we may reorder (if necessary) and combine two adjacent transpositions with a common element. If the transpositons are identical, they cancel. If they have exactly one element in common, then we get a 3-cycle in which all three elements are related under a. This process continues until we reach a product of disjoint cycles in which all elements in each cycle are related under a. In other words, the unique (up to order) cyclic decomposition of τ has the property that all elements in a given cycle are related under a.

In particular, any transposition $(u\,v)$ in S_a has the property that uav and so a transposition $(u\,v)$ is in S_a if and only if uav, which implies that the map $a \mapsto S_a$ is injective. Also, $\tau \in S_a \cap S_b$ if and only if the cycles in its unique cycle decomposition have the property that every element in each cycle is related by a and by b, and so by $a \wedge b$. Thus, $S_a \cap S_b = S_{a \wedge b}$. Finally, $S_a \subseteq S_{a \vee b}$ and $S_b \subseteq S_{a \vee b}$ imply that $S_a \vee S_b \subseteq S_{a \vee b}$. For the reverse inclusion, if $\tau \in S_{a \vee b}$, then

$$\tau = (x_1\,y_1)\cdots(x_n\,y_n)$$

where x_iay_i or x_iby_i. Thus, each transposition $(x_i\,y_i)$ is in S_a or S_b and therefore in $S_a \vee S_b$. Hence, $\tau \in S_a \vee S_b$, which shows that $S_a \vee S_b = S_{a \vee b}$ and so the map $a \mapsto S_a$ is a lattice embedding from L into $\mathcal{S}(G)$.□

Distributive Lattices

Let us explore some of the basic properties of distributive lattices. First, in a distributive lattice, maximal ideals are prime.

Theorem 4.25 *In a distributive lattice L, all maximal ideals are prime.*
Proof. Let M be a maximal ideal of L. Suppose that $a \wedge b \in M$ but $a \notin M$. Then $\downarrow a \vee M = L$ and so $b \leq a \vee m$ for some $m \in M$. Hence,

$$b = b \wedge (a \vee m) = (b \wedge a) \vee (b \wedge m) \in M$$

which shows that M is prime.□

In a distributive lattice, elements are join-irreducible if and only if they are join-prime.

Theorem 4.26 *Let L be a distributive lattice.*
1) *The following are equivalent for $x \in L$:*
   a) *x is join-irreducible*
   b) *x is join-prime*
   c) *$\uparrow x$ is a prime filter, or equivalently, $(\uparrow x)^c$ is a prime ideal.*
2) *Dually, the following are equivalent for $x \in L$:*
   a) *x is meet-irreducible*
   b) *x is meet-prime*
   c) *$\downarrow x$ is a prime ideal.*

Proof. If x is join-irreducible, then

$$x \leq a \vee b \quad \Leftrightarrow \quad x \wedge (a \vee b) = x \quad \Leftrightarrow \quad (x \wedge a) \vee (x \wedge b) = x$$

and so $x \vee a = x$ or $x \wedge b = x$, that is, $x \leq a$ or $x \leq b$. Hence, x is join-prime. Conversely, if x is join-prime and $x = a \vee b$, then $x \leq a \vee b$ and so we may assume that $x \leq a$, in which case $x \leq a \leq x$, that is, $x = a$. Hence, x is join-irreducible. The rest follows from Theorem 3.37.□

We have seen that the down map $\phi_\downarrow$ is an order embedding and so its restriction $\phi_\downarrow\colon \mathcal{M}(L) \hookrightarrow \operatorname{Spec}(L)$ is an order embedding of meet-irreducibles into prime ideals. Similarly, the up map is an anti-embedding and so the *not-up map* $\phi_{\neg\uparrow}\colon \mathcal{J}(L) \to \operatorname{Spec}(L)$ defined by

$$\phi_{\neg\uparrow}(a) = (\uparrow a)^c$$

is an order embedding of join-irreducibles into prime ideals. More can be said if L has a chain condition.

Theorem 4.27 *Let L be a distributive lattice.*
1) *The not-up map $\phi_{\neg\uparrow}\colon \mathcal{J}(L) \to \operatorname{Spec}(L)$ defined by*

$$\phi_{\neg\uparrow}(a) = (\uparrow a)^c$$

is an order embedding. If L has the DCC, then all prime ideals have the form $(\uparrow a)^c$ *for* $a \in \mathcal{J}(L)$ *and so*

$$\phi_{\neg\uparrow}: \mathcal{J}(L) \approx \mathrm{Spec}(L)$$

is an order isomorphism.

2) *The down map* $\phi_\downarrow: \mathcal{M}(L) \to \mathrm{Spec}(L)$ *defined by*

$$\phi_\downarrow(m) = \downarrow m$$

is an order embedding. If L has the ACC, then

$$\phi_\downarrow: \mathcal{M}(L) \approx \mathrm{Spec}(L)$$

is an order isomorphism.

3) *If L is finite, then*

$$\phi_\downarrow \circ \phi_{\neg\uparrow}: \mathcal{J}(L) \approx \mathcal{M}(L)$$

and so $\mathcal{J}(L)$ *and* $\mathcal{M}(L)$ *are isomorphic.*

Proof. For part 1), if L has the DCC, then the prime filters are principal and so have the form $\uparrow a$, where a is join-irreducible. Hence, the prime ideals (being complements of the prime filters) have the form $(\uparrow a)^c$, where a is join-irreducible. Thus, $\phi_{\neg\uparrow}$ is surjective. Part 2) is similar.□

Distributive lattices also have nice properties with respect to finiteness.

Theorem 4.28 *For a distributive lattice L, the following are equivalent:*

1) *L is finite*
2) *L has finite length*
3) *L has no infinite chains (equivalently, L has both chain conditions).*

Proof. Clearly a finite lattice has finite length and a lattice with finite length has no infinite chains. Suppose that L has no infinite chains. Since L has the DCC, Theorem 3.29 implies that the set $\mathcal{J}(L)$ of join-irreducibles in L is finite-join-dense in L. Hence, it is sufficient to show that $\mathcal{J}(L)$ is a finite set.

Certainly, $\mathcal{J}(L)$ has no infinite chains and so Theorem 1.14 implies that it is sufficient to show that $\mathcal{J}(L)$ has no infinite antichains. If $A = (a_1, a_2, \ldots)$ is an infinite antichain in $\mathcal{J}(L)$, the ACC on L implies that the sequence of joins

$$a_1, a_1 \vee a_2, a_1 \vee a_2 \vee a_3, \ldots$$

is eventually constant and so there is an $n > 0$ for which

$$a_1 \vee \cdots \vee a_n \vee a_{n+1} = a_1 \vee \cdots \vee a_n$$

that is,

$$a_{n+1} \le a_1 \vee \cdots \vee a_n$$

Since a_{n+1} is join-irreducible, it is join-prime and so $a_{n+1} \le a_k$ for some $k \le n$, contradicting the fact that A is an antichain. Hence, $\mathcal{J}(L)$ has no infinite antichains.□

Irredundant Join-Irreducible Representations

Theorem 3.29 implies that if a lattice L has the DCC, then $\mathcal{J}(L)$ is finite-join-dense in L. For $a \in L$, an expression of the form

$$a = u_1 \vee \cdots \vee u_n$$

where the u_i's are *distinct* join-irreducibles is **irredundant** if a is not the join of any proper subset of $U = \{u_1, \ldots, u_n\}$. Clearly, if the join is irredundant, then U must be an antichain. Moreover, if L is distributive, then the join is irredundant if and only if U is an antichain. Let us refer to such a representation of a as an **irredundant join-irreducible representation** of a.

In the presence of modularity, we can say some things about the uniqueness of irredundant join-irreducible representations of elements.

Theorem 4.29 (Kurosh–Ore theorem, [40], [48]) *Let L be a modular lattice. Suppose that $a \in L$ has two irredundant join-irreducible representations*

$$a = x_1 \vee \cdots \vee x_n = y_1 \vee \cdots \vee y_m$$

1) **(Kurosh–Ore replacement property)** *For any x_i, we can substitute some y_j to get another join-irreducible representation of a.*
2) $n = m$
Thus, if L has the DCC, then every element $a \in L$ has an irredundant join-irreducible representation

$$a = x_1 \vee \cdots \vee x_n$$

and all such representatioms have the same number of terms.
Proof. For part 1), it is sufficient to prove the theorem for x_1. Let

$$\widehat{x}_1 = x_2 \vee \cdots \vee x_n \quad \text{and} \quad c_k = y_k \vee \widehat{x}_1$$

Then we must find an index k for which $c_k = a$. Now, it is true that

$$a = c_1 \vee \cdots \vee c_m$$

However, a is join-irreducible in the interval $[\widehat{x}_1, a]$, which contains each c_k, because the modularity of L implies that

$$[x_1 \wedge \widehat{x}_1, x_1] \approx [\widehat{x}_1, x_1 \vee \widehat{x}_1] = [\widehat{x}_1, a]$$

and because x_1 is join-irreducible in the interval $[x_1 \wedge \widehat{x}_1, x_1]$. Thus, $a = c_k$ for some k.

For part 2), suppose that $n < m$. Then we can replace all of the x's by y's to get a shorter join-irreducible representation of a using only y's, which contradicts the irredundancy of the original representation of a in terms of the y's.□

When L is distributive, there is uniqueness, since

$$x_1 \vee \cdots \vee x_n = y_1 \vee \cdots \vee y_n$$

implies that

$$y_i \leq x_1 \vee \cdots \vee x_n$$

and so $y_i \leq x_j$ for some j. By a symmetric argument, $y_i \leq x_j \leq y_k$ for some k and since $\{y_1, \ldots, y_n\}$ is an antichain, equality must hold and so

$$\{x_1, \ldots, x_n\} = \{y_1, \ldots, y_n\}$$

Theorem 4.30 *Let L be a distributive lattice.*

1) *Every $a \in L$ has at most one irredundant join-irreducible representation*

$$a = x_1 \vee \cdots \vee x_n$$

2) *If L has the DCC, then every $a \in L$ has exactly one irredundant join-irreducible representation.* □

Exercises

1. Prove that the modular law is self-dual.
2. Find an infinite lattice with no infinite chains.
3. Show that the lattice of normal subgroups of the noncyclic four-group

$$V = \{1, a, b, ab\}$$

is isomorphic to M_3, which is not distributive.

4. If G is a group, prove that $\mathcal{S}(G)$ need not be modular. *Hint*: Consider the alternating group A_4 of order 12. Let

$$A = \langle(1\,2)(3\,4)\rangle, B = \langle(1\,2)(3\,4), (1\,3)(2\,4)\rangle \text{ and } C = \langle(1\,2\,3)\rangle$$

5. Prove that the family $\mathcal{N}(G)$ of normal subgroups of a group G is modular.
6. Prove that a lattice is distributive if and only if

$$a \vee (b \wedge c) \geq (a \vee b) \wedge c$$

for all $a, b, c \in L$.

7. Let L be a distributive lattice and let A and B be ideals in L. Prove that

$$A \vee B = \{a \vee b \mid a \in A, b \in B\}$$

8. Prove that a lattice is distributive if and only if the map $\phi_a: L \to L \times L$ defined by

$$\phi_a(x) = (x \wedge a, x \vee a)$$

is a lattice embedding, for all $a \in L$.

9. Prove that a lattice is distributive if and only if for all $a, b, c \in L$, the self-dual **median property** holds

$$(a \wedge b) \vee (a \wedge c) \vee (b \wedge c) = (a \vee b) \wedge (a \vee c) \wedge (b \vee c)$$

10. If L is a distributive lattice, show that an expression

$$a = u_1 \vee \cdots \vee u_n$$

with u_i join-irreducible is irredundant if and only if the terms are distinct and form an antichain.

11. Prove that a finite lattice L is modular if and only if for all $a, b \in L$, the opposite sides of the quadrilateral in Figure 4.4 are isomorphic, that is,

$$[a \wedge b, a] \approx [b, a \vee b]$$

12. Is a relatively complemented lattice complemented?
13. Show that the subspaces of a vector space V form a complemented modular lattice under set inclusion. Describe the relative complements of elements. Is the lattice distributive?
14. Prove that the lattice $(\mathbb{N}, \mid)$ is distributive.
15. Prove that the lattice of subgroups of a finite cyclic group G of order n is distributive.
16. Using the fact that if the lattice $\mathcal{S}(G)$ is distributive then G is cyclic, describe all finite groups whose lattice of subgroups is isomorphic to a power set lattice $\wp(X)$.
17. Let L be a bounded distributive lattice. Prove that the set S of complemented elements of L form a sublattice of L.
18. Let X be a nonempty set. Let $\mathcal{P}$ be the family of all partitions of X. Order $\mathcal{P}$ as follows: $\lambda \leq \sigma$ if λ is a refinement of σ, that is, if every block of σ is a union of blocks of λ.
    a) Show that $(\mathcal{P}, \leq)$ is a complete lattice. Describe the meet and join.
    b) Describe the largest and smallest elements of $\mathcal{P}$.
    c) Describe the covering relation.
    d) Show that the height of a partition λ is equal to $|X| - |\lambda|$.
    e) Prove that if $n \geq 4$ then $\mathcal{P}$ is not modular.
19. Let L be a distributive lattice. If a and a' are complementary then $a \wedge a' = 0$. Prove that a' is the largest x for which $a \wedge x = 0$.
20. Let L be a distributive lattice in which $\mathcal{J}(L) = \mathcal{A}(L)$ and $\mathcal{A}(L)$ is join-dense. Suppose that $a \in L$ has a complement a'. Let

$$\mathcal{A}_{\downarrow}(a) = \{x \in \mathcal{A}(L) \mid x \leq a\}$$

and

$$\mathcal{A}_{\downarrow}(a') = \{x \in \mathcal{A}(L) \mid x \leq a'\}$$

Prove that $\mathcal{A}(L) = \mathcal{A}_{\downarrow}(a) \cup \mathcal{A}_{\downarrow}(a')$.

21. Let L be a distributive lattice and let I be a proper ideal of L and let F be a proper filter in L. Suppose that M is a maximal element of the set of all ideals of L that contain I and are disjoint from F. Prove that M is prime in L. *Hint*: Suppose that $a \wedge b \in M$ but that $a, b \notin M$ and consider the ideals

$$J_x = \downarrow\{x \vee m \mid m \in M\}$$

for $x = a$ and $x = b$.

22. Let L be a lattice and let $A \subseteq L$ be a nonempty subset of L. Then the **lattice generated** by A, denoted by $\langle A \rangle$, is defined to be the intersection of all sublattices of L containing A. If L is distributive, show that $\langle A \rangle$ can be constructed as follows: First, take the meets of all nonempty finite subsets of A. Then $\langle A \rangle$ is the set of joins of all nonempty finite subsets of these meets.
23. Prove part 2) of Theorem 4.30 using duality.

The Partition Lattice

24. Let G be a group. Show that there is a very well known representation of the lattice $\mathcal{N}(G)$ of normal subgroups of G in $\mathrm{Part}(G)$. Show that this representation is of type 1 and so $\mathcal{N}(G)$ is Arguesian.
25. Prove that $\mathrm{Part}(X)$ is strongly atomic.
26. Prove that $\mathrm{Part}(X)$ is semimodular.
27. Prove that $\mathrm{Part}(X)$ is relatively complemented.
28. Prove that an Arguesian lattice is modular by showing that N_5 is not Arguesian. *Hint*: let

$$\begin{aligned} x_0 = y_1 &= y \\ x_1 = x_2 = y_0 &= z \\ y_2 &= x \end{aligned}$$

Pseudocomplements

Let L be a lattice with a smallest element. For $a \in L$, an element $x \in L$ such that

$$a \wedge x = 0$$

is called a **semicomplement** of a. The set $\mathcal{S}(a)$ of semicomplements of a is clearly a down-set. If $\mathcal{S}(a)$ is a principal ideal, that is, if $\mathcal{S}(a)$ has a greatest element, this element is called a **pseudocomplement** of a. A lattice in which every element has a pseudocomplement is called a **pseudocomplemented lattice**.

More generally, if $a, b \in L$ and if the set

$$\{x \in L \mid a \wedge x \leq b\}$$

has a greatest element, it is called a **relative pseudocomplement** of a in b. We denote this element by $a \curvearrowright b$. (There does not appear to be a generally accepted notation for relative pseudocomplement.) If every pair of elements has a relative pseudocomplement, then L is said to be **relatively pseudocomplemented**. A relatively pseudocomplemented lattice is also called a **Heyting lattice** or a **Heyting algebra**.

The dual of a Heyting algebra is a lattice L that satisfies the following condition: For every $a, b \in L$, the set

$$\{x \in L \mid a \vee x \geq b\}$$

has a smallest element, which we will denote by $a \curvearrowleft b$. If $a \curvearrowleft b$ exists for every pair of elements in L, then L is called a **Brouwerian lattice**. (Birkhoff refers to Heyting lattices as Brouwerian lattices but others refer to the dual of a Heyting algebra as a Brouwerian lattice.)

29. Prove that every Brouwerian or Heyting lattice is distributive.
30. Show that any finite distributive lattice is Brouwerian and Heyting.
31. Let L be a distributive lattice. Suppose that a and a' are complementary in L.
    a) Prove that $a \curvearrowright b = a' \vee b$.
    b) Show that the ordinary complement a' is a special type of relative pseudocomplement.
32. Let $L = (\mathbb{N}, \mid)$. Show that $a \curvearrowleft b$ exists for all $a, b \in L$ but that the relative pseudocomplement $a \curvearrowright b$ does not exist. Thus, L is Brouwerian but not Heyting.
33. Let P be a finite poset. Let $L = \mathcal{O}(P)$ be the lattice of down-sets in P. Show that L is pseudocomplemented. *Hint*: What does it imply to say that a down-set D is disjoint from a down-set A?

Chapter 5
Boolean Algebras

Boolean Lattices

Complemented distributive lattices have particularly nice properties.

Definition *A complemented distributive lattice is called a* **Boolean lattice**.□

Since a distributive lattice has the property that the sets $a_{[u,v]}$ are empty or singletons, Boolean lattices are *uniquely* complemented. Some simple properties of Boolean lattices follow.

Theorem 5.1 *Let L be a Boolean lattice. Let $a, b \in L$. Then*
1)

$$0' = 1,\ 1' = 0 \quad \textit{and} \quad a'' = a$$

2) **(de Morgan's laws)**

$$(a \wedge b)' = a' \vee b'$$
$$(a \vee b)' = a' \wedge b'$$

3) **(Order and complements)**

$$a \leq b \quad \Leftrightarrow \quad a \wedge b' = 0 \quad \Leftrightarrow \quad a' \vee b = 1$$

In particular,
3a) **(Order-reversing)**

$$a \leq b \quad \Leftrightarrow \quad b' \leq a'$$

3b) **(Atoms)** *If $a \in \mathcal{A}(L)$, then for any $x \in L$,*

$$a \leq x \quad \textit{or} \quad a \leq x' \quad \textit{but not both}$$

Proof. For part 3), if $a \leq b$, then $a \wedge b'$ is less than or equal to both b and b' and so $a \wedge b' = 0$. Conversely, if $a \wedge b' = 0$, then

S. Roman (ed.), *Lattices and Ordered Sets*, doi: 10.1007/978-0-387-78901-9_5,

$$b = b \vee (a \wedge b') = (b \vee a) \wedge (b \vee b') = b \vee a$$

and so $a \leq b$. The other part is dual. We leave the rest of the proof to the reader.□

Theorem 3.45 (projecting a complement into an interval to get a relative complement) implies that a Boolean lattice is relatively complemented. Therefore, distributivity implies that L is uniquely relatively complemented.

Theorem 5.2 *A Boolean lattice L is uniquely relatively complemented. Hence, the following hold in L:*

1) For any $a \in [u, v]$, we have

$$a_{[u,v]} = (a' \vee u) \wedge v = (a' \wedge v) \vee u$$

2) The join-irreducible elements of L are precisely the atoms of L, that is,

$$\mathcal{J}(L) = \mathcal{A}(L)$$

3) A Boolean lattice is atomic if and only if it is atomistic.□

We have seen that in a distributive lattice, all maximal ideals are prime. In a Boolean lattice, the converse is also true.

Theorem 5.3 *Let L be a Boolean lattice and let I be a proper ideal of L. The following are equivalent:*

1) I is maximal.
2) I is prime.
3) For all $a \in L$, exactly one of a or a' is in I.

Proof. We have seen that maximal ideals are prime. If I is prime, then $a \wedge a' \in I$ and so $a \in I$ or $a' \in I$. However, if both are in I, then so is $a \vee a' = 1$, which is false since I is proper. Finally, if 3) holds and $I \subset J$, then for $a \in J \setminus I$, we have $a' \in I$ and so $1 = a \vee a' \in J$, whence $J = L$. This shows that I is maximal.□

Boolean Algebras

Just as a lattice can be defined algebraically as a nonempty set with two operations $\wedge$ and $\vee$ satisfying certain requirements, with no explicit mention of the underlying partial order, Boolean lattices can also be defined algebraically as a set with operations. In this context, the term *Boolean algebra* is usual.

Definition *A* **Boolean algebra** $(B, \wedge, \vee, 0, 1, ')$ *is a sextuple, where B is a nonempty set with two binary operations $\wedge$ and $\vee$, one unary operation $'$ and two nullary operations (constants) 0 and 1, having the following properties:*

1) $\wedge$ *and* $\vee$ *are associative, commutative, idempotent and satisfy the absorption laws, that is, the triple* $(B, \wedge, \vee)$ *is a lattice as defined algebraically in Theorem 3.17.*
2) $\wedge$ *and* $\vee$ *satisfy the distributive laws.*
3) 0 *is the identity for join and* 1 *is the identity for meet, that is,*

$$a \vee 0 = a \quad \textit{and} \quad a \wedge 1 = a$$

for all $a \in L$.
4) The unary operation $'$ *satisfies the definition of complement:*

$$a \vee a' = 1 \quad \textit{and} \quad a \wedge a' = 0$$

□

Thus, a Boolean algebra $(B, \wedge, \vee, ', 0, 1)$ is a Boolean lattice with smallest element 0 and largest element 1. Conversely, any Boolean lattice is a Boolean algebra.

Definition *A nonempty subset* S *of a Boolean algebra* B *is a* **Boolean subalgebra** *if* S *inherits the bounds* 0 *and* 1 *as well as finite meets, joins and complements from* B.□

Example 5.4

1) The power set $\wp(S)$ of a set S is a Boolean algebra, called the **power set algebra**, or just the power set of S. In fact, these are very general algebras: We will prove in a later chapter that every finite Boolean algebra is isomorphic to $\wp(S)$ for some finite set S and, more generally, every complete atomic Boolean algebra is isomorphic to a power set $\wp(S)$. A Boolean subalgebra B of a power set algbera $\wp(S)$ is called a **power set subalgebra** or an **algebra of sets** or a **field of sets**. Put another way, a family $\mathcal{F}$ of sets is an algebra of sets if it has finite unions, finite intersections and complements.
2) Let S be a nonempty set. The **finite-cofinite algebra** $FC(S)$ is the family of all subsets of S that are either finite or cofinite. (A subset is **cofinite** if its complement is finite.) For a countably infinite set S, this Boolean algebra is not isomorphic to any power set $\wp(X)$ because $|FC(S)| = \aleph_0$ whereas $|\wp(X)|$ is either finite or uncountable. Alternatively, we can observe that $FC(S)$ is not complete for any infinite set S, but any lattice of the form $\wp(X)$ is complete.□

Boolean Rings

There is yet another way to view Boolean algebras algebraically.

Definition *A* **Boolean ring** *is a ring* $(R, +, \cdot, 0, 1)$ *with identity* 1 *in which every element is* **idempotent**, *that is,*

$$r^2 = r$$

for all $r \in R$.□

We leave proof of the following to the reader.

Theorem 5.5 *A Boolean ring is commutative and* $2r = 0$ *for all* $r \in R$.□

Theorem 5.6

1) *Let* $(R, +, \cdot, 0, 1)$ *be a Boolean ring. Define binary operations* $\wedge$ *and* $\vee$ *and a unary operation* $'$ *in* R *by*

$$\begin{aligned} r \wedge s &= rs \\ r \vee s &= r + s + rs \\ r' &= 1 + r \end{aligned}$$

Then $(R, \wedge, \vee, 0, 1, ')$ *is a Boolean algebra.*

2) *Conversely, let* $(B, \wedge, \vee, 0, 1, ')$ *be a Boolean algebra. Define binary operations by*

$$\begin{aligned} rs &= r \wedge s \\ r + s &= (r \vee s) \wedge (r' \vee s') = (r \wedge s') \vee (r' \wedge s) \end{aligned}$$

Then $(B, +, \cdot, 0, 1)$ *is a Boolean ring.*

The maps

$$\alpha\colon (R, +, \cdot, 0, 1) \mapsto (R, \wedge, \vee, 0, 1, ')$$

and

$$\rho\colon (R, \wedge, \vee, 0, 1, ') \mapsto (R, +, \cdot, 0, 1)$$

described above are inverses of each other. Moreover, a nonempty subset $S \subseteq R$ *is a lattice ideal if and only if it is a ring ideal.*

Proof. For part 1), it is clear that meet is associative, commutative and idempotent and that join is commutative and idempotent. Verification of associativity of join is straightforward, and is left to the reader. Finally, the absorption laws hold since

$$\begin{aligned} a \wedge (a \vee b) &= a(a + b + ab) = a \\ a \vee (a \wedge b) &= a + ab + aab = a \end{aligned}$$

Note that the absorption laws do not hold if join is defined simply as addition.

For part 2), let B be a Boolean algebra. As to addition, a straightforward but tedious calculation shows that addition is associative. The zero of B is 0, since

$$r + 0 = (r \vee 0) \wedge (r' \vee 1) = r$$

It is clear that addition is commutative and so B is an abelian group. It is clear that multiplication is associative and the identity of B is 1. We leave verification of distributivity

$$r(s+t) = rs + rt$$

for the reader. Finally,

$$r^2 = r \wedge r = r$$

and so B is a Boolean ring.

Given a Boolean algebra $\mathcal{B} = (B, \wedge, \vee, 0, 1, ')$, the Boolean ring $\rho(\mathcal{B}) = (B, +, \cdot, 0, 1)$ satisfies

$$rs = r \wedge s$$
$$r + s = (r \vee s) \wedge (r' \vee s')$$

In the Boolean algebra $\alpha(\rho(\mathcal{B})) = (B, \overline{\wedge}, \overline{\vee}, \overline{0}, \overline{1}, \overline{'})$, we have

$$r^{\overline{'}} = 1 + r = (1 \vee r) \wedge (0 \vee r') = r'$$

and so complements in $\mathcal{B}$ and $\alpha(\rho(\mathcal{B}))$ are the same. Meets are also the same, since

$$r \overline{\wedge} s = rs = r \wedge s$$

For the join, first note that

$$(r+s)' = [(r \vee s) \wedge (r' \vee s')]' = (r \wedge s) \vee (r' \wedge s')$$

and so

$$\begin{aligned} r \overline{\vee} s &= r + s + rs \\ &= [(r \vee s) \wedge (r' \vee s')] + (r \wedge s) \\ &= \{[(r \vee s) \wedge (r' \vee s')] \vee (r \wedge s)\} \wedge \{[(r \wedge s) \vee (r' \wedge s')] \vee (r' \vee s')\} \\ &= (r \vee s) \wedge [(r' \vee s') \vee (r \wedge s)] \wedge [(r \wedge s) \vee (r' \vee s')] \\ &= (r \vee s) \wedge [r' \vee s' \vee (r \wedge s)] \\ &= (r \vee s) \wedge [(r \wedge s)' \vee (r \wedge s)] \\ &= r \vee s \end{aligned}$$

Thus, meet, join and complement are the same in $\mathcal{B}$ and $\alpha(\rho(\mathcal{B}))$, which implies that the order is the same and so $\overline{0} = 0$ and $\overline{1} = 1$. Thus, $\alpha(\rho(\mathcal{B})) = \mathcal{B}$ as Boolean algebras and so $\alpha \circ \rho = \iota$.

Given a Boolean ring $\mathcal{R} = (R, +, \cdot, 0, 1)$, the Boolean algebra $\alpha(\mathcal{R})$ satisfies

$$r \wedge s = rs$$
$$r \vee s = r + s + rs$$
$$r' = 1 + r$$

In the corresponding Boolean ring $\rho(\sigma(\mathcal{R})) = (B, \overline{+}, \overline{\cdot}, \overline{0}, \overline{1})$, we have

$$r \,\overline{\cdot}\, s = r \wedge s = rs$$

and

$$\begin{aligned} r \,\overline{+}\, s &= (r \wedge s') \vee (r' \wedge s) \\ &= rs' \vee r's \\ &= rs' + r's + rs'r's \\ &= r(1+s) + (1+r)s + r(1+s)(1+r)s \\ &= r + rs + s + rs + rs(1+s+r+rs) \\ &= r + rs + s + rs + rs + rs + rs + rs \\ &= r + s \end{aligned}$$

Since addition and multiplication are the same in $\mathcal{R}$ and $\alpha(\rho(\mathcal{R}))$, we have $\mathcal{R} = \alpha(\rho(\mathcal{R}))$ as rings and so $\rho \circ \alpha = \iota$.

Finally, let $(R, +, \cdot, 0, 1)$ be a Boolean ring and let $(R, \wedge, \vee, 0, 1)$ be the corresponding Boolean algebra. To see that a ring ideal $I \subseteq R$ is a lattice ideal, let $a, b \in I$. Then

$$a \vee b = a + b + ab \in I$$

and if $r \leq a$, then $r = r \wedge a = ra \in I$ and so I is a down-set. Thus, I is a lattice ideal. Conversely, if $I \subseteq R$ is a lattice ideal, then for $a, b \in I$,

$$a + b = (a \wedge b') \vee (a' \wedge b) \in I$$

Also, if $r \in R$ and $a \in I$, then

$$ra = r \wedge a \in I$$

Hence, I is a ring ideal.□

Boolean Homomorphisms

Boolean homomorphisms preserve the Boolean algebra structure.

Definition *Let L and M be Boolean algebras. A* **Boolean homomorphism** *from L to M is a function $f: L \to M$ with the following properties:*

1) f preserves meets and joins, that is,

$$f(a \wedge b) = f(a) \wedge f(b)$$
$$f(a \vee b) = f(a) \vee f(b)$$

2) f preserves 0 and 1, that is,

$$f0 = 0 \quad \textit{and} \quad f1 = 1$$

3) f preserves complements, that is,

$$f(a') = (fa)'$$

□

It is not hard to see that in a uniquely complemented lattice, preservation of bounds (0 and 1) is equivalent to preservation of complements.

Theorem 5.7 *The following are equivalent for a lattice homomorphism $f: L \to M$ between uniquely complemented lattices:*

1) *f preserves complements, that is,*

$$f(a') = (fa)'$$

for all $a \in L$.

2) *f is a $\{0,1\}$-homomorphism, that is,*

$$f(0) = 0 \quad \text{and} \quad f(1) = 1$$

Proof. This follows easily from the equations

$$f(0) = f(a \wedge a') = f(a) \wedge f(a')$$
$$f(1) = f(a \vee a') = f(a) \vee f(a')$$

□

Note that a Boolean homomorphism is just a lattice $\{0,1\}$-homomorphism between Boolean algebras.

One place to find Boolean homomorphisms is in induced maps. If $f: A \to B$ is a function between nonempty sets, the **induced inverse map** $f^{-1}: \wp(B) \to \wp(A)$ is defined by

$$f^{-1}(V) = \{a \in A \mid f(a) \in V\}$$

If f is bijective, the symbol f^{-1} usually denotes the ordinary inverse map defined on the elements of B, rather than the induced inverse map defined on the power set $\wp(B)$.

We leave proof of the following to the reader.

Theorem 5.8 *Let $f: A \to B$ be a function between nonempty sets. The induced inverse map $f^{-1}: \wp(B) \to \wp(A)$ is a Boolean homomorphism. In fact, for any subset U of B and any family $\{U_i \mid i \in I\}$ of subsets of B, we have*

$$f^{-1}(\bigcup U_i) = \bigcup f^{-1}(U_i)$$
$$f^{-1}(\bigcap U_i) = \bigcap f^{-1}(U_i)$$
$$f^{-1}(U') = [f^{-1}(U)]'$$

□

Note that if B is a Boolean lattice, then its dual B^∂ is also a Boolean lattice, with meet in B^∂ equal to join in B and join in B^∂ equal to meet in B. However, complement is self-dual: The complement in B^∂ is the same as the complement in B.

In a Boolean lattice B, the complement map $\sigma: x \mapsto x'$ is order-reversing and bijective and so is a lattice-isomorphism from B to its dual B^∂. Also, de Morgan's laws say precisely that σ is a Boolean algebra isomorphism.

Theorem 5.9 *Let B be a Boolean lattice. The complement map $x \mapsto x'$ is a Boolean isomorphism from B onto the dual B^∂.* □

Characterizing Boolean Lattices

We have discussed the fact that if a lattice L is uniquely *relatively* complemented, then it is distributive, but that being uniquely complemented it not sufficient to imply distributivity. This follows from a theorem of Robert Dilworth.

Theorem 5.10 (Dilworth, 1945 [16]) *Every lattice is a sublattice of a uniquely complemented lattice.* □

Thus, we are prompted to look at conditions that imply that a uniquely complemented lattice is distributive, and therefore Boolean. One such condition is the following special form of modularity.

Definition *Let L be a complemented lattice.*

1) *L satisfies* **SMP1** *(for special modular property) if*

$$a \geq c \quad \Rightarrow \quad a \wedge (a' \vee c) = c$$

for all $a, c \in L$.

2) *Dually, L satisfies* **SMP2** *if*

$$c \geq a \quad \Rightarrow \quad a \vee (a' \wedge c) = c$$

If L satisfies both SMP1 and SMP2, we say that L satisfies **SMP.** □

We can now describe conditions that characterize distributivity (and therefore Booleanness) among uniquely complemented lattices.

Theorem 5.11 *Let L be a uniquely complemented lattice. The following are equivalent:*

1) *L is distributive (and therefore a Boolean lattice).*
2) *L is modular (due to Birkhoff and von Neumann).*
3) *L satisfies SMP.*
4) *L satisfies de Morgan's laws.*

Proof. It is clear that 1) $\Rightarrow$ 2) $\Rightarrow$ 3). To see that 3) $\Rightarrow$ 1), we show that the sets $a_{[u,v]}$ are empty or singletons, which implies distributivity by Theorem 4.8. Actually, SMP is precisely what is needed to prove that any relative complement is also a complement and is therefore unique by assumption.

Specifically, we will show that if $b \in a_{[u,v]}$, then b is the complement of the element

$$e = (a \wedge u') \vee v'$$

which is formed by projecting a into the interval $[v', u']$. First, SMP1 and the fact that $v \geq a \wedge u'$ imply that

$$e \wedge b \leq e \wedge v = [(a \wedge u') \vee v'] \wedge v = a \wedge u'$$

and so $e \wedge b$ is contained in a, b and u' and since $a \wedge b = u$, it follows that $e \wedge b$ is contained in u and u', whence $e \wedge b = 0$. Dually, SMP2 implies that

$$e \vee b \geq e \vee u = v' \vee (a \wedge u') \vee u = v' \vee a$$

Hence, $e \vee b$ contains v', a and b and since $a \vee b = v$, it follows that $e \vee b$ contains v as well, and so $e \vee b = 1$. Thus, $b = e'$ and 1)–3) are equivalent.

We leave it to the reader to show that the distributive law implies de Morgan's laws. Finally, we show that de Morgan's laws imply SMP. For SMP2, it is sufficient to show that

$$x \geq a \quad \Rightarrow \quad [(x \wedge a') \vee a]' = x'$$

which, using de Morgan's laws, is

$$x \geq a \quad \Rightarrow \quad (x \wedge a')' \wedge a' = x'$$

The consequent is equivalent to

$$x \wedge [(x \wedge a')' \wedge a'] = 0$$
$$x \vee [(x \wedge a')' \wedge a'] = 1$$

The first of these equations is true. For the second equation, since $x' \leq a'$, we have

$$x \vee [(x \wedge a')' \wedge a'] = x \vee [(x' \vee a) \wedge a'] \geq x \vee [(x' \vee a) \wedge x'] = x \vee x' = 1$$

Hence, SMP2 holds. Condition SMP1 follows by duality.□

We also have the following result.

Theorem 5.12

1) **(Birkhoff and Ward [9])** *A uniquely complemented atomic lattice is distributive.*

2) A uniquely complemented lattice of finite width is distributive.

Proof. For part 1), let L be a uniquely complemented lattice and let $\mathcal{A} = \mathcal{A}(L)$ be the set of atoms of L. We first show that the $\mathcal{A}$-down map $\phi_{\mathcal{A},\downarrow} : L \to \wp(L)$ preserves strict order, that is,

$$x < y \quad \Rightarrow \quad \mathcal{A}_{\downarrow}(x) \subset \mathcal{A}_{\downarrow}(y)$$

It then follows from Theorem 3.32 that the $\mathcal{A}$-down map is injective. Now, if $x < y$ and x' is the complement of x, then $x' \vee y = 1$. It follows that $x' \wedge y > 0$, since otherwise, y and x would be distinct complements of x'. Hence, there is an atom $a \leq x' \wedge y$. In particular, $a \leq y$ and $a \leq x'$. Hence, $a \not\leq x$ and so $a \in \mathcal{A}_{\downarrow}(y) \setminus \mathcal{A}_{\downarrow}(x)$. Thus, the $\mathcal{A}$-down map is injective.

The $\mathcal{A}$-down map is a meet-homomorphism (as are all B-down maps). To show that the $\mathcal{A}$-down map is a join-homomorphism, we show that

$$\mathcal{A}_{\downarrow}(x \vee y) \subseteq \mathcal{A}_{\downarrow}(x) \cup \mathcal{A}_{\downarrow}(y)$$

since the reverse inclusion is clear. This is equivalent to

$$a \leq x \vee y \quad \Rightarrow \quad a \leq x \quad \text{or} \quad a \leq y$$

for all atoms a. If a is not contained in x, then Theorem 3.43, which says that every atom is contained in the complement of all atoms other than itself, implies that the atoms contained in x are also contained in a' and so

$$\mathcal{A}_{\downarrow}(x) = \mathcal{A}_{\downarrow}(a' \wedge x)$$

which implies that $x = a' \wedge x$, that is, $x \leq a'$. Similarly, if $a \not\leq y$, then $y \leq a'$. But then $a \leq x \vee y \leq a'$, which is false. Hence, either $a \leq x$ or $a \leq y$.

Thus, the $\mathcal{A}$-down map is a lattice embedding of L into the distributive lattice $\wp(L)$ and so L is distributive (in fact, L is isomorphic to a power set sublattice).

To prove part 2), we show that L has the DCC, in which case it is atomic and so part 1) completes the proof. If L does not have the DCC, then L has a strictly descending sequence

$$a_1 > a_2 > \cdots$$

If $b_k = a_k \wedge a'_{k+1}$, then $a_k \vee a'_{k+1} \geq a_{k+1} \vee a'_{k+1} = 1$ and so b_k cannot be equal to 0, since otherwise, we have $a'_k = a'_{k+1}$, which implies that $a_k = a_{k+1}$. Thus, $b_k \neq 0$ for all k. Moreover, for $k < j$,

$$b_k \wedge b_j \leq b_k \wedge b_{k+1} = (a_k \wedge a'_{k+1}) \wedge (a_{k+1} \wedge a'_{k+2}) = 0$$

and so the infinite family $\{b_k\}$ is an antichain, which is not possible. Thus, L has the DCC.□

Complete and Infinite Distributivity

We now wish to look at various generalizations of distributivity to infinite meets or joins.

Infinite Distributivity

Perhaps the simplest generalization is given by the following definition.

Definition *Let L be a complete lattice.*
1) *L is* **join-infinitely distributive** *if for any $x \in L$ and any $\{a_i \mid i \in I\} \subseteq L$,*

$$x \wedge \left(\bigvee_{i \in I} a_i \right) = \bigvee_{i \in I} (x \wedge a_i)$$

2) *L is* **meet-infinitely distributive** *if for any $x \in L$ and any $\{a_i \mid i \in I\} \subseteq L$,*

$$x \vee \left(\bigwedge_{i \in I} a_i \right) = \bigwedge_{i \in I} (x \vee a_i)$$

We say that L is **infinitely distributive** *if both 2) and 3) hold.*□

We leave it as an exercise to show that the join-infinite distributive law implies that

$$\left(\bigvee_{i \in I} a_i \right) \wedge \left(\bigvee_{j \in J} b_j \right) = \bigvee_{(i,j) \in I \times J} (a_i \wedge b_j)$$

and similarly for meets.

Example 4.5 shows that if A is the set of odd natural numbers in the complete lattice $(\mathbb{N}, \mid)$, then

$$2 \wedge \left(\bigvee_{a \in A} a \right) \neq \bigvee_{a \in A} (2 \wedge a)$$

and so a complete distributive lattice need not be infinitely distributive. However, the infinite distributive laws hold in a Boolean lattice, whenever the relevant meets and joins exist.

Theorem 5.13 *In a Boolean algebra, the infinite meet and join distributive laws hold whenever the relevant meets and joins exist. In particular, a complete Boolean algebra is infinitely distributive.*
Proof. Let B be a Boolean algebra and let $\{a_i \mid i \in I\} \subseteq B$. To establish join-infinite distributivity, it is clear that for any $x \in B$,

$$\bigvee (x \wedge a_i) \leq x \wedge \left(\bigvee a_i \right)$$

For the reverse inequality, note that in any Boolean algebra we have

$$a \leq x' \vee a = x' \vee (a \wedge x)$$

and so

$$\bigvee a_i \leq \bigvee (x' \vee (a_i \wedge x)) \leq x' \vee \left(\bigvee (x \wedge a_i)\right)$$

Hence, (finite) distributivity gives

$$\begin{aligned} x \wedge \left(\bigvee a_i\right) &\leq x \wedge \left[x' \vee \left(\bigvee (x \wedge a_i)\right)\right] \\ &= x \wedge \left(\bigvee (x \wedge a_i)\right) \\ &\leq \bigvee (x \wedge a_i) \end{aligned}$$

A dual argument will establish meet-infinite distributivity.□

Complete Distributivity

Consider now an arbitrary meet of arbitrary joins in a lattice L. Let us index the joins by an index set T and so, for each $t \in T$, we have

$$x_t = \bigvee_{i \in I_t} a_{t,i}$$

where $a_{t,i} \in L$. Thus, the full expression is

$$x = \bigwedge_{t \in T} \left(\bigvee_{i \in I_t} a_{t,i}\right)$$

The most general form of distributivity can be described in words as follows: Choose a term from each join x_t, take the meet of these terms and then take the join of all such meets. Moreover, the choice can be accomplished by a *choice function*, which can be thought of as a function f that maps each $t \in T$ to an element of the index set I_t, thus choosing the member $x_{t,f(t)}$ from x_t. The maps $f\colon T \to \bigcup I_t$ for which $f(t) \in I_t$ are precisely the elements of the cartesian product $\prod I_t$. Hence, this form of distributivity can be expressed as follows:

$$\bigwedge_{t \in T} \left(\bigvee_{i \in I_t} a_{t,i}\right) = \bigvee_{f \in \prod I_t} \left(\bigwedge_{t \in T} a_{t,f(t)}\right)$$

Definition *Let L be a complete lattice. Let $\{I_t \mid t \in T\}$ be any family of nonempty sets and let $a_{t,i} \in L$ for each $t \in T$ and $i \in I_t$. Then L is* **completely distributive** *if the following hold:*

$$\bigwedge_{t\in T}\left(\bigvee_{i\in I_t} a_{t,i}\right) = \bigvee_{f\in\prod I_t}\left(\bigwedge_{t\in T} a_{t,f(t)}\right) \tag{5.14}$$

and, dually,

$$\bigvee_{t\in T}\left(\bigwedge_{i\in I_t} a_{t,i}\right) = \bigwedge_{f\in\prod I_t}\left(\bigvee_{t\in T} a_{t,f(t)}\right) \tag{5.15}$$ □

It is not hard to see that (5.14) implies join-infinite distributivity and (5.15) implies meet-infinite distributivity.

Let B be a complete Boolean algebra and assume that (5.14) holds. Consider the special case of the meet of the joins $b \vee b'$, for all $b \in B$:

$$\bigwedge_{b\in B}(b \vee b')$$

Of course, this is equal to 1. If we index the set $\{b, b'\}$ by setting $b = b_0$ and $b' = b_1$, then $I_b = \{0,1\}$ for all $b \in B$ and so we do not need double indexing on the elements of B. Complete distributivity gives

$$1 = \bigwedge_{b\in B}(b \vee b') = \bigvee_{f\in\prod I_b}\left(\bigwedge_{b\in B} b_{f(b)}\right)$$

Each term

$$x_f = \bigwedge_{b\in B} b_{f(b)}$$

is the meet of the set of elements described as follows: For each $b \in B$, choose b or b'. If $x_f \neq 0$, then it must be an atom, for if $b < x_f$, then none of the terms in x_f is equal to b and so one of the terms is b', whence $b < x_f \leq b'$ and so $b = 0$. Thus, $x_f = 0$ or x_f is an atom. It follows that 1 is the join of atoms, say

$$1 = \bigvee a_i$$

Then by join-infinite distributivity, for any nonzero $b \in B$, we have

$$b = b \wedge \left(\bigvee a_i\right) = \bigvee (b \wedge a_i)$$

and so b is the join of atoms. Thus, B is atomistic (and atomic).

The converse is also true. Indeed, we will see in a later chapter that a complete atomic Boolean algebra is isomorphic to a power set algebra, which is completely distributive.

Theorem 5.16 *Let B be a complete Boolean algebra.*
1) B satisfies (5.14) if and only if B is atomic.
2) B is completely distributive if and only if it is atomic. □

Exercises

1. Prove that for an infinite set X, the finite-cofinite algebra $FC(X)$ is not complete.
2. Prove that if B is a Boolean lattice then its dual B^{∂} is also a Boolean lattice, with meet in B^{∂} equal to join in B and join in B^{∂} equal to meet in B. Prove that complement is self-dual: the complement in B^{∂} is the same as the complement in B.
3. If B is a Boolean algebra show that $a \in B$ is an atom if and only if a' is a coatom.
4. Let B be a Boolean lattice. If A is a nonempty subset of B, describe the Boolean sublattice $\langle A \rangle$ generated by A, that is, the smallest Boolean sublattice of B containing A. (Thus, $\langle A \rangle$ contains all finite joins, meets and complements formed in B using elements of A.)
5. Prove that a Boolean lattice with the DCC also has the ACC.
6. Prove that a Boolean algebra B is finite if and only if 1 is the join of a finite number of atoms.
7. Prove that a Boolean ring is commutative and that $2r = 0$ for all $r \in R$.
8. In the correspondence between ring ideals of a Boolean ring R and lattice ideals in the Boolean lattice R, what can be said about the properties of being principal, prime or maximal?
9. Those who enjoy arithmetic with meets and joins can show that the operation
$$r + s = (r \wedge s') \vee (r' \wedge s)$$
in a Boolean algebra is associative.
10. Prove that the join-infinite distributive law implies that
$$\left(\bigvee_{i \in I} a_i\right) \wedge \left(\bigvee_{j \in J} b_j\right) = \bigvee_{(i,j) \in I \times J} (a_i \wedge b_j)$$
11. Let B be the collection of all finite unions of intervals of the real line $\mathbb{R}$ of the form
$$(-\infty, a), [a, b), [b, \infty)$$
together with the empty set. Show that B is a Boolean algebra under set inclusion but has no atoms.
12. Prove that every Boolean lattice is Brouwerian and Heyting.
13. A lattice L is **disjunctive** if $a \not\leq b$ implies that there is an $x \in L$ such that
$$x \wedge a > 0 \quad \text{and} \quad x \wedge b = 0$$
   a) Formulate the dual concept of being **conjunctive**.

b) Show that a Boolean algebra is both disjunctive and conjunctive.

For the following exercises, let $f: A \to B$ and $g: B \to C$ be functions between nonempty sets. Let $f^*: \wp(A) \to \wp(B)$ be the induced map defined by

$$f(S) = \{f(s) \mid s \in S\}$$

for $S \in \wp(A)$ and let $f^{-1}: \wp(B) \to \wp(A)$ be the induced inverse map defined by

$$f^{-1}(S) = \{a \in A \mid f(a) \in S\}$$

for all $S \in \wp(B)$, and similarly for g. Prove the following statements.

14. For all $T \in \wp(B)$,

$$f \circ f^{-1}(T) \subseteq T$$

with equality if and only if f is surjective.

15. For all $S \in \wp(A)$,

$$f^{-1} \circ f(S) \supseteq S$$

with equality if and only if f is injective.

16. a) f is injective (surjective) if and only if f^{-1} is surjective (injective).
    b) f is injective (surjective) if and only if $(f^{-1})^{-1}$ is injective (surjective).
17. $(g \circ f)^{-1} = f^{-1} \circ g^{-1}$
18. f^{-1} is a Boolean homomorphism, that is, for subsets U, U_i of B,

$$f^{-1}\left(\bigcup U_i\right) = \bigcup f^{-1}(U_i)$$
$$f^{-1}\left(\bigcap U_i\right) = \bigcap f^{-1}(U_i)$$
$$f^{-1}(A \setminus U) = B \setminus f^{-1}(U)$$

19. The induced map $f^*: \wp(A) \to \wp(B)$ preserves union, that is, for $U_i \subseteq A$,

$$f^*\left(\bigcup U_i\right) = \bigcup f^*(U_i)$$

20. If $f: A \to B$ is a bijection then

$$(f^{-1})^{-1} = f^{**}$$

Frink's Condition and Distributivity

Definition *A poset* P *has* **Frink's condition** *if it has finite meets and if there is a function* $f: P \to P$ *and an element* $e \in P$ *for which*

$$u \le v \quad \Leftrightarrow \quad u \wedge f(v) = e$$

21. Assume that P is a poset satisfying Frink's condition. Prove the following statements about P:
   a) e is the smallest element of P, so we will denote it by 0.
   b) $x \wedge f(x) = 0$ for all x.
   c) $f^2(x) \leq x$ for all x.
   d) f is order-reversing:
   $$x \leq y \quad \Leftrightarrow \quad f(y) \leq f(x)$$
   e) $f(0)$ is the largest element of P, so we write it as 1.
   f) $f^2(x) = x$ for all x.
   g) f is unique, that is, if any function $g\colon P \to P$ has the same property as f then $g = f$.
   h) P has finite joins. Hence P is a lattice. Look at $f(f(a) \wedge f(b))$.
   i) de Morgan's laws hold in P.
   j) The function f is a complement hence P is a complemented lattice.
   k) Complements are unique, hence P is a uniquely complemented lattice and we can write $f(x) = x'$.
   l) Conclude that P is a uniquely complemented lattice in which de Morgan's laws hold and so P is a Boolean lattice.

Chapter 6
The Representation of Distributive Lattices

We now turn to the issue of representing distributive lattices as power set sublattices.

The Representation of Distributive Lattices with DCC

The B-down map $\phi_{B,\downarrow}: L \to \wp(B)$ is the key to several important representation theorems for distributive lattices. The B-down map is a meet-homomorphism, that is,

$$B_{\downarrow}(x \wedge y) = B_{\downarrow}(x) \cap B_{\downarrow}(y)$$

for all $x, y \in L$. Furthermore, $\phi_{B,\downarrow}$ is a join-homomorphism if and only if

$$B_{\downarrow}(x \vee y) \subseteq B_{\downarrow}(x) \cup B_{\downarrow}(y)$$

for all $x, y \in L$, since the reverse inclusion is always true. But this is equivalent to the statement that every $b \in B$ is join-prime:

$$b \leq x \vee y \quad \Rightarrow \quad b \leq x \quad \text{or} \quad b \leq y$$

Thus, $\phi_{B,\downarrow}$ is a lattice homomorphism if and only if every element of B is join-prime. Also, Theorem 3.32 implies that $\phi_{B,\downarrow}$ is injective if and only if B is join-dense in L.

Theorem 6.1 *Let B be a nonempty subset of a lattice L and let*

$$\phi_{B,\downarrow}: L \to \wp(B)$$

be the B-down map. Then the following hold:

1) *$\phi_{B,\downarrow}$ is a lattice homomorphism if and only if every element of B is join-prime in L.*
2) *$\phi_{B,\downarrow}$ is injective if and only if B is join-dense in L.*
3) *$\phi_{B,\downarrow}$ is a lattice embedding if and only if B is a join-dense set of join-prime elements.* □

S. Roman (ed.), *Lattices and Ordered Sets*, doi: 10.1007/978-0-387-78901-9_6,

Now, if L is distributive, then the join-prime elements are the join-irreducible elements $\mathcal{J}(L)$. Moreover, if L also has the DCC, then Theorem 3.29 implies that $\mathcal{J}(L)$ is join-dense in L. Thus, we have the following representation theorem for distributive lattices with the DCC.

Theorem 6.2 (Representation theorem for distributive lattices with DCC) *Let L be a distributive lattice with the DCC and let*

$$\phi_{\mathcal{J},\downarrow}: L \hookrightarrow \mathcal{O}(\mathcal{J})$$

be the $\mathcal{J}$-down map, where $\mathcal{J} = \mathcal{J}(L)$.

1) $\phi_{\mathcal{J},\downarrow}$ *is a lattice embedding and so L is isomorphic to a power set sublattice of $\wp(\mathcal{J})$.*

2) **(Birkhoff [6], 1933)** *If L is also finite, then $\phi_{\mathcal{J},\downarrow}$ is a lattice isomorphism and so $L \approx \mathcal{O}(\mathcal{J})$.*

Proof. For part 2), to see that $\phi_{\mathcal{J},\downarrow}$ is surjective, we have $\phi_{\mathcal{J},\downarrow}(0) = \emptyset$. Let

$$D = \{d_1, \ldots, d_n\} \in \mathcal{O}(\mathcal{J})$$

be a nonempty down-set of join-irreducibles in L and let $a = \bigvee d_i$. If $j \in \mathcal{J}$ satisfies $j \le a$, then $j \le d_k$ for some k and so $j \in D$. Thus, $D = \mathcal{J}_\downarrow(a)$.□

Note that by duality, a distributive lattice with the ACC is also isomorphic to a power set sublattice.

The Representation of Atomic Boolean Algebras

If L is a bounded lattice, then the $\mathcal{J}$-down map is a $\{0,1\}$-homomorphism, since

$$\phi_{\mathcal{J},\downarrow}(0) = \emptyset \quad \text{and} \quad \phi_{\mathcal{J},\downarrow}(1) = \mathcal{J}(L)$$

Hence, if B is a Boolean algebra, then the $\mathcal{J}$-down map is a Boolean homomorphism. Moreover, Theorem 5.2 implies that

$$\mathcal{J}(B) = \mathcal{A}(B)$$

and that $\mathcal{A}(B)$ is join-dense (that is, B is atomistic) if and only if B is atomic. Also, since $\mathcal{A} = \mathcal{A}(B)$ is an antichain, it follows that

$$\mathcal{O}(\mathcal{A}) = \wp(\mathcal{A})$$

This leads to the following representation theorem for Boolean algebras.

Theorem 6.3 (Representation theorem for atomic Boolean algebras) *Let B be an atomic Boolean algebra and let*

$$\phi_{\mathcal{A},\downarrow}: B \to \wp(\mathcal{A})$$

be the $\mathcal{A}$-down map, where $\mathcal{A} = \mathcal{A}(B)$.

1) *$\phi_{\mathcal{A},\downarrow}$ is a Boolean algebra embedding and so B is isomorphic to a power set subalgebra of $\wp(\mathcal{A})$.*
2) **(Lindenbaum and Tarski**, see [62]) *If B is complete, then $\phi_{B,\downarrow}$ is a Boolean algebra isomorphism and so $B \approx \wp(\mathcal{A})$. Conversely, any power set is complete and atomic. Thus, up to isomorphism, the complete atomic Boolean algebras are precisely the power set algebras.*

Proof. For part 2), to see that $\phi_{\mathcal{A},\downarrow}$ is surjective, we have $\phi_{\mathcal{A},\downarrow}(0) = \emptyset$. Let $A \subseteq \mathcal{A}$ be nonempty. We wish to show that $A = \mathcal{A}_\downarrow(\bigvee A)$. Clearly, $A \subseteq \mathcal{A}_\downarrow(\bigvee A)$. For the reverse inclusion, if $\alpha \in \mathcal{A}_\downarrow(\bigvee A)$, then infinite join-distributivity implies that

$$\alpha = \alpha \wedge \left(\bigvee A\right) = \bigvee_{a \in A} (\alpha \wedge a)$$

Since α is an atom, it follows that $\alpha = \alpha \wedge a$ for some $a \in A$, that is, $\alpha \leq a$. But a is also an atom and so $\alpha = a \in A$. We leave proof of the converse to the reader.□

The Representation of Arbitrary Distributive Lattices

Garrett Birkhoff also showed that an arbitrary distributive lattice L is isomorphic to a power set sublattice, using the family of prime ideals as a carrier set. For $a \in L$, let

$$\mathcal{P}_a = \{P \in \operatorname{Spec}(L) \mid a \in P\}$$

and let

$$\mathcal{P}_{\neg a} = (\mathcal{P}_a)^c = \{P \in \operatorname{Spec}(L) \mid a \notin P\}$$

The first order of business is to show that $\mathcal{P}_{\neg a}$ is nonempty for $a > 0$.

Theorem 6.4 *Let L be a distributive lattice.*
1) *If $a \not\leq b$, then there is a prime ideal containing b but not a.*
2) *If $a > 0$, then $\mathcal{P}_{\neg a}$ is nonempty.*
3) *If $a \neq b$, then $\mathcal{P}_{\neg a} \neq \mathcal{P}_{\neg b}$.*

Proof. For part 1), since $a \notin {\downarrow} b$, the family $\mathcal{F}$ of all ideals containing b but not a is nonempty. Since the union of any chain in $\mathcal{F}$ is in $\mathcal{F}$, Zorn's lemma implies that $\mathcal{F}$ has a maximal ideal M. But M is also prime, for if $x \wedge y \in M$ but $x, y \notin M$, then $M \vee {\downarrow} x = L$ and $M \vee {\downarrow} y = L$. Thus, $a \leq m \vee x$ and $a \leq n \vee y$ for some $m, n \in M$. Setting $p = m \vee n \in M$ gives $a \leq p \vee x$ and $a \leq p \vee y$ and so

$$a \leq (p \vee x) \wedge (p \vee y) = p \vee (x \wedge y) \in M$$

whence $a \in M$, a contradiction. Thus, M is prime. The other parts follow from part 1).□

It is easy to see that

$$\mathcal{P}_a \cap \mathcal{P}_b = \mathcal{P}_{a \vee b}, \quad \mathcal{P}_a \cup \mathcal{P}_b = \mathcal{P}_{a \wedge b} \quad \text{and} \quad (\mathcal{P}_a)^c = \mathcal{P}_{a'}$$

the latter when a has a complement. Thus

$$\mathcal{P}_{\neg a} \cap \mathcal{P}_{\neg b} = \mathcal{P}_{\neg(a \wedge b)}, \quad \mathcal{P}_{\neg a} \cup \mathcal{P}_{\neg b} = \mathcal{P}_{\neg(a \vee b)} \quad \text{and} \quad (\mathcal{P}_{\neg a})^c = \mathcal{P}_{\neg(a')}$$

the latter when a has a complement. In other words, the set

$$\mathcal{S} = \{\mathcal{P}_{\neg a} \mid a \in L\}$$

is a sublattice of the power set $\wp(\mathrm{Spec}(L))$ and if L is complemented, then $\mathcal{S}$ is a subalgebra of $\wp(\mathrm{Spec}(L))$.

Also, the map $\rho: L \to \mathcal{O}(\mathrm{Spec}(L))$ defined by

$$\rho(a) = \mathcal{P}_{\neg a}$$

which we will call the **rho map** for L, is a lattice homomorphism that preserves existing complements. Moreover, Theorem 6.4 shows that the rho map is a lattice embedding.

We can now summarize.

Theorem 6.5 *Let L be a lattice and let*

$$\mathcal{S} = \{\mathcal{P}_{\neg a} \mid a \in L\}$$

1) *$\mathcal{S}$ is a sublattice of the power set lattice $\wp(\mathrm{Spec}(L))$, in fact,*

$$\mathcal{P}_{\neg a} \cap \mathcal{P}_{\neg b} = \mathcal{P}_{\neg(a \wedge b)} \quad \textit{and} \quad \mathcal{P}_{\neg a} \cup \mathcal{P}_{\neg b} = \mathcal{P}_{\neg(a \vee b)}$$

   Also, if L is complemented, then $\mathcal{S}$ is a subalgebra of $\wp(\mathrm{Spec}(L))$, that is,

$$(\mathcal{P}_{\neg a})^c = \mathcal{P}_{\neg(a')}$$

2) *The rho map $\rho: L \to \mathcal{O}(\mathrm{Spec}(L))$ defined by*

$$\rho(a) = \mathcal{P}_{\neg a}$$

   is a lattice embedding if and only if L is distributive. $\square$

If L is a finite distributive lattice, then the rho map is also surjective. To see this, Theorem 4.27 says that all prime ideals have the form $(\uparrow j)^c$ for $j \in \mathcal{J}(L)$. Let

$$D = \{(\uparrow j_1)^c, \ldots, (\uparrow j_n)^c\} \in \mathcal{O}(\mathrm{Spec}(L))$$

We show that $D = \mathcal{P}_{\neg a}$, where $a = \bigvee j_i$. Now, if j is join-irreducible, then since

D is a down-set, the following are equivalent:

$$\begin{aligned}(\uparrow j)^c &\in \mathcal{P}_{\neg a}\\ a &\notin (\uparrow j)^c\\ j \le a &= \bigvee j_k\\ j &\le j_k \text{ for some } k\\ (\uparrow j)^c &\subseteq (\uparrow j_k)^c\\ (\uparrow j)^c &\in D\end{aligned}$$

Thus, $D = \mathcal{P}_{\neg a}$ and so ρ is surjective and hence a lattice isomorphism.

Theorem 6.6 *Let L be a nontrivial distributive lattice and let*

$$\rho: L \to \mathcal{O}(\mathrm{Spec}(L))$$

be the rho map for L.

1) **(Representation theorem for distributive lattices, Birkhoff, 1933 [6])** *The rho map ρ is an embedding of L into the power set $\wp(\mathrm{Spec}(L))$ and so L is isomorphic to a power set sublattice (ring of sets). Moreover, if L is finite, then ρ is an isomorphism and $L \approx \mathcal{O}(\mathrm{Spec}(L))$.*
2) **(Representation theorem for Boolean algebras)** *If L is a Boolean algebra, then ρ is a Boolean algebra embedding of L into the power set $\wp(\mathrm{Spec}(L))$ and so L is isomorphic to a power set subalgebra (field of sets). Moreover, if L is finite, then ρ is an isomorphism.* □

In the case of a finite distributive lattice L, the rho map

$$\rho: L \approx \mathcal{O}(\mathrm{Spec}(L))$$

has image $\mathcal{O}(\mathrm{Spec}(L))$. However, to describe the image of the rho map for an arbitrary distributive lattice requires some topological notions and so we postpone this matter until the chapter on duality theory, where the requisite topological notions are developed.

Summary

Let us summarize our results on lattice representations.

Theorem 6.7 (Representation theorems)

1) **(Arbitrary lattices)** *Any lattice L is isomorphic to the lattice $\mathcal{P}(L)$ of principal ideals of L, via the down map $\phi_\downarrow: L \hookrightarrow \mathcal{I}(L)$.*
2) **(Distributive lattices)** *Any distributive lattice L is isomorphic to a power set sublattice (ring of subsets), via the rho map $\rho: L \to \mathcal{O}(\mathrm{Spec}(L))$. If L has a chain condition, then the $\mathcal{J}$-down map $\phi_{\mathcal{J},\downarrow}: L \to \mathcal{O}(\mathcal{J})$ is also an embedding and is an isomorphism if L is finite.*
3) **(Boolean algebras)** *Any Boolean algebra B is isomorphic to a power set subalgebra (field of subsets), via the rho map $\rho: L \to \mathcal{O}(\mathrm{Spec}(B))$.*

4) **(Complete atomic Boolean algebras)** *Any complete atomic Boolean algebra B is isomorphic to the power set $\wp(\mathcal{A})$, via the $\mathcal{A}$-down map $\phi_{\mathcal{A},\downarrow}: B \to \wp(\mathcal{A})$. Conversely, any power set is complete and atomic.* $\square$

Exercises

1. Let L be a nontrivial lattice and let

$$\mathcal{S} = \{\mathcal{P}_{\neg a} \mid a \in L\}$$

Prove that $\mathcal{S}$ is a sublattice of the power set sublattice $\wp(\mathrm{Spec}(L))$, in particular,

a) $\mathcal{P}_{\neg(a \wedge b)} = \mathcal{P}_{\neg a} \cap \mathcal{P}_{\neg b}$

b) $\mathcal{P}_{\neg(a \vee b)} = \mathcal{P}_{\neg a} \cup \mathcal{P}_{\neg b}$

Prove that if L is complemented then $\mathcal{S}$ is a subalgebra of $\wp(\mathrm{Spec}(L))$, in particular,

c) $\mathcal{P}^c_{\neg a} = \mathcal{P}_{\neg(a')}$

2. Let L be a distributive lattice with the DCC. Show that L can be embedded in a Boolean lattice B. If $n = |L|$ is finite, then show that B can be chosen with $|B| \leq 2^{n-1}$.
3. Find a lattice L for which $\mathcal{J}(L) = \mathcal{A}(L)$ but L is not a Boolean algebra.
4. Let P be a finite poset.
   a) Describe the covering relation in the lattice $\mathcal{O}(P)$.
   b) What is the length of $\mathcal{O}(P)$.
   c) Show that the length of a finite distributive lattice L is equal to $|\mathcal{J}(L)|$.
5. Let L be a lattice with the DCC. Show that

$$\rho = \phi_{\neg\uparrow} \circ \phi_{\mathcal{J},\downarrow}$$

where ρ is the rho map, $\phi_{\neg\uparrow}$ is the not-up map and $\mathcal{J} = \mathcal{J}(L)$.

6. Show that the map $\rho^\partial: L \to \mathcal{U}(\mathrm{Spec}(L))$ defined by

$$\rho^\partial(a) = \mathcal{P}_a = \{P \in \mathrm{Spec}(L) \mid a \in P\}$$

is a lattice anti-homomorphism, that is,

$$\rho^\partial(a \wedge b) = \rho^\partial(a) \cup \rho^\partial(b)$$
$$\rho^\partial(a \vee b)\rho^\partial(a) \cap \rho^\partial(b)$$

and if L is Boolean then ρ^∂ is a Boolean algebra anti-homomorphism, that is,

$$\rho^\partial(a') = \rho^\partial(a)^c$$

7. Let L be a lattice and let

$$\mathcal{P}_a := (\mathcal{P}_{\neg a})^c = \{P \in \mathrm{Spec}(L) \mid a \in P\}$$

and

$$\mathcal{P}_{\neg a,b} := \mathcal{P}_{\neg a} \cap \mathcal{P}_b$$

Let

$$\mathcal{S} = \{\mathcal{P}_{\neg a} \mid a \in L\} \quad \text{and} \quad \mathcal{B} = \{\mathcal{P}_{\neg a,b} \mid a, b \in L\}$$

Prove that $\mathcal{B}$ contains $\mathcal{S}$ and is closed under intersection and complement, in particular,

a) $\mathcal{P}_{\neg a} = \mathcal{P}_{\neg a,0}$ and $\mathcal{P}_b = \mathcal{P}_{\neg 1,b}$

b) $\mathcal{P}_{\neg a_1,b_1} \cap \mathcal{P}_{\neg a_2,b_2} = \mathcal{P}_{\neg(a_1 \wedge a_2),(b_1 \vee b_2)}$

c) $\mathcal{P}^c_{\neg a,b} = \mathcal{P}_{\neg b,a}$

Prove that if L is complemented then

$$\mathcal{P}_{\neg a,b} = \mathcal{P}_{\neg(a \wedge b')}$$

and so $\mathcal{B} = \mathcal{S}$.

8. Let L be a lattice. We define, for any ideal I of L,

$$\mathcal{P}_{\neg I} := \{P \in \operatorname{Spec}(L) \mid I \not\subseteq P\}$$

Prove that if $I, J \trianglelefteq L$ then

$$I \subseteq J \quad \Leftrightarrow \quad \mathcal{P}_{\neg I} \subseteq \mathcal{P}_{\neg J}$$

Also, for any $a \in L$,

$$\mathcal{P}_{\neg(\downarrow a)} = \mathcal{P}_{\neg a}$$

and for any nonempty subset $X \subseteq L$,

$$\bigcup_{x \in X} \mathcal{P}_{\neg x} = \mathcal{P}_{\neg(X]}$$

Chapter 7
Algebraic Lattices

Motivation

We have seen in Theorem 3.8 that the meet-structures in a complete lattice L are the families of closed sets of the closure operators on L. Specifically, the intersection-structures on a nonempty set X are the families of closed sets of the closure operators on X. We have also seen that there are two types of closure operators on X: finitary and infinitary. A *finitary closure operator* is one for which the closure operator is completely determined by its behavior on finite sets, in the sense that

$$\text{cl}(S) = \bigcup\{\text{cl}(F) \mid F \subseteq S, F \text{ finite}\}$$

The closure operators one meets (no pun intended) in algebra tend to be finitary, but as we remarked earlier, topological closure operators on T_1 spaces are finitary if and only if the topology is discrete. In a sense that we will make more precise, the property of being finitary is the dividing line between closure operators that come from algebraic structures and other types of closure operators.

To explain this more clearly, we define the concept of a "general algebraic structure," which encompasses some of the common algebraic structures, such as groups, rings and lattices (but not fields).

Definition

1) *If n is a positive integer, an n-ary* **operation** *on a nonempty set A is a map $f\colon A^n \to A$. A 0-ary* **operation** *on A is a distinguished element of A. The 0-ary, 1-ary and 2-ary operations are also called* **nullary**, **unary** *and* **binary** *operations, respectively. An operation that is n-ary for some $n \geq 0$ is said to be a* **finitary operation**. *(One can also define* **infinitary** *κ-ary operations $f\colon A^\kappa \to A$, where κ is any infinite cardinal.)*

S. Roman (ed.), *Lattices and Ordered Sets*, doi: 10.1007/978-0-387-78901-9_7,

2) *An* Ω*-***algebra** *(or just an* **algebra***), also called a* **universal algebra***, is a set* A*, together with a set* Ω *of finitary operations on* A*. We denote an algebra by* $\langle A;\Omega\rangle$*, or simply by* A *when the set of operators is understood.*
3) *Let* $\langle A;\Omega\rangle$ *be an algebra. If a subset* $Y\subseteq A$ *is invariant under* $f\in\Omega$*, that is, if*

$$y_1,\ldots,y_n\in Y \quad\Rightarrow\quad f(y_1,\ldots,y_n)\in Y$$

then the restriction $f|_Y\colon Y^n\to Y$ *is an* n*-ary operation on* Y*. (A set* Y *is invariant under a nullary operation* $x\in A$ *if* $x\in Y$*.) If* Y *is* f*-invariant for all* $f\in\Omega$*, then the set* Y*, together with the restrictions* $\{f|_Y\mid f\in\Omega\}$ *is a* **subalgebra** *of* $\langle A;\Omega\rangle$*. When the operations are understood, we simply refer to* Y *as a subalgebra of* A*. Let* $\operatorname{Sub}(\langle A;\Omega\rangle)$ *denote the family of subalgebras of an algebra* $\langle A;\Omega\rangle$*.*□

To clarify an earlier remark, a field is not an algebra, since the inverse operation is only a *partial operation*: 0 has no inverse.

The family $\operatorname{Sub}(A)$ of subalgebras of an algebra A is easily seen to be an $\cap$-structure, called a **subalgebra lattice**. Note also that the empty set is a subalgebra of A if and only if Ω has no nullary operations. If A is an algebra and $S\subseteq A$, then the **subalgebra generated by** S is the smallest subalgebra of A containing S, which is the intersection of all subalgebras containing S. The subalgebra generated by a finite nonempty set S is said to be **finitely-generated**.

Our ultimate goal is to characterize the $\cap$-structures that are subalgebra lattices. As we will see, these $\cap$-structures can be characterized in several ways. In terms of the $\cap$-structure itself:

1) They are the $\cap$-structures that inherit *directed* unions from $\wp(X)$. For example, the directed union of subgroups is a subgroup. Note that this is not true for nondirected (even finite) unions. While these $\cap$-structures are most commonly referred to as *algebraic* $\cap$*-structures*, reflecting their usual origin, we will use the more descriptive name $\cap\overrightarrow{\cup}$ -structures.
2) They are the $\cap$-structures for which the *compact* subsets (defined later) are join-dense. For example, every subgroup H of a group is the join of all finitely-generated subgroups of H.

In terms of the corresponding closure operator:

3) They are the families of closed sets of *finitary* closure operators. For example, the "subgroup generated by" operator is finitary.

4) They are the families of closed sets of *join-continuous* closure operators, that is, closure operators with the property that for a directed set D,

$$\mathrm{cl}\left(\bigcup_{d\in D} d\right) = \bigvee_{d\in D} \mathrm{cl}(d)$$

A word on notation: The symbols

$$\overrightarrow{\bigvee} D \quad \text{and} \quad \bigvee^{\rightarrow} D$$

denote the join of a *directed* subset D, referred to as a **directed join**. Note that a *finite* directed set has a maximum element. Hence, the join of a finite directed set is equal to the maximum element.

Algebraic Lattices

The finitely-generated subalgebras of an algebra A have a special property, namely, if a finitely-generated subalgebra S is contained in the join of an arbitrary family $\mathcal{F}$ of subalgebras, then S is contained in the join of a finite subfamily of $\mathcal{F}$, that is,

$$S \subseteq \bigvee \mathcal{F} \quad \Rightarrow \quad S \subseteq \bigvee \mathcal{F}_0$$

for some finite subfamily $\mathcal{F}_0$ of $\mathcal{F}$. This motivates the following definitions.

Definition *Let L be a complete lattice and let $a \in L$.*

1) A ***join-cover*** *of a is a subset B of L for which*

$$a \leq \bigvee B$$

2) The element a is ***compact*** *if every join-cover of a has a finite join-subcover, that is,*

$$a \leq \bigvee B \quad \Rightarrow \quad a \leq \bigvee B_0 \textit{ for some finite subset } B_0 \subseteq B$$

The set of compact elements of L is denoted by $\mathcal{K}(L)$.□

Theorem 7.1 *Let L be a complete lattice. An element $a \in L$ is compact if and only if every directed join-cover of a has a finite (therefore singleton) join-subcover, that is,*

$$a \leq \overrightarrow{\bigvee} D \quad \Rightarrow \quad a \leq d \textit{ for some } d \in D$$

Proof. It is clear that compact elements have this property. Conversely, suppose that this property holds and $a \leq \bigvee B$ for some set B, then

$$a = \bigvee B = \bigvee \left\{ \bigvee S \,\middle|\, S \subseteq B,\, S \text{ finite} \right\}$$

But the set $\{\ S \mid S \subseteq B,\ S$ finite$\}$ is directed and so $a \leq\ \ S$ for some finite subset $S \subseteq B$.□

Example 7.2

1) All elements of a finite lattice are compact.
2) The compact elements of $\wp(X)$ are the finite subsets of X.
3) The compact elements of the subgroup lattice $\mathcal{S}(G)$ of a group G are the finitely-generated subgroups.
4) The compact elements of the subspace lattice $\mathcal{S}(V)$ of a vector space V are the finite-dimensional subspaces.□

Theorem 7.3 *Let L be a complete lattice.*

1) $0 \in L$ *is compact.*
2) The compact elements $\mathcal{K}(L)$ *inherit finite joins from* L *and so* $\mathcal{K}(L)$ *is a join-semilattice with* 0.
3) The ideals of $\mathcal{K}(L)$ *are precisely the sets of the form* $\mathcal{K}_{\downarrow}(a)$ *for* $a \in L$.

Proof. For part 3), let $I \in \mathcal{I}$ and let $a = \bigvee^{\rightarrow} I$. Then $I \subseteq \mathcal{K}_{\downarrow}(a)$. On the other hand, if $x \in \mathcal{K}_{\downarrow}(a)$, then x is compact and $x \leq \bigvee^{\rightarrow} I$ and so $x \leq i$ for some $i \in I$, whence $x \in I$. Thus, $I = \mathcal{K}_{\downarrow}(a)$.□

Definition *A complete lattice L is* **algebraic**, *or* **compactly-generated** *if the compact elements of L are join-dense in L.*□

Note that not all authors require completeness in the definition of an algebraic lattice.

All finite lattices are algebraic. We have just seen that the subalgebra lattice Sub(A) of an algebra is algebraic and we will show that every algebraic lattice is the subalgebra lattice for some algebra.

$\cap\overrightarrow{\cup}$-Structures

Subalgebra lattices are $\cap$-structures but, in general, they do not inherit arbitrary unions from $\wp(X)$. However, they do inherit *directed* unions. This follows from the finitary nature of the operations in an algebra. For if $\mathcal{F} = \{S_i \mid i \in I\}$ is a directed subfamily of $\mathcal{S}$ and if $U = \bigcup^{\rightarrow} \mathcal{F}$, then

$$x_1, \ldots, x_n \in U \quad \Rightarrow \quad x_1, \ldots, x_n \in F \text{ for some } F \in \mathcal{F}$$

and so $f(x_1, \ldots, x_n) \in F \subseteq U$ for all $f \in \Omega$. Hence, U is a subalgebra of X. The following definition is motivated by this example.

Definition *If a nonempty subset* $\mathcal{M}$ *of a power set* $\wp(X)$ *inherits arbitrary meets and directed joins, we call* $\mathcal{M}$ *an* $\cap\overrightarrow{\cup}$**-structure.**□

Some authors refer to $\cap\overrightarrow{\cup}$-structures as **algebraic intersection-structures**, but we prefer to avoid this terminology, since intersection-structures are complete lattices and the term *algebraic* has a different definition when applied to complete lattices (as we will see a bit later).

It is not hard to characterize the compact subsets of an $\cap\overrightarrow{\cup}$-structure $\mathcal{M}$ in X and to show that an $\cap\overrightarrow{\cup}$-structure is an algebraic lattice. Recall that the $\mathcal{M}$-closure of a subset S of X is defined by

$$\langle S\rangle = \bigcap\{M \in \mathcal{M} \mid S \subseteq M\}$$

Definition *Let $\mathcal{M}$ be an $\cap\overrightarrow{\cup}$-structure in X. A set in $\mathcal{M}$ that is the $\mathcal{M}$-closure $\langle S\rangle$ of a finite subset S of X is called a* **finitely-generated** *element of $\mathcal{M}$.*$\square$

Any finitely-generated element $\langle F\rangle$ of $\mathcal{M}$ is compact, for if

$$\langle F\rangle \subseteq \overrightarrow{\bigcup}\{M_i \mid M_i \in \mathcal{M}, i \in I\}$$

then F is also contained in this directed union and so $F \subseteq M_j$ for some $j \in I$. Hence, $\langle F\rangle \subseteq M_j$. Conversely, suppose that K is compact in $\mathcal{M}$. The family of all $\mathcal{M}$-closures $\langle F\rangle$ of finite subsets of K is directed, for if F_1 and F_2 are finite, then $\langle F_1\rangle, \langle F_2\rangle \subseteq \langle F_1 \cup F_2\rangle$. Hence,

$$K \subseteq \overrightarrow{\bigcup}\{\langle F\rangle \mid F \subseteq K, F \text{ finite}\}$$

and so there is a finite subset F of K for which $K \subseteq \langle F\rangle$. Since the reverse inclusion is clear, we have $K = \langle F\rangle$.

Theorem 7.4 *Let $\mathcal{M}$ be an $\cap\overrightarrow{\cup}$-structure in $\wp(X)$.*
1) *The compact elements of $\mathcal{M}$ are the finitely-generated elements of $\mathcal{M}$.*
2) *$\mathcal{M}$ is an algebraic lattice.*

Proof. The last statement follows from the fact that for any $M \in \mathcal{M}$,

$$M = \overrightarrow{\bigcup}\{\langle F\rangle \mid F \subseteq M, F \text{ finite}\}$$

$\square$

Thus, $\cap\overrightarrow{\cup}$-structures are algebraic lattices. On the other hand, every algebraic lattice is isomorphic to an $\cap\overrightarrow{\cup}$-structure. To see this, we show that every algebraic lattice is isomorphic to the ideal lattices of a join-semilattice with 0 and that these lattices are $\cap\overrightarrow{\cup}$-structures. Note that the notion of an ideal makes sense in a join-semilattice L: An ideal is a nonempty down-set that inherits binary joins from L.

Theorem 7.5 *An algebraic lattice L is isomorphic to the ideal lattice of a join-semilattice with* 0. *In particular, if $\mathcal{K} = \mathcal{K}(L)$, then the $\mathcal{K}$-down map*

$$\phi_{\mathcal{K},\downarrow}: L \approx \mathcal{I}(\mathcal{K}(L))$$

defined by

$$\phi_{\mathcal{K},\downarrow}(a) = \mathcal{K}_{\downarrow}(a)$$

is an isomorphism. Moreover, $\mathcal{I}(\mathcal{K}(L))$ *is an* $\cap\vec{\cup}$ *-structure in* $\wp(\mathcal{K}(L))$ *and so every algebraic lattice is isomorphic to an* $\cap\vec{\cup}$ *-structure.*

Proof. Theorem 7.3 implies that $\mathcal{K}(L)$ is a join-semilattice with 0 and that $\mathcal{I} = \mathcal{I}(\mathcal{K}(L))$ is an $\cap$-structure whose elements are the sets $\mathcal{K}_{\downarrow}(a)$ for $a \in L$. Hence, the $\mathcal{K}$-down map $\phi_{\mathcal{K},\downarrow}: L \to \mathcal{I}$ is surjective. If L is algebraic, then $\mathcal{K}(L)$ is join-dense in L and so Theorem 3.32 implies that $\phi_{\mathcal{K},\downarrow}$ is an order embedding and therefore a lattice isomorphism.

To see that $\mathcal{I}$ has directed unions, let $\{\mathcal{K}_{\downarrow}(d) \mid d \in D\}$ be a directed family in $\mathcal{I}$. Since $\phi_{\mathcal{K},\downarrow}$ is an order isomorphism, it follows that D is also directed. Moreover,

$$\begin{aligned}\vec{\bigcup}\{\mathcal{K}_{\downarrow}(d) \mid d \in D\} &= \{k \in \mathcal{K}(L) \mid k \leq d \text{ for some } d \in D\} \\ &= \{k \in \mathcal{K}(L) \mid k \leq \vec{\bigvee} D\} \\ &= \mathcal{K}_{\downarrow}\left(\bigvee D\right)\end{aligned}$$

and so $\mathcal{I}$ has directed unions and is an $\cap\vec{\cup}$ -structure.□

Algebraic Closure Operators

As we mentioned earlier, $\cap\vec{\cup}$-structures correspond to finitary closure operators. Such operators have another name.

Definition *A finitary closure operator on a set X is called an* **algebraic closure operator**.

Algebraic closure operators are continuous in the following sense.

Definition *Let P and Q be posets that have directed joins. A function* $f: P \to Q$ *is* **continuous** (*also called* **join-continuous**) *if for any directed subset D in P, the set* $f(D)$ *is also directed and*

$$f\left(\vec{\bigvee}_{d \in D} d\right) = \vec{\bigvee}_{d \in D} f(d)$$ □

A continuous map is monotone, for if $a \leq b$, then the set $\{a, b\}$ is directed and so

$$f(b) = f(a \vee b) = f(a) \vee f(b)$$

which implies that $f(a) \leq f(b)$. The converse is not true, however, but we can say that any monotone map sends directed sets to directed sets and that

$$\overrightarrow{\bigvee}_{d \in D} f(d) \leq f\left(\overrightarrow{\bigvee}_{d \in D} d\right)$$

Example 7.6 Let $f: \wp(\mathbb{N}) \to \wp(\mathbb{N})$ be defined by $f(S) = \emptyset$ for S a finite set, $f(S) = \mathbb{N}$ otherwise. Then f is monotone but not continuous, since if $\wp_0(\mathbb{N})$ is the family of all finite subsets of $\mathbb{N}$, then

$$f(\bigcup \wp_0(\mathbb{N})) = f(\mathbb{N}) = \mathbb{N}$$

but

$$\bigcup f(\wp_0(\mathbb{N})) = \bigcup \emptyset = \emptyset$$

□

Theorem 7.7 *A closure operator on a set X is algebraic if and only if it is continuous.*

Proof. Suppose that cl is algebraic and $D = \{d_i \mid i \in I\}$ is directed in $\wp(X)$. Then $\{\text{cl}(d_i) \mid i \in I\}$ is also directed, since $\text{cl}(d_i), \text{cl}(d_j) \leq \text{cl}(d_i \vee d_j)$. Also,

$$\begin{aligned} \text{cl}\left(\overrightarrow{\bigcup} d_i\right) &= \overrightarrow{\bigcup}\left\{\text{cl}(x) \,\middle|\, x \text{ finite}, x \subseteq \overrightarrow{\bigcup} d_i\right\} \\ &= \overrightarrow{\bigcup}\left\{\text{cl}(x) \,\middle|\, x \subseteq d_i \text{ for some } d_i \in D\right\} \\ &\subseteq \overrightarrow{\bigcup}\{\text{cl}(d_i) \mid d_i \in D\} \end{aligned}$$

and since the reverse inclusion is clear, we have equality and so cl is join-continuous. Conversely, if cl is join-continuous, then for any $a \in L$,

$$a = \overrightarrow{\bigcup}\{k \mid k \text{ finite}, k \subseteq a\}$$

and so join-continuity gives

$$\text{cl}(a) = \overrightarrow{\bigcup}\{\text{cl}(k) \mid k \text{ finite}, k \subseteq a\}$$

which shows that cl is algebraic.□

The Main Correspondence

If L is a complete lattice, then we have seen in an earlier chapter that the maps

$$\Gamma: \text{cl} \to \text{Cl}(X) \quad \text{and} \quad \Pi: C \to C\text{-closure}$$

are inverse bijections between closure operators on a set X and $\cap$-structures on $\wp(X)$. As pictured in Figure 7.1, under these bijections, the algebraic closure operators correspond to the $\cap\overrightarrow{\bigcup}$-structures.

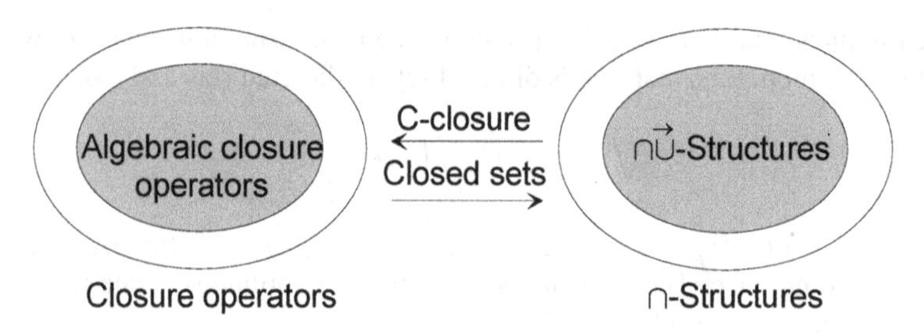

Figure 7.1

Theorem 7.8 *Let X be a nonempty set. The maps*

$$\Gamma\colon \mathrm{cl} \to \mathrm{Cl}(X) \quad \text{and} \quad \Pi\colon C \to C\text{-closure}$$

are inverse bijections between the closure operators on X and the $\cap$-structures on $\wp(X)$. These bijections map algebraic closure operators onto $\cap\overrightarrow{\cup}$-structures.

Proof. Suppose that cl is algebraic and that $D = \{d_i \mid i \in I\}$ is directed in $\mathrm{Cl}(X)$. Since cl is join-continuous,

$$\mathrm{cl}\left(\overrightarrow{\bigcup} d_i\right) = \overrightarrow{\bigcup} \mathrm{cl}(d_i) = \overrightarrow{\bigcup} d_i$$

Hence, $\bigcup^{\rightarrow} D \in \mathrm{Cl}(X)$ and so $\mathrm{Cl}(X)$ is an $\cap\overrightarrow{\cup}$-structure. Conversely, if cl is join-continuous and if $a \in \wp(X)$, then

$$a = \overrightarrow{\bigcup} \{x \mid x \text{ finite}, x \subseteq a\} \subseteq \overrightarrow{\bigcup} \{\mathrm{cl}(x) \mid x \text{ finite}, x \subseteq a\} \subseteq \mathrm{cl}(a)$$

But the element

$$b = \overrightarrow{\bigcup} \{\mathrm{cl}(x) \mid x \text{ finite},\ x \subseteq a\}$$

is closed by the join-continuity of the closure operator and so taking closures gives $\mathrm{cl}(a) = b$ and so cl is algebraic.□

Subalgebra Lattices

We can now characterize subalgebra lattices.

Theorem 7.9 (Birkhoff–Frink, 1948 [8]) *A lattice L is isomorphic to the subalgebra lattice* $\mathrm{Sub}(\langle A, \Omega\rangle)$ *of some algebra $\langle A, \Omega\rangle$ if and only if L is algebraic.*

Proof. Since a subalgebra lattice is an $\cap\overrightarrow{\cup}$-structure, it is algebraic. Let us give two proofs that algebraic lattices are isomorphic to subalgebra lattices. First, Theorem 7.5 implies that L is isomorphic to the ideal lattice $\mathcal{I}(J)$ of the join-semilattice $J = \mathcal{K}(L)$ with 0. Thus, all we need to do is find operations Ω on J so that the subalgebras of the algebra $\langle J; \Omega \rangle$ are precisely the ideals of J.

At the start, J is an algebra with no operations and so every subset is a subalgebra. To reduce this family to the ideals of J, we first include the binary join operation and so $\text{Sub}(J)$ is now the family of subsets of J that inherit binary joins. Ideals are also down-sets and to restrict the subalgebras to down-sets, that is, to force

$$x \in S, a \leq x \quad \Rightarrow \quad a \in S$$

for every subalgebra S, we include a unary operation

$$f_a(x) = \begin{cases} a & \text{if } a \leq x \\ x & \text{otherwise} \end{cases}$$

for each $a \in J$. Then $x \in S$ and $a \leq x$ implies that $a = f_a(x) \in S$. Now $\text{Sub}(J)$ is the family of all down-sets that inherit binary joins. These are precisely the ideals of J with one exception. Our definition of ideal requires that ideals be nonempty. Thus, we include the nullary operation $0 \in J$ to remove the empty set from $\text{Sub}(J)$. Now we have

$$\text{Sub}(\langle J, \Omega \rangle) = \mathcal{I}(J) \approx L$$

As a second proof, L is isomorphic to an $\cap\overrightarrow{\cup}$ -structure $\mathcal{M}$ on a set X and so $\mathcal{M}$ is the family $\text{Cl}(X)$ of closed sets of a finitary closure operator cl on X. Thus, for any $S \subseteq X$,

$$\text{cl}(S) = \bigcup \{\text{cl}(F) \mid F \subseteq S, F \text{ finite}\}$$

and so S is closed if and only if it has the property that

$$F \subseteq S, F \text{ finite} \quad \Rightarrow \quad \text{cl}(F) \subseteq S \tag{7.10}$$

We seek operations on X for which this characterizes the subalgebras as well.

Given a finite set $F \subseteq X$ of size $n \geq 0$ and an $a \in \text{cl}(F) \setminus F$, we can force a subalgebra S of X that contains F to also contain a by requiring that S be invariant under any n-ary operation $f_{(a,F)}$ satisfying

$$f_{(a,F)}(x_1, \ldots, x_n) = a \quad \text{if} \quad F = \{x_1, \ldots, x_n\}$$

when $n \geq 1$ and under the nullary operation $f_{(a,\emptyset)} = a$ when $n = 0$.

But we also need S to be $f_{(a,F)}$-invariant even if S does not contain F. Of course, this is only an issue when F is nonempty, that is, when $n \geq 1$. In this case, $f_{(a,F)}(x_1, \ldots, x_n)$ can be any member of S, so we set

$$f_{(a,F)}(x_1,\ldots,x_n) = \begin{cases} a & \text{if } F = \{x_1,\ldots,x_n\} \\ x_1 & \text{otherwise} \end{cases}$$

Then a nonempty subset S of X is Ω-invariant, that is, S is a subalgebra of X if and only if (7.10) holds.□

Congruence Lattices

We have seen (Theorem 7.9) that the *subalgebra lattices* of algebras are precisely the algebraic lattices. There has been much work done on the question of characterizing the *congruence lattices* of an algebra.

Definition *A* **congruence relation** *on an algebra $\langle X;\Omega\rangle$ is an equivalence relation $\equiv$ on X with the property that for each n-ary operation $f \in \Omega$ and for all $x_i, y_i \in X$,*

$$x_i \equiv y_i \text{ for } 1 \le i \le n \quad \Rightarrow \quad f(x_1,\ldots,x_n) \equiv f(y_1,\ldots,y_n)$$

The family of all congruence relations on $\langle X;\Omega\rangle$ is denoted by $\mathrm{Con}(\langle X;\Omega\rangle)$ *or* $\mathrm{Con}(X)$.□

The meet and join of congruence relations is again a congruence relation and so $\mathrm{Con}(X)$ is a complete sublattice of the lattice of equivalence relations on X. The lattice $\mathrm{Con}(X)$ is called the **congruence lattice** of X. We will study congruence lattices in some detail in a later chapter.

In 1963, Grätzer and Schmidt proved that the congruence lattices of algebras are also precisely the algebraic lattices.

Theorem 7.11 (Grätzer and Schmidt, 1963 [27]) *A lattice L is isomorphic to the congruence lattice* $\mathrm{Con}(\langle A,\Omega\rangle)$ *of some algebra $\langle A,\Omega\rangle$ if and only if L is algebraic.*□

The proof of Theorem 7.11 is rather difficult and several proofs have appeared as time has gone by (some published and others not). However, as Grätzer [26] points out in his recent survey article, the proof is somehow inherently complex, as witnessed by the work of Freese, Lampe and Taylor.

Theorem 7.12 (Freese, Lampe, and Taylor, 1979 [20]) *Let V be an infinite-dimensional vector space whose base field has uncountable cardinality κ and let* $\mathrm{Sub}(V)$ *be the subalgebra lattice of V, which is algebraic. If* $\mathrm{Sub}(V)$ *is isomorphic to a congruence lattice of an algebra $\langle A;\Omega\rangle$, then $|\Omega| \ge \kappa$, that is, $\langle X,\Omega\rangle$ has at least κ operations.*□

It is also natural to ask in particular whether every algebraic lattice is isomorphic to the congruence lattice $\mathrm{Con}(L)$ of some lattice L. (A latttice is an algebra.) In a later chapter, we will study the congruence lattice $\mathrm{Con}(L)$ of a lattice L. We will show that $\mathrm{Con}(L)$ is distributive and this brings us to one of the most important and longstanding problems in lattice theory: The **congruence lattice problem (CLP)** is the problem of whether all distributive, algebraic lattices are isomorphic to $\mathrm{Con}(L)$ for some lattice L. For the finite case, we have the following positive result.

Theorem 7.13

1) **(Dilworth)** *Every finite distributive lattice is isomorphic to* $\mathrm{Con}(L)$ *for some finite lattice* L.

2) **(Grätzer and Schmidt, 1963 [27])** *Every finite distributive lattice is isomorphic to* $\mathrm{Con}(L)$ *for some finite sectionally complemented lattice* L.□

There is a modestly amusing story behind the theorem of Dilworth stated above. Apparently, Dilworth did not publish this theorem, but it did appear as an *exercise* (marked with an asterisk to indicate that it is hard) in Birkhoff's book [5]. As Grätzer states in [26]:

> E. T. Schmidt and I got really interested in the result and inquired from G. Birkhoff where the result came from, but he did not know and encouraged us to write to R. P. Dilworth. Unfortunately, Dilworth was busy editing the proceedings of a lattice theory meeting, but eventually we got a response. Yes, he proved the result, and the proof was in his lecture notes. No, copies of his lecture notes were no longer available.

On the other hand, the general congruence lattice problem has very recently (2006) been solved in the negative by Wehrung [64].

Meet-Representations

We have defined a meet-irreducible element $m \in L$ to be a nonunit element with the property that

$$m = a \wedge b \quad \Rightarrow \quad m = a \text{ or } m = b$$

We can define an analogous notion for arbitrary meets.

Definition *Let* L *be a complete lattice.*

1) *An element* $a \in L$ *is* **completely meet-irreducible** *if* $a = \bigwedge S$ *for a subset* $S \subseteq L$ *implies that* $a = s$ *for some* $s \in S$.

2) *An element* $a \in L$ *is* **strictly meet-irreducible** *if* $a = 1$ *or if the set* $\uparrow a \setminus \{a\}$ *of elements properly containing* a *has a smallest element.*□

We leave it as an exercise to show that these two concepts are equivalent.

Theorem 7.14 *An element of a complete lattice is completely meet-irreducible if and only if it is strictly meet-irreducible.* □

Example 7.15 To illustrate the concept, let S be a proper subspace of a vector space V and let $S^{\subset} = (\uparrow S) \setminus \{S\}$. If $v \notin S$, then the subspace $\langle S, v \rangle$ covers S and so is a minimal element of $S^{\subset}$. Hence, S is strictly meet-irreducible if and only if $S = V$ or S has codimension 1. Note also that any subspace of V is the intersection of strictly meet-irreducible subspaces. □

Example 7.16 We can also find strictly meet-irreducible subgroups of a group G as follows. Let H be a proper subgroup of G and let $k \notin H$. Let $\mathcal{F}$ be the family of all subgroups of G that contain H but not k. Since the union of any chain in $\mathcal{F}$ is also in $\mathcal{F}$, Zorn's lemma implies that $\mathcal{F}$ has a maximal member M.

It is easy to see that $M \vee \langle k \rangle$ is the smallest element of $M^{\subset} = (\uparrow M) \setminus \{M\}$, for if $M \subset N \leq G$, then $H \subseteq N$ and the maximality of M implies that $k \in N$ and so $M \vee \langle k \rangle \subseteq N$. Hence, M is strictly meet-irreducible. We have shown that given any proper subgroup H of G, for each $k \notin H$ there is a strictly meet-irreducible subgroup M_k for which $H \subseteq M_k$ and $k \notin M_k$. Thus,

$$H = \bigcap \{M_k \mid k \notin H\}$$

that is, the strictly meet-irreducible subgroups are meet-dense in the lattice $\mathcal{S}(G)$ of subgroups of G. □

Definition *Let L be a lattice. A representation of an element $a \in L$ of the form*

$$a = \bigwedge \{q_i \mid i \in I\}$$

where each q_i is strictly meet-irreducible is called a **meet-representation** *of a and is said to be* **irredundant** *if a is not the meet of any proper subfamily of $\{q_i \mid i \in I\}$.* □

The existence of meet-representations does not guarantee the existence of irredundant meet-representations, as the following example shows.

Example 7.17 The set $L = \{0, 1, 1/2, 1/3, \ldots\} \subseteq \mathbb{Q}$ is a chain under the usual order. It is clear that every positive element of L is strictly meet-irreducible, but 0 is not. Also, 0 has a meet-representation but no irredundant meet-representation. □

Theorem 7.18 *If L is an algebraic lattice, then every element of L has a meet-representation.*

Proof. The proof mimics the case of groups described earlier, where the role of the element $k \in G$ is played by a compact element of L. Let $a, b \in L$ satisfy $b \not\leq a$. Since $\mathcal{K}(L)$ is join-dense, there is a $k \in \mathcal{K}(L)$ for which $k \leq b$ but

$k \not\leq a$. Let

$$X = \{x \in L \mid a \leq x, k \not\leq x\}$$

which is nonempty since $a \in X$. Any directed subset D of X has a join $j = \bigvee D$, which is also in X since $a \leq j$ and if $k \leq j$, then the compactness of k implies that $k \leq d$ for some $d \in D$, which is false. Thus, Zorn's lemma implies that X has a maximal element m. It follows that $a \leq m$ but $k \not\leq m$ and so $m < m \vee k$. Moreover, if $m < n$, then the maximality of m implies that $k \leq n$ and so $m \vee k \leq n$. It follows that $m \vee k$ is the least element of $(\uparrow m) \setminus \{m\}$ and so m is strictly meet-irreducible.

Now, if $a \in L$, then for each $b \in L$ with the property that $b \not\leq a$, there is a strictly meet-irreducible element m_b for which $a \leq m_b$ but $b \not\leq m_b$. Let $\mathcal{M}$ be the family of all strictly meet-irreducible elements of L that contain a and let

$$b = \bigwedge \mathcal{M}$$

Then $a \leq b$ but if $a < b$, then there is a strictly meet-irreducible element m_b for which $a \leq m_b$ but $b \not\leq m_b$. But $m_b \in \mathcal{M}$ implies that $b \leq m_b$, a contradiction. Hence, $a = b$ is the meet of strictly meet-irreducible elements of L.□

The book by Crawley and Dilworth [11] contains many results relating the existence of irredundant meet-representations to strong atomicity. Here is a sampling.

Theorem 7.19 *Let L be an algebraic lattice.*
1) *If L is strongly atomic, then every element of L has an irredundant meet-representation.*
2) *By way of converse, if L is modular and every element of L has an irredundant meet-representation, then L is strongly atomic.*□

It is interesting to note that the statement

1) If L is a strongly atomic algebraic lattice, then every element of L has an irredundant meet-representation

is actually equivalent to the axiom of choice. Crawley and Dilworth's proof of this statement uses the Hausdorff maximality principle, which is equivalent to the axiom of choice.

On the other hand, suppose that 1) holds and let P be a poset. We may assume that P is not a chain. Let L be the family of all chains in P, along with the empty set and P. Then L is an $\cap\overrightarrow{\cup}$-structure and therefore an algebraic lattice. Also, the strictly meet-irreducible elements in L are P, the maximal chains and the chains that are covered by a maximal chain. If $C \subset D$ in L, then there is a $d \in D \setminus C$ and so $C \cup \{d\}$ is a chain that covers C. Hence, L is strongly atomic and so 1) implies that every proper chain in P is the meet of maximal chains and thus contained in a maximal chain. Thus, the Hausdorff maximality principle holds.

Exercises

1. Prove that $C(p)$ is the smallest closed element containing p and that 1 is closed.
2. Prove that the compact elements $\mathcal{K}(L)$ in a complete lattice L inherit finite joins from L and so $\mathcal{K}(L)$ is a join-semilattice with 0.
3. Is the meet of compact elements necessarily compact?
4. Prove that if L is a complete lattice with ACC, then L is algebraic.
5. Prove that the compact elements of $\wp(X)$ are the finite subsets of X.
6. Find the compact elements of the chain $[0,1] \subseteq \mathbb{R}$.
7. Find the compact elements of the lattice $(\mathbb{N}, \mid)$.
8. Prove that the compact elements of the subgroup lattice of a group G are the finitely-generated subgroups.
9. Let L be an algebraic lattice. Show that if $a < k$ and $k \in \mathcal{K}(L)$, then k is a cover for some element b for which $a \leq b < k$.
10. Let L be an algebraic lattice and let S be a complete sublattice of L. Show that S is also algebraic. Conclude that any closed interval of L is algebraic.
11. Prove that any algebraic lattice is weakly atomic.
12. Let L be an algebraic lattice.
    a) Prove that L is **meet-continuous**, that is, meet distributes over arbitrary directed joins, that is,
    $$a \wedge \overrightarrow{\bigvee} \{b_i \mid i \in I\} = \overrightarrow{\bigvee} \{a \wedge b_i \mid i \in I\}$$
    b) Prove that if L is distributive, then meet distributes over arbitrary join.
13. Prove that an element $a \in L$ is completely meet-irreducible if and only if it is strictly meet-irreducible.
14. Let L be a lattice. An element $a \in L$ is **join-inaccessible** if for any subset B of L,
    $$a = \bigvee B \quad \Rightarrow \quad a = \bigvee B_0 \text{ for some finite subset } B_0 \subseteq B$$
    Prove the following for a complete lattice L.
    a) The compact elements of L are join-inaccessible.
    b) If L is algebraic, then the compact elements are precisely the join-inaccessible elements.
15. Let L be a finite lattice in which every element has a unique irredundant meet-representation. Let $S = \{s_i \mid i \in I\}$ be the family of strictly meet-irreducibe elements of L. Thus, for each $a \in L$, there is a unique subset $S(a) \subseteq S$ for which

$$a = \bigwedge\nolimits_{i \in S(a)} s_i$$

is an irredundant meet-representation of a.

a) Show that $a = b$ if and only if $S(a) = S(b)$.
b) Show that $a < b$ if and only if $S(b) \subset S(a)$.
c) Show that $a \sqsubset b$ if and only if $S(b) \sqsubset S(a)$.
d) Show that

$$c = \bigwedge\nolimits_{i \in S(a) \cap S(b)} s_i$$

is an irredundant meet-representation of $a \vee b$.

e) Show that L is upper semimodular.

16. Prove the following (unpublished?) result of Bill Hanf: For any nontrivial algebraic lattice L, the following are equivalent:
    a) If $k \in \mathcal{K}(L)$ is compact, then $\mathcal{K}_{\downarrow}(k)$ is countable.
    b) L is isomorphic to the subalgebra lattice $\mathrm{Sub}(\mathcal{A})$ of an algebra $\mathcal{A} = \langle X, \Omega \rangle$ that has only a countable number of operations (including nullary operations).
    c) L is isomorphic to the subalgebra lattice $\mathrm{Sub}(\mathcal{A})$ of an algebra $\mathcal{A} = \langle X, \Omega \rangle$ that has only a single operation, which is binary.

Chapter 8
Prime and Maximal Ideals; Separation Theorems

Separation Theorems

The existence of maximal and prime ideals is an important topic in lattice theory. In discussing the existence of such ideals, it is of considerable interest to know not just when a lattice L has such an ideal, but when L has such an ideal that "separates" certain elements or subsets of L. Theorems to this effect are called **separation theorems**. The following notation will prove convenient.

Definition *Let L be a lattice. If $A, X \subseteq L$ are disjoint subsets of L, then we denote the set of all proper ideals of L that contain A and are disjoint from X by $\mathcal{I}(L; A, X)$, that is,*

$$\mathcal{I}(L; A, X) = \{J \in \mathcal{I}(L) \setminus \{L\} \mid A \subseteq J, J \cap X = \emptyset\}$$

If I is a proper ideal and F is a proper filter, we denote this set by

$$\mathcal{I}(L; I \lhd L, F \rhd L) \qquad \square$$

An element $J \in \mathcal{I}(L; A, X)$ represents a type of *separation* of A and X and it will be convenient to establish some general terminology to describe this separation.

Definition *Let $\mathcal{F}$ and $\mathcal{G}$ represent families of subsets of lattices, such as the family of proper subsets, proper ideals or proper filters, ordered by set inclusion.*

1) *A lattice L has the* **maximal $(\mathcal{F}, \mathcal{G})$-separation property** *if for any disjoint $F \in \mathcal{F}$ and $G \in \mathcal{G}$, the set $\mathcal{I}(L; F, G)$ has a maximal element.*
2) *A lattice L has the* **prime $(\mathcal{F}, \mathcal{G})$-separation property** *if for any disjoint $F \in \mathcal{F}$ and $G \in \mathcal{G}$, the set $\mathcal{I}(L; F, G)$ contains a prime ideal of L.* $\square$

There are some special cases of these definitions that are of interest.

S. Roman (ed.), *Lattices and Ordered Sets*, doi: 10.1007/978-0-387-78901-9_8,

Definition

1) $(\mathcal{G} = \{\emptyset\})$ *A lattice* L *has the* **maximal (prime)** $\boldsymbol{\mathcal{F}}$**-extension property** *if any* $F \in \mathcal{F}$ *is contained in a maximal (prime) ideal of* L.

2) $(\mathcal{F} = \{\emptyset\})$ *A lattice* L *has the* **maximal (prime)** $\boldsymbol{\mathcal{G}}$**-avoidance property** *if for any* $G \in \mathcal{G}$, *the set* $\mathcal{I}(L; \emptyset, G)$ *contains a maximal element (prime ideal).*

3) $(\mathcal{F} = \mathcal{G} = \{\emptyset\})$ *A lattice* L *has the* **maximal (prime) existence property** *if* L *contains a maximal (prime) ideal.* □

It is clear that the separation property implies the other properties and all properties imply the existence property.

The families of interest to us with regard to separation are the families $\mathcal{I}^*$, $\mathcal{F}^*$ and $\wp^*$ of proper ideals, proper filters and proper subsets, respectively and the separations of interest are

$$(\mathcal{I}^*, \wp^*)\text{-separation} \quad \text{and} \quad (\mathcal{I}^*, \mathcal{F}^*)\text{-separation}$$

and those cases formed by replacing one or more of $\mathcal{I}^*$, $\mathcal{F}^*$ or $\wp^*$ by the family $\{\emptyset\}$. In particular, for a given class $\mathcal{C}$ of lattices, we consider the following six separation statements:

$$\text{Every nontrivial } \mathcal{C}\text{-lattice has the maximal} \left\{ \begin{array}{l} (\mathcal{I}^*, \wp^*)\text{-separation} \\ (\mathcal{I}^*, \mathcal{F}^*)\text{-separation} \\ \mathcal{I}^*\text{-extension} \\ \wp^*\text{-avoidance} \\ \mathcal{F}^*\text{-avoidance} \\ \text{existence} \end{array} \right\} \text{property} \quad (8.1)$$

Maximal Separation Theorems for Lattices with 1

We will prove that if $\mathcal{C}$ is a class of lattices with 1 that contains the class of complete distributive lattices, then each of the six statements in (8.1) is equivalent to the axiom of choice.

It is easy to see that the axiom of choice, or equivalently Zorn's lemma, implies the strongest of these statements, and therefore all of these statements:

1) Every nontrivial lattice with 1 has the maximal $(\mathcal{I}^*, \wp^*)$-separation property.

For if $I \in \mathcal{I}^*$ and $X \in \wp^*$ are disjoint, then $\mathcal{I} = \mathcal{I}(L; I, X)$ is nonempty, since it contains I. Since the union of any chain $\mathcal{C}$ in $\mathcal{I}$ is again in $\mathcal{I}$, Zorn's lemma implies that $\mathcal{I}$ has a maximal element. Hence, 1) holds.

Actually, there has been much work done proving that various forms of the six statements in (8.1) are equivalent to the axiom of choice. Here are some examples:

- Every nontrivial lattice with 1 has the maximal $\mathcal{I}^*$-extension property (Mrówka, 1956 [46]).
- Every nontrivial distributive lattice with 1 has the maximal $\mathcal{I}^*$-extension property (Klimowsky, 1958 [38]).
- Every nontrivial lattice with 1 has a maximal ideal (Scott, 1954 [55]).
- Every nontrivial distributive lattice with 1 has a maximal ideal (Klimowsky, 1958 [38]; see also Rubin and Rubin, 1985 [54], page 101).
- Every nontrivial complete lattice has a maximal ideal (Banaschewski, 1961 [2]).
- For each nonempty set X, every complete sublattice of $\wp(X)$ has a maximal ideal (Bell and Fremlin, 1972 [3]).

Of course, the weakest statement of the form (8.1) is

2) Every nontrivial complete distributive lattice has a maximal ideal.

In 2003, Herrlich [33] proved that the axiom of choice is equivalent to the following statement:

3) The lattice of closed sets of any topological space has a maximal filter.

But it is easy to see that the dual of 2),

2^{∂}) Every nontrivial complete distributive lattice has a maximal filter

which is equivalent to 2), implies 3). For the family Γ of closed subsets of a topological space is a $\cap$-structure containing $\emptyset$ and therefore a complete lattice. Also, since meet is intersection and *finite* join is union, it follows that Γ is distributive. Thus, Γ is a nontrivial complete distributive lattice and so 2^{∂}) implies 3). Thus, 2) implies the axiom of choice and each of the six statements in (8.1) implies the axiom of choice.

The following theorem uses a few basic notions from topology, some of which are covered briefly in an appendix to this book.

Theorem 8.2 (Herrlich, 2003 [33])

1) The following statement is equivalent to the axiom of choice: The lattice of closed sets of any topological space has a maximal filter.

2) Each of the six statements in (8.1) is equivalent to the axiom of choice.

Proof. We have seen that the axiom of choice implies statement 2) and therefore also this statement. For the converse, let $\{X_i \mid i \in I\}$ be a nonempty family of nonempty sets. We wish to find a choice function $f \in \prod X_i$. We may use the axioms of set theory without the axiom of choice. This allows us to take unions and cartesian products and to use the *axiom schema of comprehension*, which says that if S is a set and if P is a property, then we may form the subset

$$A = \{x \in S \mid P(x) \text{ holds}\}$$

Now, let ∞ be a symbol not in $\bigcup X_i$ and let $Y_i = X_i \cup \{\infty\}$ for each $i \in I$. Define a topology on Y_i by taking the closed sets to be the finite subsets of X_i, along with the whole space Y_i. The cartesian product

$$Y = \prod_{i \in I} Y_i$$

is nonempty, since it contains the constant function $f: I \to \bigcup Y_i$ defined by $f(i) = \infty$ for all $i \in I$, or in more formal notation,

$$f = I \times \{\infty\} \subseteq I \times \bigcup Y_i$$

For each $i \in I$, the *projection map* $\pi_i: Y \to Y_i$ is defined by $\pi_i(f) = f(i)$ for all $f \in Y$. Also, for each $B \subseteq Y_i$, the *cylinder* with *base* B is the set

$$\pi_i^{-1}(B) = \{f \in Y \mid f(i) \in B\}$$

We give Y the product topology τ, that is, the set of all cylinders with open bases is a *subbasis* for a topology on Y. Since the projection maps are open, a cylinder is an open subset in the product topology if and only if its base is open. Such cylinders are called *open cylinders*.

The family Γ of closed subsets of Y is a nontrivial complete distributive lattice and so Γ has a maximal filter $\mathcal{F}$, which is also a prime filter by Theorem 4.25. We refer to a cylinder in $\mathcal{F}$ as an **$\mathcal{F}$-cylinder** and the base of an $\mathcal{F}$-cylinder as an **$\mathcal{F}$-base**.

We wish to show that for each coordinate $i \in I$, there is an $\mathcal{F}$-base that is a finite subset of X_i. For this, it is sufficient to show that there is an $\mathcal{F}$-base that is a *proper* subset of Y_i, since any $\mathcal{F}$-base is closed and so is either Y_i or a finite subset of X_i.

However, if there is *any* $F \in \mathcal{F}$ for which $\pi_i(F)$ is a proper subset of Y_i, then F is contained in the cylinder $\pi_i^{-1}(\pi_i(F))$ and since $\mathcal{F}$ is a filter and

$$\pi_i^{-1}(\pi_i(F)) \supseteq F \in \mathcal{F}$$

it follows that $\pi_i^{-1}(\pi_i(F))$ is an $\mathcal{F}$-cylinder whose base is a proper subset of Y_i.

This leaves only the possibility that $\pi_i(F) = Y_i$ for all $F \in \mathcal{F}$. But this contradicts the maximality of $\mathcal{F}$. To see this, let $x \in X_i$ and let

$$\mathcal{G} = [\mathcal{F}, \pi_i^{-1}(x))$$

be the filter generated by $\mathcal{F}$ and the closed cylinder $\pi_i^{-1}(x)$. The filter $\mathcal{G}$ consists of all closed sets $K \supseteq F \cap \pi_i^{-1}(x)$, for some $F \in \mathcal{F}$. But $\mathcal{G}$ is strictly larger than $\mathcal{F}$, since $\pi_i^{-1}(x)$ is not in $\mathcal{F}$. Also, $\mathcal{G}$ is a proper filter, since $\emptyset \in \mathcal{G}$ implies that

$$\emptyset = F \cap \pi_i^{-1}(x)$$

for some $F \in \mathcal{F}$, which is false since $\pi_i(F) = Y_i$ and so there is an $f \in F$ for which $\pi_i(f) = x$.

Thus, we have shown that for each coordinate $i \in I$, there is an $\mathcal{F}$-base that is a finite subset of X_i. Next, we show that there is a *unique singleton* $\mathcal{F}$-base $\{x_i\}$ in each X_i. If $\pi_i^{-1}(B)$ is an $\mathcal{F}$-cylinder with base $B = \{b_1, \ldots, b_n\}$ in X_i, then

$$\pi_i^{-1}(b_1) \cup \cdots \cup \pi_i^{-1}(b_n) = \pi_i^{-1}(B) \in \mathcal{F}$$

and since $\mathcal{F}$ is prime, at least one $\pi_i^{-1}(b_i)$ is in $\mathcal{F}$ and so $\{b_i\}$ is also an $\mathcal{F}$-base in X_i. On the other hand, if $\{c_i\}$ is also an $\mathcal{F}$-base in X_i with $c_i \neq b_i$, then

$$\emptyset = \pi_i^{-1}(b_i) \cap \pi_i^{-1}(c_i) \in \mathcal{F}$$

which contradicts the maximality of $\mathcal{F}$. Hence, there is a unique singleton $\mathcal{F}$-base $\{b_i\}$ in each X_i and so the function $f: I \to \bigcup X_i$ defined by

$$f(i) = b_i$$

is a choice function for the family $\{X_i \mid i \in I\}$.□

The fact that lattices have the maximal $(\mathcal{I}^*, \wp^*)$-separation property is sometimes referred to as the **maximal separation theorem** for lattices. Similarly, lattices satisfy the **maximal extension theorem** for proper ideals and the **maximal ideal theorem**.

Maximal Separation Theorems for Boolean Algebras

Of course, the class of Boolean algebras does not contain the class of complete distributive lattices and so the previous results do not apply to Boolean algebras. With respect to Boolean algebras, we confine ourselves (for now) to the following two statements:

1) Every nontrivial Boolean algebra has the maximal $(\mathcal{I}^*, \wp^*)$-separation property.
2) Every nontrivial Boolean algebra has the maximal $\wp^*$-avoidance property.

We will prove that each of these statements is equivalent to the axiom of choice. On the other hand, the statement that every nontrivial Boolean algebra has the maximal $\mathcal{I}^*$-extension property is strictly weaker than the axiom of choice. (We will revisit this issue a bit later.)

Since a Boolean algebra is a lattice with 1, we have seen that the axiom of choice implies statement 1), which implies statement 2).

Theorem 8.3 *Each of statements 1) and 2) is equivalent to the axiom of choice.*
Proof. It is sufficient to show that statement 2) implies Tukey's lemma. We begin with a few remarks about families of finite character. First, if $\mathcal{F}$ has finite character and if $A \in \mathcal{F}$, then any subset S of A is also in $\mathcal{F}$, since all finite subsets of S are finite subsets of A and so belong to $\mathcal{F}$. Hence, $A \in \mathcal{F}$ implies $\downarrow A \subseteq \mathcal{F}$.

Second, let $\mathcal{I}$ be an ideal of $\wp(X)$ contained in $\mathcal{F}$. Then $\bigcup \mathcal{I} \in \mathcal{F}$, for if $F \subseteq \bigcup \mathcal{I}$ is finite, then F is contained in a union $\bigcup \mathcal{I}_0$ of a finite subset of $\mathcal{I}$, which is in $\mathcal{I}$, whence $F \in \mathcal{I}$. Thus, $\bigcup \mathcal{I} \in \mathcal{F}$. Moreover, if $\mathcal{I}$ is maximal among all ideals of $\wp(X)$ contained in $\mathcal{F}$, then $\bigcup \mathcal{I}$ is a maximal element of $\mathcal{F}$. For if $\bigcup \mathcal{I} \subset N \in \mathcal{F}$, then

$$\mathcal{I} \subseteq \downarrow\left(\bigcup \mathcal{I}\right) \subset \downarrow N \subseteq \mathcal{F}$$

Hence, the maximality of $\mathcal{I}$ implies that $\downarrow N = \mathcal{F}$. Thus, $\mathcal{F}$ is an ideal of $\wp(X)$ and so $\mathcal{I} = \mathcal{F}$, which is false.

Now, to see that 2) implies Tukey's lemma, let $\mathcal{F} \subseteq \wp(X)$ be a nonempty family of finite character. Then $\wp(X) \setminus \mathcal{F}$ is a proper subset of $\wp(X)$ and so statement 2) implies that there is an ideal $\mathcal{I}$ of $\wp(X)$ that is maximal with respect to avoiding $\wp(X) \setminus \mathcal{F}$, that is, maximal with respect to being contained in $\mathcal{F}$. Hence, by the previous remarks, $\bigcup \mathcal{I}$ is maximal in $\mathcal{F}$ and Tukey's lemma holds.□

Prime Separation Theorems

There are a variety of prime separation theorems that follow from the aciom of choice. Here are some examples:

1) (**The distributive prime ideal theorem**) Every nontrivial distributive lattice has a prime ideal.
2) (**The Boolean prime ideal theorem**) Every nontrivial Boolean algebra has a prime (i.e., maximal) ideal.
3) (**The Boolean prime $\mathcal{I}^*$-extension theorem**) Any proper ideal of a Boolean algebra B can be extended to a prime ideal.

It is known that these statements are strictly weaker than the axiom of choice. However, we will not pursue this line of inquiry further. The interested reader may want to consult Rubin and Rubin [53].

Our interest centers upon the fact that, assuming the axiom of choice, separation by prime ideals is equivalent to distributivity. First, we show that for a distributive lattice, a maximal element of $\mathcal{I}(L; I \lhd L, F \rhd L)$ is a prime ideal in L.

Theorem 8.4 *If L is a distributive lattice, then any maximal element of $\mathcal{I}(L; I \lhd L, F \rhd L)$ is prime in L.*

Proof. Let M be maximal in $\mathcal{I}(L; I, F)$. Suppose that $a \wedge b \in M$ but that $a, b \notin M$. Hence, the ideal

$$J_a = M \vee \downarrow a$$

properly contains M and so intersects F. Thus, $a \vee m \in F$ for some $m \in M$ and similarly, $b \vee n \in F$ for some $n \in M$. Hence, if $p = n \vee m \in M$, then $a \vee p$ and $b \vee p$ are in F, whence

$$(a \wedge b) \vee p = (a \vee p) \wedge (b \vee p) \in F$$

But $a \wedge b$ and p are in M and so $M \cap F \neq \emptyset$, a contradiction. Thus, M is prime.□

We can now show that separation by prime ideals is equivalent to distributivity.

Theorem 8.5 *The following are equivalent for a nontrivial lattice L (assuming the axiom of choice):*

1) *L is distributive.*
2) **(The prime $(\mathcal{I}^*, \mathcal{F}^*)$-separation property)** *The set $\mathcal{I}(L; I \lhd L, F \rhd L)$ contains a prime ideal of L.*
3) **(The prime point-separation property)** *If $b \not\leq a$ in L, then there is a prime ideal containing a but not b. In particular, if $a \neq b$, then there is a prime ideal containing one of a or b but not the other.*

The fact that 1) implies 2) is sometimes called the **prime separation theorem**.

Proof. Assume that L is distributive. Zorn's lemma implies that the set $\mathcal{I}(L; I \lhd L, F \rhd L)$ has a maximal ideal M, which is prime by Theorem 8.4. It is clear that 2) implies 3). Finally, if L is not distributive, then it has a sublattice isomorphic to M_3 or N_5, as shown in Figure 8.1.

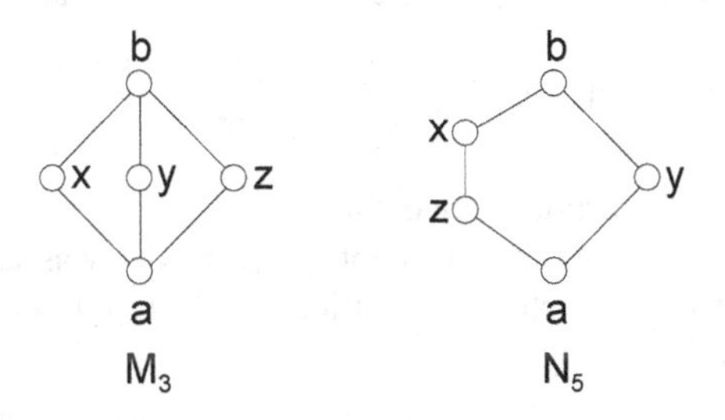

Figure 8.1

Now, suppose that P is a prime ideal containing z. Then $x \wedge y = a \in P$ and so one of x or y is in P. But if $y \in P$, then $b = z \vee y \in P$ and so $x \in P$. Thus, $z \in P$ implies $x \in P$. But $x \not\leq z$ and so 3) fails.□

Exercises

1. Let L be a distributive lattice. Prove that $\mathcal{P}_a = \mathcal{P}_{\neg b}$ if and only if a and b are complements.
2. Let L be a nontrivial bounded lattice. Show that L is distributive if and only if every proper ideal of L is the intersection of prime ideals.
3. Let M be a distributive lattice and let N be a sublattice of M. Show that if $Q \in \operatorname{Spec}(N)$ then there is a $Q' \in \operatorname{Spec}(M)$ such that $N \cap Q' = Q$.
4. Let L be a Boolean lattice. Prove that a proper filter F is an ultrafilter if and only if for each $a \in L$, we have $a \in F$ or $a' \in F$.
5. Describe the principal ultrafilters of the power set lattice $\wp(S)$.
6. Let $\mathcal{F}$ be a proper filter in the power set lattice $\wp(S)$. Prove that the following are equivalent:
   a) F is an ultrafilter.
   b) If $\mathcal{P} = \{A_1, \ldots, A_n\}$ is a partition of S then exactly one of the blocks A_i is in $\mathcal{F}$.
   c) For every $A \subseteq S$, either $A \in \mathcal{F}$ or $A^c \in \mathcal{F}$.
   d) If a subset A of S intersects every subset in $\mathcal{F}$ nontrivially then $A \in \mathcal{F}$.
7. True or false: Distributive lattices with 1 have the property that if $I \lhd L$ and $F \rhd L$ then $\mathcal{I}(L; I, F)$ has an element that is maximal in L.
8. Let L be a distributive lattice and let M be a proper sublattice of L. Prove that there are prime ideals P and Q of L for which $P \cap M \subseteq Q$ but $P \not\subseteq Q$.
9. For a Boolean lattice L, prove that L has the maximal separation property for proper ideals and proper filters, that is, the set

   $$\mathcal{I}(L; I \lhd L, F \rhd L)$$

   has a maximal element if and only if this set has an element that is maximal in L.
10. Show that if every nontrivial lattice with 1 has a maximal ideal then Tukey's lemma holds as follows. Let $\mathcal{F} \subseteq \wp(X)$ be a nonempty family of finite character and let $\mathcal{G} = \mathcal{F} \cup \{X\}$. Let $A \wedge B = A \cap B$ and let

    $$A \vee B = \begin{cases} A \cup B & \text{if } A \cup B \in \mathcal{F} \\ X & \text{otherwise} \end{cases}$$

    Show that $\mathcal{G}$ is a nontrivial lattice with 1.
11. In an exercise from an earlier chapter, the reader was asked to prove the following: Let L be a distributive lattice and let $I \trianglelefteq L$ and $F \trianglerighteq L$. Define a binary relation θ on L by

    $$x\theta y \quad \text{if} \quad (x \vee u) \wedge v = (y \vee u) \wedge v \text{ for some } u \in I \text{ and } v \in F$$

a) θ is a congruence relation on L.
b) If L is bounded then $I \subseteq [0]$ and $F \subseteq [1]$.
c) θ is proper ($\theta \subset L \times L$) if and only if $I \cap F = \emptyset$.
Use this to prove that the prime existence property implies that prime $(\mathcal{I}^*, \mathcal{F}^*)$-separation property. *Hint*: Apply the prime ideal theorem to the quotient L/θ, which is also a distributive lattice with 1.

Chapter 9
Congruence Relations on Lattices

If S is an algebraic structure, such as a lattice, group, ring or module, then an equivalence relation $\equiv$ on S that also preserves the algebraic operations of S is called a **congruence relation** on S. For example, a congruence relation $\equiv$ on a group G is an equivalence relation for which

1) $a \equiv b \Rightarrow a^{-1} \equiv b^{-1}$
2) $a \equiv b, a' \equiv b' \Rightarrow aa' \equiv bb'$

Now, in elementary algebra, one teaches that there is a correspondence between certain special types of substructures and quotient structures. In the case of groups, for example, there is a correspondence between normal subgroups and quotient groups. A more complete story for groups must include the fact that a subgroup H of a group G is normal if and only if equivalence modulo H:

$$a \equiv b \quad \Leftrightarrow \quad aH = bH$$

is a *congruence relation* on G. For if H is normal in G and $aH = a'H$ and $bH = b'H$, then

$$abH = aHb' = a'Hb' = a'b'H$$

Also, if $aH = bH$, then

$$a^{-1}H = Ha^{-1} = Hb^{-1} = b^{-1}H$$

Conversely, if equivalence modulo H is a congruence relation, then

$$aH = a'H, \quad bH = b'H \quad \Rightarrow \quad abH = a'b'H$$

or equivalently,

$$a' \in aH, \quad b' \in bH \quad \Rightarrow \quad a'b' \in abH$$

which is equivalent to the equation

$$aHbH = abH$$

S. Roman (ed.), *Lattices and Ordered Sets*, doi: 10.1007/978-0-387-78901-9_9,

Setting $b = a^{-1}$ gives $aHa^{-1}H = H$, which implies that $aHa^{-1} \subseteq H$ for all $a \in G$ and so H is normal in G.

Similar statements can be made for rings or modules. However, in most treatments of elementary algebra, the role of the congruence relation is underplayed in favor of the role of the special substructure (normal subgroup, ideal, submodule).

On the other hand, in algebras in general, there is no special class of substructures that plays the roles of normal subgroups for groups and ideals for rings and we are left only with the relationship between quotient structures and congruence relations. In particular, this is the case for lattices. Only in distributive, sectionally complemented lattices is there a bijection between quotient lattices and ideals.

Congruence Relations on Lattices

We begin with the definition of a congruence relation on a lattice.

Definition *An equivalence relation θ on a lattice L is a* **congruence relation** *on L if for all $a, b, x, y \in L$,*

$$a\theta x \quad \text{and} \quad b\theta y \quad \Rightarrow \quad (a \wedge b)\theta(x \wedge y) \quad \text{and} \quad (a \vee b)\theta(x \vee y)$$

The equivalence classes under a congruence relation θ are called **congruence classes**. *The congruence class containing $a \in L$ is denoted by $[a]$ or $[a]_\theta$. The set of all congruence classes for θ is denoted by L/θ. The set of all congruence relations on L is denoted by* $\mathrm{Con}(L)$.□

We will use the notations

$$a\theta b, \quad a \equiv_\theta b \quad \text{and} \quad a \overset{\theta}{\equiv} b$$

interchangeably throughout the chapter and write $a \equiv b$ when the specific congruence is understood. Also, the notation

$$a \overset{\theta}{\equiv} b \overset{\sigma}{\equiv} c$$

is shorthand for $a \overset{\theta}{\equiv} b$ and $b \overset{\sigma}{\equiv} c$.

Note that if $u \le v$ and $u \equiv v$, then all elements in the interval $[u, v]$ are congruent, for if $x \in [u, v]$, then

$$x = v \wedge x \equiv u \wedge x = u$$

and so every element of $[u, v]$ is congruent to u. Thus, congruence classes are convex subsets of L. Moreover, if $a \equiv b$, then

$$a \wedge b \equiv b \wedge b = b \quad \text{and} \quad a \vee b \equiv b \vee b = b$$

and so all elements of the interval $[a \wedge b, a \vee b]$ are congruent.

Theorem 9.1 *Let θ and σ be congruence relations on a lattice L.*
1) *The congruence classes of θ are convex sets.*
2) *For all $a, b \in L$,*

$$a\theta b \quad \Leftrightarrow \quad (a \wedge b)\theta(a \vee b)$$

in which case every element of the interval $[a \wedge b, a \vee b]$ is congruent to a.

3) *$\theta = \sigma$ if and only if*

$$a\theta b \quad \Leftrightarrow \quad a\sigma b$$

for all $a < b$ in L.

4) *If μ is a binary relation on L satisfying*

$$a\mu b \quad \Leftrightarrow \quad (a \wedge b)\mu(a \vee b)$$

for all $a, b \in L$, then $\theta = \mu$ if and only if

$$a\theta b \quad \Leftrightarrow \quad a\mu b$$

for all $a \leq b$ in L.

Proof. For part 4),

$$\begin{aligned} a\theta b \quad &\Leftrightarrow \quad (a \wedge b)\theta(a \vee b) \\ &\Leftrightarrow \quad (a \wedge b)\mu(a \vee b) \\ &\Leftrightarrow \quad a\mu b \end{aligned}$$

□

Characterizing Congruence Relations

The following simple result is very useful.

Theorem 9.2 *An equivalence relation $\equiv$ on a lattice L is a congruence relation if and only if for all $a, b, x \in L$,*

$$a \equiv b \quad \Rightarrow \quad a \wedge x \equiv b \wedge x \quad \textit{and} \quad a \vee x \equiv b \vee x$$

Proof. If the stated property holds, then

$$\begin{aligned} a \equiv b, x \equiv y &\Rightarrow a \wedge x \equiv b \wedge x, b \wedge x \equiv b \wedge y \\ &\Rightarrow a \wedge x \equiv b \wedge y \end{aligned}$$

and similarly for joins.□

Example 9.3 Let L be a distributive lattice and let $t \in L$. Then the binary relations defined by

$$a\delta b \quad \text{if} \quad a \vee t = b \vee t$$

and

$$a\mu b \quad \text{if} \quad a \wedge t = b \wedge t$$

are both congruence relations on L. It is easy to see that these relations are equivalence relations. For δ, it is clear that

$$a\delta b \quad \Rightarrow \quad (a \vee x)\delta(b \vee x)$$

and for meet, the distributivity of L gives

$$\begin{aligned} a\delta b \quad &\Rightarrow \quad a \vee t = b \vee t \\ &\Rightarrow \quad (a \vee t) \wedge (x \vee t) = (b \vee t) \wedge (x \vee t) \\ &\Rightarrow \quad (a \wedge x) \vee t = (b \wedge x) \vee t \\ &\Rightarrow \quad (a \wedge x)\delta(b \wedge x) \end{aligned}$$

A similar argument can be made for μ.□

There is another characterization of congruence relations, due to Grätzer and Schmidt, that assumes only a reflexive binary relation.

Theorem 9.4 *A reflexive binary relation* $\equiv$ *on a lattice* L *is a congruence relation on* L *if and only if it satisfies the following properties:*
1) *For all* $a, b \in L$,

$$a \equiv b \quad \Leftrightarrow \quad a \wedge b \equiv a \vee b$$

2) **(Comparable transitivity)** *If* $a \leq b \leq c$ *in* L, *then*

$$a \equiv b, \quad b \equiv c \quad \Rightarrow \quad a \equiv c$$

3) *If* $a \leq b$ *and* $a \equiv b$, *then*

$$a \wedge x \equiv b \wedge x \quad \textit{and} \quad a \vee x \equiv b \vee x$$

for all $x \in L$.

Proof. First note that $\equiv$ is symmetric by 1). Next, we show that if $a \leq b$ and $a \equiv b$, then all elements in $[a, b]$ are congruent. It is easy to see that any $x \in [a, b]$ is congruent to both a and b, for we have

$$a = a \wedge x \equiv b \wedge x = x$$

and

$$x = a \vee x \equiv b \vee x = b$$

Hence, if $x, y \in [a, b]$, then $x \vee y \equiv a$. But $x \wedge y \in [a, x \vee y]$ and so $x \wedge y \equiv x \vee y$, which implies that $x \equiv y$.

To establish transitivity, let $a \equiv b \equiv c$. Then $a \wedge b \equiv a \vee b$ and so

$$a \wedge b \equiv a \vee b \quad \Rightarrow \quad b \equiv a \vee b \quad \Rightarrow \quad b \vee c \equiv a \vee b \vee c$$

Similarly,

$$a \wedge b \equiv a \vee b \quad \Rightarrow \quad a \wedge b \equiv b \quad \Rightarrow \quad a \wedge b \vee c \equiv b \wedge c$$

Hence,

$$a \wedge b \wedge c \equiv b \wedge c \equiv b \vee c \equiv a \vee b \vee c$$

and so

$$a \wedge b \wedge c \equiv a \vee b \vee c$$

Therefore, all elements in the interval $[a \wedge b \wedge c, a \vee b \vee c]$, including a and c, are congruent. Thus, $\equiv$ is an equivalence relation.

We complete the proof using Theorem 9.2. Suppose that $a \equiv b$. If a and b are comparable, then 3) and the symmetry of the relation imply that

$$a \wedge x \equiv b \wedge x \quad \text{and} \quad a \vee x \equiv b \vee x$$

for all $x \in L$. If $a \parallel b$, then $a \wedge b \equiv a \vee b$ and 3) gives

$$(a \wedge b) \wedge x \equiv (a \vee b) \wedge x \quad \text{and} \quad (a \wedge b) \vee x \equiv (a \vee b) \vee x$$

for all $x \in L$. But

$$a \wedge x, b \wedge x \in [(a \wedge b) \wedge x, (a \vee b) \wedge x]$$

and so $a \wedge x \equiv b \wedge x$. Similarly,

$$a \vee x, b \vee x \in [(a \wedge b) \vee x, (a \vee b) \vee x]$$

and so $a \vee x \equiv b \vee x$.□

Congruence Relations and Partitions

It is useful to look at the congruence relations on a lattice L in three ways (we use the same notation for all three viewpoints, relying on the context to make the necessary distinctions):

1) As special types of binary relations on L.
2) As special types of partitions of L.
3) As special subsets of the lattice product $L^2 = L \times L$ (with coordinatewise meet and join).

The first viewpoint provides the definition. For the others, we have the following.

Theorem 9.5

1) A partition θ on a lattice L defines a congruence relation on L if and only if the following hold:

a) Each block of the partition is a convex sublattice of L.

b) *The* **quadrilateral property** *holds: If two adjacent elements of a quadrilateral* $Q = (a \wedge b, a, b, a \vee b)$ *belong to the same block, then so do the complementary adjacent elements. This is shown in Figure 9.1.*

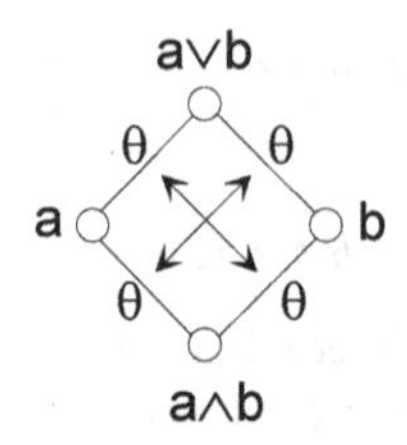

Figure 9.1

2) *A subset* θ *of* L^2 *is a congruence relation on* L *if and only if the following hold:*
   a) θ *is an equivalence relation on* L.
   b) θ *is a sublattice of the lattice* L^2 *(under coordinatewise meet and join).*

Proof. For part 1), one direction has been proved. Assume that 1a) and 1b) hold. We prove that $a \equiv b$ implies that $a \wedge x \equiv b \wedge x$ for all $x \in L$. Since 1a) and 1b) are self-dual, it follows that $a \vee x = b \vee x$ for all $x \in L$.

First suppose that $u \equiv v$ and $u \leq v$. Then we can project x into the interval $[u, v]$ to get $u \vee (v \wedge x)$, as shown in Figure 9.2.

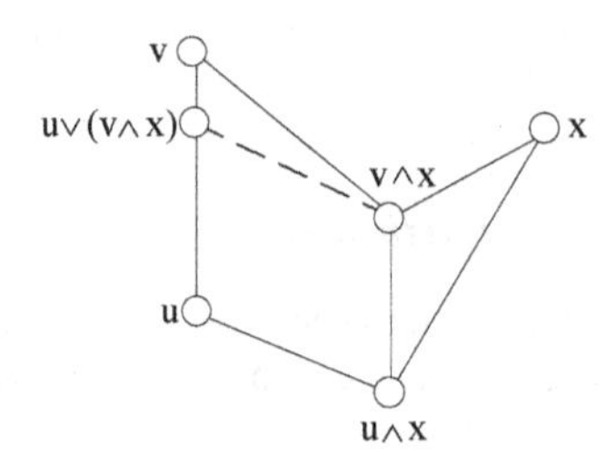

Figure 9.2

The convexity of blocks implies that

$$u \equiv u \vee (v \wedge x)$$

and the quadrilateral property then implies that $u \wedge x \equiv v \wedge x$.

For the general case, if $a \equiv b$, then letting $u = a \wedge b$ and $v = a \vee b$, we have $a \wedge b \equiv a \vee b$ and $a \wedge b \leq a \vee b$. Hence, the previous remarks show that

$$(a \wedge b) \wedge x \equiv (a \vee b) \wedge x$$

and since $a \wedge x$ and $b \wedge x$ lie in the interval between $(a \wedge b) \wedge x$ and

$(a \vee b) \wedge x$, convexity implies that $a \wedge x \equiv b \wedge x$, as desired. We leave proof of part 2) as an exercise.□

The Lattice of Congruence Relations

We have seen in an earlier chapter that if X is a nonempty set, then the poset $\mathrm{Equ}(X)$ is a complete lattice. When X is a lattice, we can say a bit more. Let us recall the definition of a witness sequence and add an additional definition.

Definition *Let $\mathcal{F}$ be a family of binary relations on a lattice L. If $a, b \in L$, then a* **witness sequence** *from a to b is a sequence*

$$a = a_1, a_2, \ldots, a_{n-1}, a_n = b$$

in L for which

$$a = a_1 \stackrel{\theta_1}{\equiv} a_2 \stackrel{\theta_2}{\equiv} a_3 \equiv \cdots \equiv a_{n-1} \stackrel{\theta_{n-1}}{\equiv} a_n = b$$

where $\theta_i \in \mathcal{F}$. When $a \le b$, an **increasing witness sequence** *from a to b is a witness sequence for which $a_i \le a_{i+1}$ for all i.*□

Theorem 9.6 *Let L be a lattice. Then* $\mathrm{Equ}(L)$ *is an $\cap\overrightarrow{\cup}$-structure and therefore an algebraic lattice.*

1) *If*

$$\theta = \bigvee \mathcal{F}$$

then $a\theta b$ if and only if there is an increasing witness sequence

$$a \wedge b = x_1 \stackrel{\theta_{i_1}}{\equiv} x_2 \stackrel{\theta_{i_2}}{\equiv} x_3 \equiv \cdots \equiv x_{m-1} \stackrel{\theta_{i_{m-1}}}{\equiv} x_m = a \vee b$$

from $a \wedge b$ to $a \vee b$ using $\mathcal{F}$.

2) $\mathrm{Con}(L)$ *is a complete sublattice of* $\mathrm{Equ}(L)$*, called the* **congruence lattice** *of L. The smallest element of* $\mathrm{Con}(L)$ *is equality and the largest is given by $a\theta b$ for all $a, b \in L$.*
3) $\mathrm{Con}(L)$ *is an $\cap\overrightarrow{\cup}$-structure and therefore an algebraic lattice.*
4) $\mathrm{Con}(L)$ *is distributive.*

Proof. We leave the proof that $\mathrm{Equ}(L)$ is an $\cap\overrightarrow{\cup}$-structure to the reader. For part 1), let $\theta = \bigvee \mathcal{F}$. If there is an increasing witness sequence from $a \wedge b$ to $a \vee b$, then $(a \wedge b)\theta(a \vee b)$ and so $a\theta b$. Conversely, if $a\theta b$ then $(a \wedge b)\theta(a \vee b)$. Suppose that

$$a \wedge b = a_1 \equiv a_2 \equiv a_3 \equiv \cdots \equiv a_{n-1} \equiv a_n = a \vee b$$

is a witness sequence using $\mathcal{F}$. Write $\alpha = a \wedge b$ and $\beta = a \vee b$. Joining the endpoints of a step $a_k \equiv a_{k+1}$ with α and then meeting with β gives

$$(a_k \vee \alpha) \wedge \beta \equiv (a_{k+1} \vee \alpha) \wedge \beta$$

where each term is in the interval $[a \wedge b, a \vee b]$. Moreover, since this process has no effect on $a \wedge b$ and $a \vee b$, we may assume to start with that $a_k \in [a \wedge b, a \vee b]$. Then replacing a_k with $a_1 \vee \cdots \vee a_k$ gives

$$\begin{aligned} a \wedge b = a_1 &\equiv a_1 \vee a_2 \equiv a_1 \vee a_2 \vee a_3 \equiv \cdots \\ \cdots &\equiv (a_1 \vee \cdots \vee a_{n-1}) \equiv (a_1 \vee \cdots \vee a_n) = a \vee b \end{aligned}$$

which is an increasing witness sequence from $a \wedge b$ to $a \vee b$.

For part 2), if $\mathcal{F} = \{\theta_i \in \mathrm{Con}(L) \mid i \in I\}$ and if $\sigma = \bigcap \mathcal{F}$, then for any $x \in L$,

$$\begin{aligned} a\sigma b &\Rightarrow a\theta_i b \text{ for all } i \\ &\Rightarrow (a \wedge x)\theta_i(b \wedge x) \text{ and } (a \vee x)\theta_i(b \vee x) \text{ for all } i \\ &\Rightarrow (a \wedge x)\sigma(b \wedge x) \text{ and } (a \vee x)\sigma(b \vee x) \end{aligned}$$

and so $\sigma \in \mathrm{Con}(L)$. For the join $\theta = \bigvee \mathcal{F}$, if

$$a = a_1, a_2, \ldots, a_{n-1}, a_n = b$$

is a witness sequence from a to b, then

$$a \wedge x = a_1 \wedge x, a_2 \wedge x, \ldots, a_{n-1} \wedge x, a_n \wedge x = b \wedge x$$

is a witness sequence from $a \wedge x$ to $b \wedge x$ and similarly for the join in place of meet. Hence, $\theta \in \mathrm{Con}(L)$.

We leave proof of part 3) as an exercise. For part 4), since

$$\theta \wedge (\tau \vee \sigma) \supseteq (\theta \cap \tau) \vee (\theta \cap \sigma)$$

it is sufficient to prove that

$$a[\theta \wedge (\tau \vee \sigma)]b \quad \Rightarrow \quad a[(\theta \cap \tau) \vee (\theta \cap \sigma)]b$$

for all $a, b \in L$. The antecedent implies that $a\theta b$ and that there is an increasing witness sequence

$$a \wedge b = a_1 \overset{\tau\cup\sigma}{\equiv} a_2 \overset{\tau\cup\sigma}{\equiv} a_3 \equiv \cdots \equiv a_{n-1} \overset{\tau\cup\sigma}{\equiv} a_n = a \vee b$$

Since $(a \wedge b)\theta(a \vee b)$ and since $a_i \in [a \wedge b, a \vee b]$, it follows that

$$a \wedge b = a_1 \overset{\theta}{\equiv} a_2 \overset{\theta}{\equiv} a_3 \equiv \cdots \equiv a_{n-1} \overset{\theta}{\equiv} a_n = a \vee b$$

and so if $\alpha = \theta \cap (\tau \cup \sigma)$, then

$$a \wedge b = a_1 \overset{\alpha}{\equiv} a_2 \overset{\alpha}{\equiv} a_3 \equiv \cdots \equiv a_{n-1} \overset{\alpha}{\equiv} a_n = a \vee b$$

But $\alpha = (\theta \cap \tau) \cup (\theta \cap \sigma)$ and so

$$(a \wedge b)[(\theta \cap \tau) \vee (\theta \cap \sigma)](a \vee b)$$

which implies that $a[(\theta \cap \tau) \vee (\theta \cap \sigma)]b$. Hence, $\mathrm{Con}(L)$ is distributive.□

Commuting Congruences and Joins

We have seen that $a(\theta \vee \sigma)b$ if and only if there is a witness sequence

$$a = a_1 \overset{\mu_1}{\equiv} a_2 \overset{\mu_2}{\equiv} a_3 \equiv \cdots \equiv a_{n-1} \overset{\mu_{n-1}}{\equiv} a_n = b$$

where $\mu_i = \theta$ or $\mu_i = \sigma$. By transitivity, we can assume that adjacent μ_i's are distinct. Let us focus on one portion of this sequence, say

$$x \overset{\theta}{\equiv} y \overset{\sigma}{\equiv} z$$

If it were true that θ and σ *commute*, even at the cost of changing the intermediate element y, that is, if

$$x \overset{\theta}{\equiv} y \overset{\sigma}{\equiv} z \quad \Rightarrow \quad x \overset{\sigma}{\equiv} w \overset{\theta}{\equiv} z$$

for some $w \in L$, then in the witness sequence, we can move all of the θ's to the front to get

$$a = a_1 \overset{\theta}{\equiv} w_2 \equiv \cdots \equiv w_{k-1} \overset{\theta}{\equiv} w_k \overset{\sigma}{\equiv} w_{k+1} \equiv \cdots \equiv w_{n-1} \overset{\sigma}{\equiv} a_n = b$$

and transitivity would imply that

$$a = a_1 \overset{\theta}{\equiv} w_k \overset{\sigma}{\equiv} a_n = b$$

Thus, in this case, the join takes the relatively simple form

$$a(\theta \vee \sigma)b \quad \Leftrightarrow \quad a \overset{\theta}{\equiv} x \overset{\sigma}{\equiv} b$$

for some $x \in L$.

Definition *Let θ and σ be congruence relations on a lattice L. Define the* **product** $\theta\sigma$ *by*

$$a \overset{\theta\sigma}{\equiv} b \quad \text{if} \quad a \overset{\theta}{\equiv} x \overset{\sigma}{\equiv} b \text{ for some } x \in L$$

We say that θ and σ **commute** *if*

$$\theta\sigma = \sigma\theta$$ □

Thus, we are led to examine conditions under which congruence relations commute. The following facts are helpful in this regard.

Lemma 9.7 *Let θ and σ be congruence relations on a lattice L.*
1) If $a \le b$ then

$$a \overset{\theta}{\equiv} x \overset{\sigma}{\equiv} b \text{ for some } x \in L \quad \Leftrightarrow \quad a \overset{\theta}{\equiv} y \overset{\sigma}{\equiv} b \text{ for some } y \in [a, b]$$

2) *If* $a, b \leq u$, *then*

$$a \overset{\theta}{\equiv} u \overset{\sigma}{\equiv} b \quad \Rightarrow \quad a \overset{\sigma}{\equiv} a \wedge b \overset{\theta}{\equiv} b$$

Proof. For part 1), since $a \leq b$, taking the join with a and the meet with b gives

$$a \overset{\theta}{\equiv} x \overset{\sigma}{\equiv} b \quad \Rightarrow \quad a \overset{\theta}{\equiv} (x \vee a) \wedge b \overset{\sigma}{\equiv} b$$

For part 2), taking the meet with a gives $a\sigma(a \wedge b)$ and taking the meet with b gives $(a \wedge b)\theta b$ and so $a\sigma(a \wedge b)\theta b$.□

Theorem 9.8 *Let* θ *and* σ *be congruence relations on a lattice* L. *The following are equivalent:*

1) θ *and* σ *commute.*
2) *For all* $a < b$,

$$a \overset{\theta\sigma}{\equiv} b \quad \Leftrightarrow \quad a \overset{\theta\sigma}{\equiv} b$$

3) $\theta\sigma$ *is a symmetric binary relation.*
4) $\theta\sigma = \theta \vee \sigma$.

Proof. It is clear that 1) implies 2). To see that 2) implies 1), if $a\theta\sigma b$ then $a\theta x\sigma b$ for some $x \in L$ and so

$$a \overset{\theta}{\equiv} (a \vee x) \overset{\sigma}{\equiv} (a \vee x \vee b)$$

Hence, Lemma 9.7 implies that

$$a \overset{\sigma}{\equiv} y \overset{\theta}{\equiv} (a \vee x \vee b)$$

where $a \leq y \leq a \vee x \vee b$. Also, the symmetry of both θ and σ imply that $b\sigma x\theta a$ and so a similar argument gives

$$b \overset{\theta}{\equiv} z \overset{\sigma}{\equiv} (a \vee x \vee b)$$

where $b \leq z \leq a \vee x \vee b$. Thus,

$$a \overset{\sigma}{\equiv} y \overset{\theta}{\equiv} (a \vee x \vee b) \overset{\sigma}{\equiv} z \overset{\theta}{\equiv} b$$

Since y and z are less than or equal to $a \vee x \vee b$, Lemma 9.7 implies that

$$a \overset{\sigma}{\equiv} y \overset{\sigma}{\equiv} (y \wedge z) \overset{\theta}{\equiv} z \overset{\theta}{\equiv} b$$

and so

$$a \overset{\sigma}{\equiv} y \wedge z \overset{\theta}{\equiv} b$$

whence $a\sigma\theta b$. This proves that $a\theta\sigma b$ implies $a\sigma\theta b$. The reverse implication follows by symmetry and so 1) and 2) are equivalent.

If θ and σ commute, then

$$a \overset{\theta\sigma}{\equiv} b \quad \Leftrightarrow \quad a \overset{\sigma\theta}{\equiv} b \quad \Leftrightarrow \quad b \overset{\theta\sigma}{\equiv} b$$

and so $\theta\sigma$ is symmetric. Conversely, if $\theta\sigma$ is symmetric, then

$$a \overset{\sigma\theta}{\equiv} b \quad \Leftrightarrow \quad b \overset{\theta\sigma}{\equiv} a \quad \Leftrightarrow \quad a \overset{\theta\sigma}{\equiv} b$$

and so θ and σ commute. Thus, 1) and 3) are equivalent. Finally, we have seen that if θ and σ commute, then $\theta \vee \sigma = \theta\sigma$ and if $\theta \vee \sigma = \theta\sigma$, then $\theta\sigma$ is symmetric.□

In 1950, Dilworth proved the following theorem.

Theorem 9.9 (Dilworth [17]) *Any two congruence relations on a relatively complemented lattice commute.*

Proof. Theorem 9.8 implies that it is sufficient to show that for any congruence relations θ and σ,

$$a \overset{\theta\sigma}{\equiv} b \quad \Rightarrow \quad a \overset{\sigma\theta}{\equiv} b$$

for all $a < b$ in L. If $a\theta\sigma b$, then $a\theta x\sigma b$, where $x \in [a, b]$. Hence, as shown in Figure 9.3,

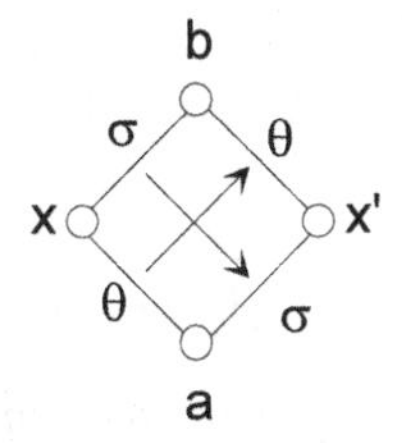

Figure 9.3

if x' is a relative complement of x in $[a, b]$, the quadrilateral property applied to (a, x, x', b) gives $a\sigma x'\theta b$ and so $a\sigma\theta b$.□

Quotient Lattices and Kernels

It is now time to take a closer look at the properties of quotient lattices L/θ, where θ is a congruence relation on L. The set L/θ of congruence classes of a congruence relation θ is also a lattice. Indeed, it is precisely because θ is a *congruence* relation that the inherited lattice operations of L are well defined on L/θ.

Theorem 9.10 *Let θ be a congruence relation on a lattice L. The set L/θ of congruence classes is a lattice, called the* **quotient lattice** *of L modulo θ, under* *the operations*

$$[a] \wedge [b] = [a \wedge b] \quad \text{and} \quad [a] \vee [b] = [a \vee b]$$

Proof. First, we must check that these operations are well defined. But if $a \equiv a'$ and $b \equiv b'$, then $a \wedge b \equiv a' \wedge b'$ and so

$$[a] = [a'] \text{ and } [b] = [b'] \quad \Rightarrow \quad [a \wedge b] = [a' \wedge b']$$

which says that the meet operation is well defined. A similar argument applies to the join operation. Finally, it is routine to check the properties of a lattice. For instance, the associativity law

$$([a] \vee [b]) \vee [c] = [a] \vee ([b] \vee [c])$$

is equivalent to

$$[(a \vee b) \vee c] = [a \vee (b \vee c)]$$

which follows from associativity in L.□

We now describe the order relation in the quotient lattice.

Theorem 9.11 *Let L be a lattice and let $\theta \in \mathrm{Con}(L)$.*

1) *$[a] \leq [b]$ if and only if some element of $[a]$ is less than or equal to some element of $[b]$. In particular,*

$$a \leq b \quad \Rightarrow \quad [a] \leq [b]$$

2) *$[a] < [b]$ if and only if*

$$\alpha < \beta \quad \text{or} \quad \alpha \parallel \beta$$

for every $\alpha \in [a]$ and $\beta \in [b]$, with strict inequality for at least one pair (α, β).

3) *$[a] \parallel [b]$ if and only if every element of $[a]$ is parallel to every element of $[b]$.*

4) *A congruence class $[x]$ is an ideal of L if and only if it is the smallest element of L/θ, in which case it is called the* **kernel** *or* **ideal kernel** *of θ and is denoted by $\ker(\theta)$. If L has a 0, then $\ker(\theta) = [0]$.*

Proof. For part 1),

$$a \leq b \Leftrightarrow a \wedge b = a \Rightarrow [a \wedge b] = [a] \Leftrightarrow [a] \wedge [b] = [a] \Leftrightarrow [a] \leq [b]$$

Conversely,

$$[a] \leq [b] \Rightarrow [a \wedge b] = [a] \Rightarrow a \wedge b \in [a]$$

and so $\alpha \leq \beta$, where $\alpha = a \wedge b \in [a]$ and $\beta = b \in [b]$.

For part 2), if $[a] < [b]$, then $[a] \leq [b]$, $[a] \neq [b]$ and $[b] \not\leq [a]$. Thus, part 1) implies that $\alpha < \beta$ for some $(\alpha, \beta) \in [a] \times [b]$ and that $\beta \leq \alpha$ cannot happen for any $(\alpha, \beta) \in [a] \times [b]$. Conversely, suppose that $\alpha < \beta$ or $\alpha \parallel \beta$ for every $(\alpha, \beta) \in [a] \times [b]$, with strict inequality for at least one pair. Then part 1) implies that $[a] \leq [b]$. However, $[a] = [b]$ implies that $(a, a) \in [a] \times [b]$, which is false. Hence, $[a] < [b]$.

Part 3) follows from part 1). For part 4), let $x \in L$. Since $[x]$ is a sublattice, it is an ideal if and only if it is a down-set. If $[x]$ is a down-set, then for any congruence class $[a]$, we have $x \wedge a \in [x]$ and $x \wedge a \leq a \in [a]$ and so $[x] \leq [a]$. Thus, $[x]$ is the smallest element of L/θ. Conversely, if $[x]$ is the smallest element of L/θ, then for any $a \leq y \in [x]$ we have $[a] \leq [y] = [x]$ and so $[a] = [x]$, that is, $a \in [x]$, which shows that $[x]$ is a down-set.□

Figure 9.4 demonstrates that we may have $[x] < [y]$ but $x \parallel y$.

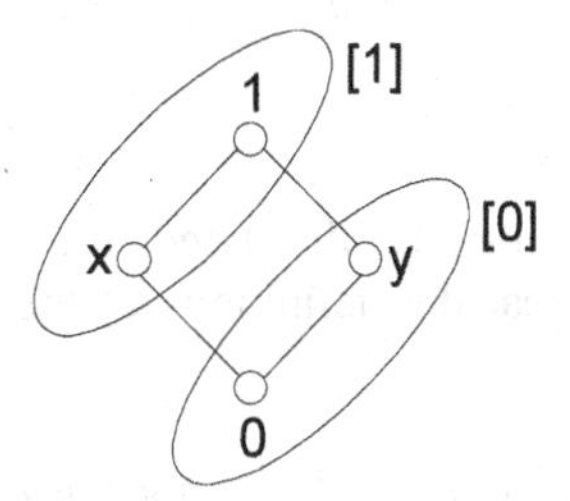

Figure 9.4

Congruence Relations and Lattice Homomorphisms

Now we want to look at the relationship between congruence relations and lattice homomorphisms.

Theorem 9.12

1) *Every lattice homomorphism $f: L \to M$ defines a congruence relation θ_f given by*

$$a\theta_f b \quad \Leftrightarrow \quad f(a) = f(b)$$

and called the **congruence kernel** *of f. The congruence classes of θ_f are the sets*

$$L/\theta_f = \{f^{-1}(x) \mid x \in \text{im}(f)\}$$

Thus, f is injective if and only if θ_f is equality.

2) *Every congruence relation θ on a lattice L defines a lattice epimorphism $\pi_\theta: L \to L/\theta$ given by*

$$\pi_\theta(a) = [a]$$

and called the **natural projection** *or* **canonical projection** *of L modulo θ.*

The congruence kernel of π_θ *is* θ, *that is,*

$$\theta_{\pi_\theta} = \theta$$

Proof. For part 1), if $a\theta_f b$ and $x\theta_f y$, then $f(a) = f(b)$ and $f(x) = f(y)$ and so

$$f(a \vee x) = f(a) \vee f(x) = f(b) \vee f(y) = f(b \vee y)$$

which implies that $(a \vee x)\theta_f(b \vee y)$. A similar argument can be made for meets. For part 2), it is clear that π_θ is surjective. Also,

$$\pi_\theta(a \wedge b) = [a \wedge b] = [a] \wedge [b] = \pi_\theta(a) \wedge \pi_\theta(b)$$

and similarly for join. Hence, π_θ is a lattice epimorphism. Finally, the congruence kernel of π_θ is θ, since

$$a\theta b \quad \Leftrightarrow \quad [a]_\theta = [b]_\theta \quad \Leftrightarrow \quad \pi_\theta(a) = \pi_\theta(b)$$

□

Kernels

Earlier in the book, we briefly discussed the ideal kernel of a lattice homomorphism. Let us repeat the definitions of the two types of kernels of a lattice homomorphism.

Definition *Let* $f: L \to M$ *be a lattice homomorphism.*

1) The **congruence kernel** *of* f *is the congruence relation* θ_f.

2) If M has a smallest element 0 *and if* $f^{-1}(0)$ *is nonempty, then the set*

$$\ker(f) = f^{-1}(0)$$

is called the **ideal kernel** *of* f.□

The ideal kernel of a homomorphism is easily seen to be an ideal. Also, as shown in Figure 9.5, the ideal kernel of the congruence kernel θ_f is the ideal kernel of f.

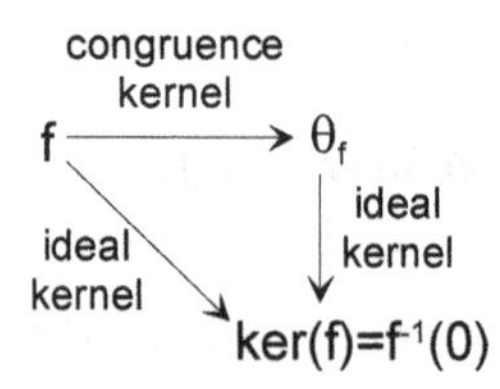

Figure 9.5

Theorem 9.13 *Let* M *be a lattice with* 0 *and let* $f: L \to M$ *be a lattice homomorphism. Then the ideal kernel* $\ker(f) = f^{-1}(0)$ *is an ideal of L and*

$$\ker(\theta_f) = \ker(f)$$

Proof. We leave proof of part 1) for the reader. For part 2), it is easy to see that $f^{-1}(0)$ is an ideal of L and so Theorem 9.11 implies that $\ker(\theta_f) = f^{-1}(0)$.□

Universality and the Lattice Homomorphism Theorem

The quotient lattice and its natural projection have a universal property.

Theorem 9.14 *Let θ be a congruence relation on a lattice L. The pair $(L/\theta, \pi_\theta)$ has the following* **universal property of quotients***: Referring to Figure 9.6,*

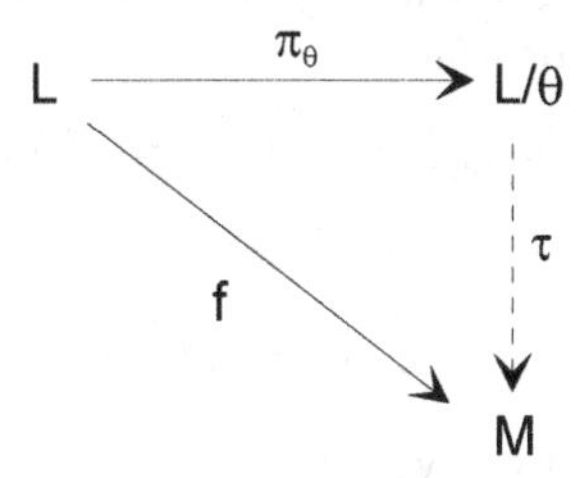

Figure 9.6

if $f: L \to M$ is any lattice homomorphism for which $\theta \subseteq \theta_f$, that is,

$$a \overset{\theta}{\equiv} b \quad \Rightarrow \quad f(a) = f(b)$$

then there is a unique lattice homomorphism $\tau: L/\theta \to M$, which we call the **mediating morphism** *for f, with the property that*

$$\tau \circ \pi_\theta = f$$

The pair $(L/\theta, \pi_\theta)$ is said to be **universal***. Also, τ and f have the same image:*

$$\operatorname{im}(\tau) = \operatorname{im}(f)$$

and the same congruence kernel modulo θ, in the sense that

$$[x]\theta_\tau[y] \quad \Leftrightarrow \quad x\theta_f y$$

Proof. The condition $\tau \circ \pi_\theta = f$ uniquely determines τ to be

$$\tau([p]) = f(p)$$

This function is well defined on L/θ since f is constant on $[p]$. Also,

$$\begin{aligned}\tau([p] \wedge [q]) &= \tau([p \wedge q]) \\ &= f(p \wedge q) \\ &= f(p) \wedge f(q) \\ &= \tau([p]) \wedge \tau([q])\end{aligned}$$

and similarly for join. Hence, τ is a lattice homomorphism. Finally,

$$\operatorname{im}(\tau) = \{\tau([p]) \mid p \in L\} = \{f(p) \mid p \in L\} = \operatorname{im}(f)$$

and

$$[x]\theta_\tau[y] \Leftrightarrow \tau[x] = \tau[y] \Leftrightarrow f(x) = f(y) \Leftrightarrow x\theta_f y$$ □

Theorem 9.14 has a familiar corollary, which is obtained by taking $\theta = \theta_f$.

Corollary 9.15 (*The* **homomorphism theorem** *or* **first isomorphism theorem**) *Let $f: L \to M$ be a lattice homomorphism. Then there is a unique lattice embedding $\tau: L/\theta_f \hookrightarrow M$ for which*

$$\tau \circ \pi_{\theta_f}(p) = \tau([p]) = f(p)$$

Thus,

$$\tau: L/\theta_f \approx \operatorname{im}(f)$$

Proof. To see that τ is injective, we have

$$[x]\theta_\tau[y] \quad \Leftrightarrow \quad x\theta_f y \quad \Leftrightarrow \quad xK_{\pi_f}y \quad \Leftrightarrow \quad [x] = [y]$$ □

There is also a correspondence theorem for lattices.

Theorem 9.16 (*The* **correspondence theorem**) *Let L be a lattice and let $\theta \in \operatorname{Con}(L)$. Then there is a bijection between the congruence relations on L/θ and the congruence relations of L that contain θ.*
Proof. If τ is a congruence on L/θ, then

$$x\tau' y \quad \text{if} \quad [x]_\theta \tau [y]_\theta$$

is a congruence on L containing θ and if σ is a congruence on L containing θ, then

$$[x]_\theta \sigma' [y]_\theta \quad \text{if} \quad x\sigma y$$

is a congruence on L/θ. We leave the details to the reader.□

Subdirect Product Representations

Definition *Let L be a lattice and let $\mathcal{F} = \{K_i \mid i \in I\}$ be a nonempty family of lattices. An embedding*

$$\sigma: L \to \prod\nolimits_{i \in I} K_i$$

is called a **subdirect representation** *of L in $\prod K_i$ if the projection maps $\pi_i: L \to K_i$ are surjective for all $i \in I$, that is, if for any $k \in K_i$, there is an $a \in L$ for which $(\sigma a)(i) = k$.*□

Theorem 9.17 *Let L be a lattice.*

1) *Let* $\mathcal{F} = \{\theta_i \mid i \in I\}$ *be a nonempty family of congruence relations on* L. *If* $\lambda = \bigwedge \theta_i$, *then* L/λ *has a subdirect representation in* $\prod L/\theta_i$. *In particular, if* $\bigwedge \theta_i$ *is equality, then* L *has a subdirect representation in* $\prod L/\theta_i$.

2) *Conversely, let* $\mathcal{F} = \{K_i \mid i \in I\}$ *be a nonempty family of lattices. If* L *has a subdirect representation in* $\prod K_i$, *with projection maps* $\pi_i \colon L \to K_i$, *then* $K_i \approx L/\ker(\pi_i)$ *and* $\bigwedge \ker(\pi_i)$ *is equality.*

Proof. For part 1), consider the map $\sigma \colon L/\lambda \to \prod L/\theta_i$ defined by

$$(\sigma[a]_\lambda)(i) = [a]_{\theta_i}$$

which is well defined since $a\lambda a'$ implies that $a\theta_i a'$ for all $i \in I$. This is easily seen to be a homomorphism. Moreover, $\sigma[a]_\lambda = \sigma[b]_\lambda$ if and only if $[a]_{\theta_i}[b]$ for all $i \in I$, which holds if and only if $a \bigwedge \theta_i b$, that is, if and only if $a\lambda b$. Hence, σ is an embedding. It is clear that π_i is surjective for all i.

For part 2), since the projection maps are surjective, the first isomorphism theorem gives $K_i \approx L/\ker(\pi_i)$ and since $a\ker(\pi_i)b$ for all i if and only if $\pi_i a = \pi_i b$ for all i, which holds if and only if $a = b$, we see that $\bigwedge \ker(\pi_i)$ is equality.□

Standard Ideals and Standard Congruence Relations

Our goal now is to describe conditions on a lattice L under which there is a bijection (in fact, a lattice isomorphism) between congruence relations on L and ideals of L. To this end, we define a special type of lattice element.

Definition *Let* L *be a lattice. An element* $i \in L$ *is* **standard** *if*

$$a \wedge (i \vee b) = (a \wedge i) \vee (a \wedge b)$$

for all $a, b \in L$.□

Note that standard elements might have been called join-standard elements, since there is a dual notion for this concept. Our primary interest is in standard elements of the ideal lattice $\mathcal{I}(L)$ of a lattice L. Here is the definition in this context. For convenience, we introduce the new term *friendly*.

Definition *Let* $\mathcal{I}(L)$ *be the ideal lattice of a lattice* L.

1) *An ideal* I *of* L *is* **standard** *if*

$$A \wedge (I \vee B) = (A \wedge I) \vee (A \wedge B)$$

for all $A, B \in \mathcal{I}(L)$. *Let* $\mathcal{SI}(L)$ *denote the set of all standard ideals of* L.

2) *An ideal* I *of* L *is* **friendly** *if*

$$I \vee A = \{i \vee a \mid i \in I, a \in A\}$$

for all $A \in \mathcal{I}(L)$. *Let* $\mathcal{FI}(L)$ *denote the set of all friendly ideals of* L.□

Theorem 9.18 *An ideal I of a lattice L is friendly if and only if for any $i \in I$ and $a, b \in L$,*

$$b \leq a \vee i \quad \Rightarrow \quad b = (b \wedge a) \vee j$$

for some $j \in I$.
Proof. Suppose that I is friendly. If $b \leq a \vee i$ for $i \in I$, then $b \in \downarrow a \vee I$ and so $b = a' \vee j$ for $a' \leq a$ and $j \in I$. But $a' \leq b$ implies that

$$b = (a' \wedge b) \vee j \leq (a \wedge b) \vee j \leq b$$

and so $b = (a \wedge b) \vee j$. The converse is clear.□

We next show that an ideal is standard if and only if it is friendly.

Theorem 9.19 *Let I be an ideal of a lattice L. The following are equivalent:*
1) I is standard
2) I is standard with respect to principal ideals, that is, for all $a, b \in L$,

$$\downarrow a \wedge (I \vee \downarrow b) = (\downarrow a \wedge I) \vee (\downarrow a \wedge \downarrow b)$$

3) I is friendly
Proof. It is clear that 1) implies 2). To see that 2) implies 3), if $b \leq a \vee i$ for some $i \in I$, then

$$b = b \wedge (a \vee i) \in \downarrow b \wedge (\downarrow a \vee I) = \downarrow (a \wedge b) \vee (\downarrow b \wedge I)$$

and so $b \leq (a \wedge b) \vee j$ for some $j \in I$, $j \leq b$. Hence, $b = (a \wedge b) \vee j$. To show that 3) implies 1), it is sufficient to show that

$$A \wedge (I \vee B) \subseteq (A \wedge I) \vee (A \wedge B)$$

since the reverse inclusion is clear. If $x \in A \wedge (I \vee B)$, then $x \in A$ and $x = i \vee b$ for $i \in I$ and $b \in B$. Thus, $i \vee b \in A$ and so $i, b \in A$, whence

$$x = i \vee b \in (A \wedge I) \vee (A \wedge B)$$ □

The standard ideals form a sublattice of the ideal lattice $\mathcal{I}(L)$.

Theorem 9.20 *The set $\mathcal{SI}(L)$ of standard ideals of a lattice L is a sublattice of $\mathcal{I}(L)$, that is, the intersection and join of standard ideals is standard.*
Proof. Let I and J be standard. As to join,

$$\begin{aligned} A \wedge [(I \vee J) \vee B] &= A \wedge [I \vee (J \vee B)] \\ &= (A \wedge I) \vee [A \wedge (J \vee B)] \\ &= (A \wedge I) \vee [(A \wedge J) \vee (A \wedge B)] \\ &= [A \wedge (I \vee J)] \vee (A \wedge B) \end{aligned}$$

and so $I \vee J$ is standard. As to intersection $I \cap J$, suppose that

$$b \le a \vee k$$

where $k \in I \cap J$. Then writing $c = a \wedge b$, we have

$$b = c \vee i = c \vee j$$

where $i \in I$ and $j \in J$ and $i \le c \vee j$ implies that $i = (c \wedge i) \vee j'$ for $j' \in J$. But $j' \le i \in I$ and so $j' \in I \cap J$. Hence,

$$b = c \vee i = c \vee (c \wedge i) \vee j' = c \vee j' = (a \wedge b) \vee j'$$

Hence, $I \cap J$ is standard.□

Standard Congruence Relations

We can now describe the relationship between standard ideals and certain types of congruence relations.

Definition *Let L be a lattice and let I be an ideal of L. If the binary relation defined by*

$$x \sigma_I y \quad \textit{if} \quad (x \wedge y) \vee i = x \vee y, \textit{ for some } i \in I$$

is a congruence relation on L, we say that σ_I is **standard.**□

Note that if $x \le y$, then $x \sigma_I y$ if and only if there is an $i \in I$ for which $x \vee i = y$.

Theorem 9.21 *Let $I \in \mathcal{I}(L)$. Then σ_I is standard if and only if I is standard. Moreover, if I is standard, then*

a) *σ_I is the smallest congruence relation on L for which all elements of I are congruent*

b) $\ker(\sigma_I) = I$.

We denote the set of all standard congruence relations on L by $\mathrm{StCon}(L)$.

Proof. If $\sigma = \sigma_I$ is standard and if $b \le a \vee i$, for $i \in I$, then

$$b = b \wedge (a \vee i) \overset{\sigma}{\equiv} b \wedge a$$

and so there exists $j \in I$ for which

$$b = (b \wedge a) \vee j$$

Hence, I is standard.

Conversely, suppose that I is standard. We show that σ is a congruence relation on L using Theorem 9.4. It is clear that σ is reflexive and it is easy to see that $x \sigma y$ if and only if $(x \wedge y) \sigma (x \vee y)$. Next, suppose that $x \le y \le z$ and $x \sigma y \sigma z$. Then there exist $i, j \in I$ for which

$$x \vee i = y \quad \text{and} \quad y \vee j = z$$

and so

$$x \vee i \vee j = y \vee j = z$$

Finally, suppose that $x \leq y$, $x\sigma y$ and $t \in L$. Then $x \vee i = y$ for some $i \in I$ and so

$$(x \vee t) \vee i = y \vee t$$

For the meet condition, since $x \vee i = y$, we have $y \wedge t \leq x \vee i$ and so the standardness of I implies that

$$y \wedge t = (y \wedge t \wedge x) \vee j = (x \wedge t) \vee j$$

for some $j \in I$. Hence, $(x \wedge t)\sigma(y \wedge t)$.

For part a), it is easy to see that all elements of I are congruent. Moreover, if a congruence relation θ has the property that all elements of I are congruent and if $x\sigma_I y$, then $(x \wedge y) \vee i = x \vee y$ for some $i \in I$ and so

$$x \wedge y = [(x \wedge y) \vee (x \wedge y \wedge i)]\theta[(x \wedge y) \vee i] = x \vee y$$

Hence, $(x \wedge y)\theta(x \vee y)$, which implies that $x\theta y$. Thus, $\sigma_I \subseteq \theta$. For part b), if $i \in I$ and $x\sigma_I i$, then $(x \wedge i) \vee j = x \vee i$ for some $j \in I$ and so $x \vee i \in I$, whence $x \in I$. Thus, I is a congruence class and so $\ker(\sigma_I) = I$.□

We have seen that $\mathcal{SI}(L)$ is a sublattice of $\mathcal{I}(L)$. It is also true that $\operatorname{StCon}(L)$ is a sublattice of $\operatorname{Con}(L)$.

Theorem 9.22 *Let L be a lattice.*

1) $\operatorname{StCon}(L)$ *is a sublattice of* $\operatorname{Con}(L)$.
2) *For all* $I, J \in \mathcal{I}(L)$,

$$\sigma_{I \cap J} = \sigma_I \cap \sigma_j \quad \textit{and} \quad \sigma_{I \vee J} = \sigma_I \vee \sigma_j$$

3) *The maps*

$$\sigma\colon \mathcal{SI}(L) \to \operatorname{StCon}(L) \quad \textit{and} \quad \ker\colon \operatorname{StCon}(L) \to \mathcal{SI}(L)$$

where

$$\sigma(I) = \sigma_I$$

are inverse lattice isomorphisms.

Proof. Since $\ker(\sigma_I) = I$, it follows that σ is bijective, with inverse equal to the kernel map. For the intersection, it is clear that $\sigma_{I \cap J} \subseteq \sigma_I \cap \sigma_J$ and so it is sufficient to show that $x(\sigma_I \cap \sigma_J)y$ implies $x\sigma_{I \cap J}y$, where $x < y$. In this case, there exist $i \in I$ and $j \in J$ for which

$$x \vee i = y \quad \text{and} \quad x \vee j = y$$

Hence, $i \leq x \vee j$ and since J is standard, there is a $j' \in J$ for which

$$i = (i \wedge x) \vee j'$$

But $j' \leq i$ implies that $j' \in I$ and so

$$y = x \vee i = x \vee j'$$

where $j' \in I \cap J$. Hence, $x\sigma_{I\cap J}y$ and so $\sigma_{I\cap J} = \sigma_I \cap \sigma_j$.

For the join, it is clear that $\sigma_I, \sigma_J \subseteq \sigma_{I\vee J}$ and so $\sigma_I \vee \sigma_J \subseteq \sigma_{I\vee J}$. To see that $x\sigma_{I\vee J}y$ implies $x(\sigma_I \vee \sigma_J)y$, again we assume that $x < y$. Then there is a $k \in I \vee J$ for which $x \vee k = y$. Since I is friendly, $k = i \vee j$ for some $i \in I$ and $j \in J$ and so $y = x \vee (i \vee j)$. Therefore,

$$x \overset{\sigma_I}{\equiv} x \vee i \overset{\sigma_J}{\equiv} x \vee i \vee j = y$$

which shows that $x(\sigma_I \vee \sigma_J)y$. It follows that σ is a lattice isomorphism, with inverse map ker.□

It is also true that standard congruence relations commute.

Theorem 9.23 *Any two standard congruence relations on a lattice L commute, that is,*

$$\sigma_I\sigma_J = \sigma_J\sigma_I$$

for all standard ideals I and J.
Proof. Let σ_I and σ_J be standard congruence relations on L. Theorem 9.8 implies that it is sufficient to show that

$$a\sigma_I\sigma_J b \quad \Rightarrow \quad a\sigma_J\sigma_I b$$

for all $a < b$ in L. But if

$$a\sigma_I x\sigma_J b$$

for some $x \in [a, b]$, then there are $i \in I$ and $j \in J$ for which

$$a \vee i = x \quad \text{and} \quad x \vee j = b$$

Hence,

$$a \overset{\sigma_J}{\equiv} (a \vee j) \overset{\sigma_I}{\equiv} (a \vee j \vee i) = b$$

and so $a\sigma_J\sigma_I b$.□

Standard Ideals in Distributive Lattices

We have seen that the maps

$$\sigma: \mathcal{SI}(L) \to \text{StCon}(L) \quad \text{and} \quad \text{ker}: \text{StCon}(L) \to \mathcal{SI}(L)$$

where $\sigma(I) = \sigma_I$ are inverse lattice isomorphisms. Therefore, it is of obvious interest to know when

$$\mathcal{SI}(L) = \mathcal{I}(L)$$

and

$$\text{StCon}(L) = \text{Con}(L)$$

If L is a distributive lattice, then it is easy to see that all ideals of L are friendly, for if $b \le a \vee i$ for $i \in I$, then

$$b = b \wedge (a \vee i) = (b \wedge a) \vee (b \wedge i)$$

where $b \wedge i \in I$. Hence, I is friendly. Moreover, if every ideal I is friendly, then every ideal I is the kernel of a congruence relation, namely, the standard congruence σ_I. Finally, if every ideal I is the kernel of a congruence relation, then L is distributive, for if L contains one of the forbidden sublattices in Figure 9.7,

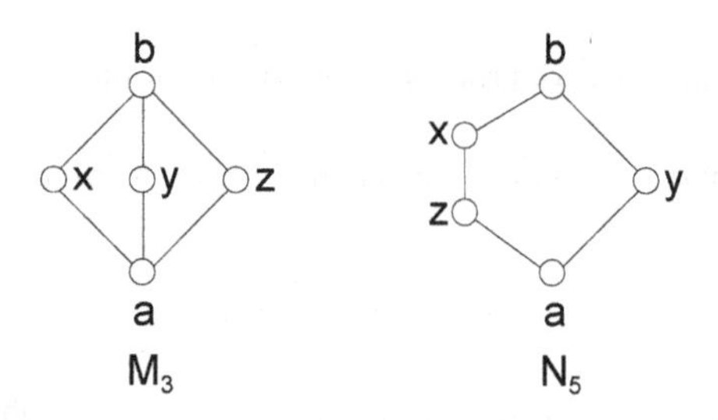

Figure 9.7

then the ideal $\downarrow z$ is not a kernel. To see this, if $\downarrow z = \ker(\theta)$, then $a\theta z$ and so the quadrilateral property implies that $b\theta y$, which in turn implies that $a\theta x$ and so $x \in \downarrow z$, which is false.

Theorem 9.24 *The following are equivalent for a lattice L:*
1) *L is distributive.*
2) *Every ideal of L is friendly (standard), that is,*

$$\mathcal{SI}(L) = \mathcal{I}(L)$$

3) *Every ideal of L is the kernel of some congruence relation on L.*□

As to the issue of when $\text{StCon}(L) = \text{Con}(L)$, note that the definition of a standard congruence is

$$a\sigma_I b \quad \text{if} \quad (a \wedge b) \vee i = a \vee b$$

for some $i \in I$ and this is one half of the requirements for i to be a relative complement of $a \wedge b$ with respect to the section $[0, a \vee b]$.

Indeed, if L is sectionally complemented, then every congruence relation θ on L is standard; in fact,

$$\theta = \sigma_{\ker(\theta)}$$

To see this, it is clear that

$$x\sigma_I y \quad \Leftrightarrow \quad (x \wedge y)\sigma_I(x \vee y)$$

for any standard congruence σ_I and since this also holds for θ, it is sufficient to show that $a\theta b \Leftrightarrow a\sigma_{\ker(\theta)}b$ for all $a \le b$. If $a\theta b$ with $a \le b$ and if y is a relative complement of a in $[0, b]$, then

$$a \vee y = b \quad \text{and} \quad a \wedge y = 0$$

But $a\theta b$ implies that $(a \wedge y)\theta(b \wedge y)$, that is, $0\theta y$. Hence $y \in \ker(\theta)$ and so $a\sigma_{\ker(\theta)}b$. Conversely, if $a\sigma_{\ker(\theta)}b$ with $a \le b$, then there is a $y \in \ker(\theta)$ for which $a \vee y = b$. But $a\theta(a \vee y)$ and so $a\theta b$. Hence $\theta = \sigma_{\ker(\theta)}$.

Thus, if L is sectionally complemented, then all congruence relations on L are standard. In looking for a converse, suppose that all congruence relations on L are standard and let $c \in [0, b]$. We wish to show that there is an $i \in L$ for which

$$c \vee i = b \quad \text{and} \quad c \wedge i = 0$$

If θ is any congruence relation on L, then $\theta = \sigma_I$ for some ideal I and so the condition $c \vee i = b$ follows from $c\theta b$. Also, since $\ker(\theta) = I$, if we can find a θ for which

$$\ker(\theta) = \mathcal{S}(c) = \{x \in L \mid c \wedge x = 0\}$$

then we are done. Now, $\mathcal{S}(c)$ is not an ideal in general, but it is an ideal if L is distributive. Moreover, it appears that one of the congruence relations of Example 9.3:

$$x\mu y \quad \text{if} \quad x \wedge c = y \wedge c$$

is just what we need, since $c\mu b$ and

$$\ker(\mu) = \{x \in L \mid x \wedge c = 0\} = \mathcal{S}(c)$$

We can now put the pieces together and add one more piece.

Theorem 9.25 *Let L be a lattice. The maps*

$$\sigma\colon \mathcal{SI}(L) \to \mathrm{StCon}(L) \quad \textit{and} \quad \ker\colon \mathrm{StCon}(L) \to \mathcal{SI}(L)$$

where $\sigma(I) = \sigma_I$ are inverse lattice isomorphisms. Moreover:

1) *L is distributive if and only if $\mathcal{I}(L) = \mathcal{SI}(L)$.*
2) *If L is sectionally complemented, then* $\mathrm{Con}(L) = \mathrm{StCon}(L)$.
3) *The following are equivalent:*
   a) *L is distributive and sectionally complemented*

b) $\mathcal{I}(L) = \mathcal{SI}(L)$ *and* $\mathrm{Con}(L) = \mathrm{StCon}(L)$
c) *The maps*

$$\sigma: \mathcal{I}(L) \to \mathrm{Con}(L) \quad \textit{and} \quad \ker: \mathrm{Con}(L) \to \mathcal{I}(L)$$

are inverse bijections, in which case these maps are inverse lattice isomorphisms.
d) *L is distributive and all congruence relations on L commute.*

Proof. We have seen that 3a) and 3b) are equivalent and it is clear that 3b) implies 3c). If 3c) holds, then all ideals I are standard, since $\sigma(I) = \sigma_I$ is a congruence relation. Also, all congruence relations θ are standard, since $\theta = \sigma(I) = \sigma_I$, for some ideal I. Hence, 3c) implies 3b).

If 3a) holds, then every section $[0, a]$ of L is modular and complemented and therefore relatively complemented. Hence, L is relatively complemented and so Theorem 9.9 implies that all congruence relations on L commute.

Finally, assume that 3d) holds. We show that L is sectionally complemented. Let $t \in [0, b]$ and consider the congruence relations

$$a\delta b \quad \text{if} \quad a \vee t = b \vee t$$

and

$$a\mu b \quad \text{if} \quad a \wedge t = b \wedge t$$

of Example 9.3. Then $0\delta t\mu b$ implies that $0\mu x\delta b$, for some $x \in [0, b]$, that is,

$$0 = t \wedge x \quad \text{and} \quad x \vee t = b$$

Hence, x is a relative complement of t in $[0, b]$ and L is sectionally complemented.□

Exercises

1. Prove that a subset θ of L^2 is a congruence relation on L if and only if the following hold:
   a) θ is an equivalence relation on L.
   b) θ is a sublattice of the lattice L^2 (under coordinatewise meet and join).
2. What is the smallest element of $\mathrm{Con}(L)$? What is the largest element?
3. Let L be a lattice and let $\theta \in \mathrm{Con}(L)$. Prove that $[a] \sqsubset [b]$ if and only if
   a) $[a] < [b]$ and
   b) $a \leq x \leq b \Rightarrow x \in [a]$ or $x \in [b]$.
4. Shows that, in general, even when L is distributive, congruence kernels need not be distinct, that is, there can be two distinct congruence relations on L that have the same kernel.
5. A lattice L is **simple** if $\mathrm{Con}(L) = \{0, 1\}$ where 0 is the smallest congruence and 1 is the largest congruence. Show that the lattice M_n shown in Figure 9.8 is simple for $n \geq 3$.

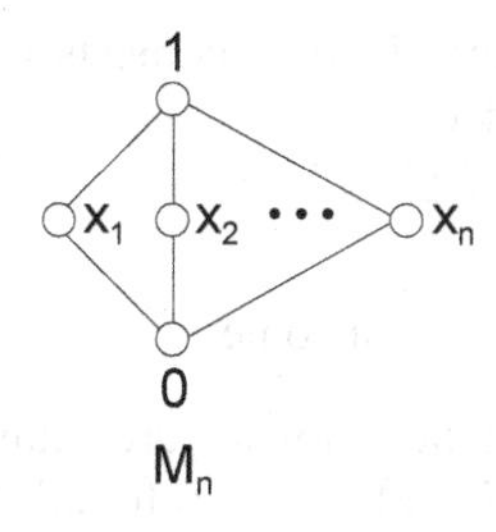

Figure 9.8

6. Let B be a Boolean lattice. Prove that a congruence relation θ on the *lattice* B is also a Boolean congruence, that is, $a\theta b$ implies $a^c\theta b^c$.
7. Let $f: L \to M$ be a lattice homomorphism where L and M have 0. Characterize the fact that f is a lattice embedding in terms of the congruence kernel θ_f. What about the ideal kernel of f (if it exists)?
8. If L is a distributive (modular) lattice, show that L/θ is also distributive (modular).
9. Let I be a standard ideal in a lattice L. Prove that the standard congruence relation σ_I is the smallest congruence relation on L for which all elements of I are congruent.
10. Let θ and σ be congruence relations on a lattice L. Show that if $\theta \le \sigma$ then there is a unique epimorphism $f: L/\theta \to L/\sigma$ such that $f \circ \pi_\theta = \pi_\sigma$.
11. Two lattice epimorphisms $f, g \in \mathrm{Epi}(L)$ are **kernel equivalent** if their congruence kernels θ_f and θ_g are equal, that is, if

$$f(a) = f(b) \quad \Leftrightarrow \quad g(a) = g(b)$$

   a) Show that this is an equivalence relation $\equiv$.
   b) Prove that the lattice epimorphisms $f : L \to M$ and $g: L \to N$ are kernel equivalent if and only if there is a lattice isomorphism $\sigma: M \to N$ for which $\sigma \circ f = g$.
   c) Let L be a bounded lattice. Show that the map $\theta \mapsto [\pi_\theta]$ sending $\theta \in \mathrm{Con}(L)$ to the equivalence class containing π_θ is bijective and so

$$\mathrm{Con}(L) \leftrightarrow \mathrm{Epi}(L)/\equiv$$

   where $\mathrm{Epi}(L)$ is the set of all epimorphisms with domain L.
12. Prove the following theorem, which Grätzer [24] refers to as the **first isomorphism theorem** for lattices. Let θ be a congruence relation on a lattice L. Let S be a set of canonical forms for θ, that is, S is a set consisting of exactly one member from each congruence class of θ. If S is a sublattice of L then $S \approx L/\theta$.
13. Let $\{\theta_i \mid i \in I\}$ be a family of congruence relations on a lattice L. Prove that if all of these congruences have kernels then

$$\ker(\bigwedge \theta_i) = \bigcap \ker(\theta_i)$$

Thus, the family $\mathcal{K}$ of kernels inherits meets from $\mathcal{I}(L)$ and so if L has 0 then $\mathcal{K}$ is a complete lattice.

14. Let L be a lattice and let $I \trianglelefteq L$.
    a) Prove that
$$x \vee i = y \vee i \text{ for some } i \quad \Leftrightarrow \quad \downarrow x \vee I = \downarrow y \vee I$$
    b) Let $i \in I$. Prove that the following are equivalent:
        i) $i \vee (x \wedge y) = (i \vee x) \wedge (i \vee y)$ for all $x, y \in L$.
        ii) The join map $j_i : x \mapsto x \vee i$ is a homomorphism.
        iii) The binary relation defined by $x\theta_i y$ if $x \vee i = y \vee i$ is a congruence relation.
15. Let L be a lattice and let $a, b \in L$. The smallest congruence relation $\theta_{a,b}$ on L for which $a\theta_{a,b}b$ is called the **principal congruence generated by** the pair (a, b).
    a) Show that
$$\theta_{a,b} = \bigwedge\{\sigma \in \text{Con}(L) \mid a\sigma b\}$$
    b) Show that for any $\tau \in \text{Con}(L)$,
$$\tau = \bigvee\{\theta_{a,b} \mid a\tau b\}$$
    Hence, the family of principal congruence relations is join-dense in $\text{Con}(L)$.
    c) Prove that the principal congruence relations are compact and so $\text{Con}(L)$ is algebraic.
    d) Find the principal congruence $\theta_{x,1}$ for the lattice N_5 in Figure 9.9.

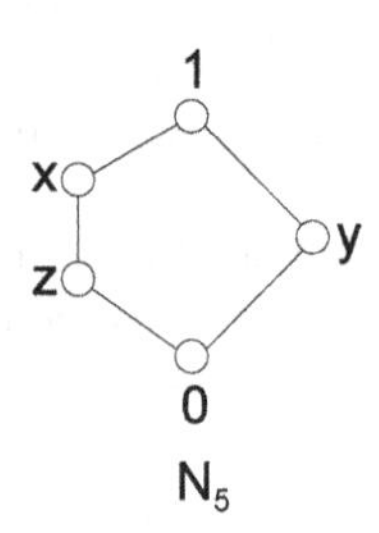

Figure 9.9

16. a) Prove that a nonempty subset S of a lattice L is a convex sublattice of L if and only if $S = I \cap F$, where I is an ideal and F is a filter.
    b) Let L be a lattice and let θ be a congruence relation on L. A filter F in L is called the **cokernel** of θ if F is the largest element in the quotient L/θ. Let I be a kernel in L and let F be a cokernel in L. Show that if $I \cap F$ is nonempty, then it is a congruence class for some congruence relation on L.

c) Prove that a lattice L is distributive if and only if every convex sublattice of L is a congruence class.

17. Let L be a distributive lattice and let $I \trianglelefteq L$ and $F \trianglerighteq L$. Define a binary relation θ on L by

$$x\theta y \quad \text{if} \quad (x \vee u) \wedge v = (y \vee u) \wedge v \text{ for some } u \in I \text{ and } v \in F$$

a) Prove that θ is a congruence relation on L.
b) If L is bounded, prove that $I \subseteq [0]$.
c) If L is bounded, prove that $F \subseteq [1]$.
d) If L is bounded, prove that θ is the smallest congruence relation on L for which $I \subseteq [0]$ and $F \subseteq [1]$.
e) Prove that θ is proper ($\theta \neq L \times L$) if and only if $I \cap F = \emptyset$.
f) Prove that θ is trivial (equality) if and only if $I = \{0\}$ and $F = \{1\}$.

Distributive Ideals

Definition *Let L be a lattice.*

1) *An element $i \in L$ is* **distributive** *if*

$$i \vee (a \wedge b) = (i \vee a) \wedge (i \vee b)$$

for $a, b \in L$.

2) *An ideal $I \in \mathcal{I}(L)$ is* **distributive** *if*

$$I \vee (A \wedge B) = (I \vee A) \wedge (I \vee B)$$

for $A, B \in \mathcal{I}(L)$. Let $\mathcal{DI}(L)$ denote the set of all distributive ideals of L.

3) *If the binary relation defined by*

$$x\delta_I y \quad \textit{if} \quad x \vee i = y \vee i, \textit{ for some } i \in I$$

is a congruence relation on L, we say that δ_I is **distributive.** $\square$

18. Let I be an ideal of a lattice L. Prove that I is standard if and only if both of the following hold:
a) I is distributive
b) Relative complements of I, if they exist, are unique, that is, for all $A, B \in \mathcal{I}(L)$,

$$I \wedge A = I \wedge B \quad \text{and} \quad I \vee A = I \vee B \quad \Rightarrow \quad A = B$$

Find a distributive ideal that is not standard. *Hint*: Look in a small nondistributive lattice.

19. Show that for a distributive congruence relation

$$x\delta_I y \quad \Leftrightarrow \quad \downarrow x \vee I = \downarrow y \vee I$$

20. Let I be a distributive ideal in a lattice L. Prove that the distributive congruence relation δ_I is the smallest congruence relation on L for which all elements of I are congruent.

21. Let I be an ideal of a lattice L. Prove that δ_I is distributive if and only if I is distributive. Moreover, if I is distributive, then
    a) δ_I is the smallest congruence relation on L for which all elements of I are congruent
    b) $\ker(\delta_I) = I$.
    Prove that if I is a standard ideal, then $\sigma_I = \delta_I$.
22. Find an example of a distributive congruence that is not standard.
23. Prove that if *every* ideal of a lattice L is distributive, then every ideal is standard.
24. Let L be a sectionally complemented lattice with the ACC. Prove that for any congruence relation θ on L there is an element $m \in L$ such that

$$a\theta b \quad \Leftrightarrow \quad a \vee m = b \vee m$$

for all $a, b \in L$.

Part II
Topics

Chapter 10
Duality for Distributive Lattices: The Priestley Topology

Let us repeat the representation theorem for distributive lattices (Theorem 6.6).

Theorem 10.1 *Let L be a nontrivial distributive lattice and let*

$$\rho_L : L \to \mathcal{O}(\mathrm{Spec}(L))$$

be the rho map for L, defined by

$$\rho_L(a) = \mathcal{P}_{\neg a} := \{P \in \mathrm{Spec}(L) \mid a \notin P\}$$

1) **(Representation theorem for distributive lattices, Birkhoff, 1933 [6])** *The rho map ρ is an embedding of L into the power set $\wp(\mathrm{Spec}(L))$ and so L is isomorphic to a power set sublattice (ring of sets). Moreover, if L is finite, then ρ is an isomorphism and $L \approx \mathcal{O}(\mathrm{Spec}(L))$.*

2) **(Representation theorem for Boolean algebras)** *If L is a Boolean algebra, then ρ is a Boolean algebra embedding of L into the power set $\wp(\mathrm{Spec}(L))$ and so L is isomorphic to a power set subalgebra (field of sets). Moreover, if L is finite, then ρ is an isomorphism.* □

There is more to the story concerning the representation of distributive lattices as power set sublattices. The family

$$\rho = \{\rho_L : L \approx \mathcal{O}(\mathrm{Spec}(L)) \mid L \text{ is a finite distributive lattice}\}$$

of lattice isomorphisms is part of a general *duality* between finite distributive lattices and finite posets. This duality also exists for arbitrary bounded distributive lattices, except that the rho maps are not surjective in this case and so we must replace the sets $\mathcal{O}(\mathrm{Spec}(L))$ with other sets that are rather more difficult to describe.

Indeed, in the early 1930s, Marshall Stone introduced the idea of topologizing the spectrum $\mathrm{Spec}(B)$ of a Boolean algebra B to help describe the image of the

S. Roman (ed.), *Lattices and Ordered Sets*, doi: 10.1007/978-0-387-78901-9_10,

rho maps. In particular, the sets $\mathcal{P}_{\neg a}$ form a basis for the *Stone topology* and are actually clopen (open and closed) in that topology. In the late 1930s, Stone generalized his topology to deal with bounded distributive lattices. Unfortunately, the resulting topology is not Hausdorff, although it is compact. Nonetheless, the introduction of topological methods led to a new connection between algebra and topology that has had a great influence in other areas of mathematics, including functional analysis, which was actually Stone's main area of interest.

In the 1970s, Hilary Priestley [52] recognized that a different topology might be a more appropriate choice for the representation theory of arbitrary bounded distributive lattices. The *Priestley topology* on the spectrum $\operatorname{Spec}(L)$ of a bounded distributive lattice has basis

$$\mathcal{B} = \{\mathcal{P}_{\neg a,b} := \mathcal{P}_{\neg a} \cap \mathcal{P}_b \mid a, b \in L\}$$

This topology is both compact and Hausdorff. As we will see, it enables us to describe the image of the rho map in elegant terms as the family of all *clopen down-sets* in $\operatorname{Spec}(L)$.

Our plan in this chapter is twofold. First, we will discuss the duality between finite distributive lattices and finite posets, since this does not require the introduction of topological notions. (More precisely, the Priestley topology is discrete when the sets are finite.) Then we will discuss the necessary topological notions so that we can present the general duality theory of bounded distributive lattices.

Both dualities are most elegantly described using the *language* of category theory. Therefore, since some readers may not be familiar with this language, it seems appropriate to include an informal discussion here and, for those who desire a more formal approach, an appendix on the subject. We suggest that all readers review at least the informal discussion.

An Informal Introduction to the Language of Category Theory

We will need three main concepts from category theory: *categories*, *functors* and *natural transformations*.

Categories

Informally speaking, a category consists simply of a class of mathematical objects and their structure-preserving maps. Examples are the category of sets, the category of groups, the category of posets and the category of lattices.

More specifically, a **category** $\mathcal{C}$ consists of a class $\mathbf{Obj}(\mathcal{C})$ whose members are called the **objects** of the category. One often writes $A \in \mathcal{C}$ as a shorthand for $A \in \mathbf{Obj}(\mathcal{C})$. Also, for each pair (A, B) of objects, there is a set $\hom_{\mathcal{C}}(A, B)$, whose elements are called **morphisms, maps** or **arrows**. Often but not always, the morphisms in $\hom_{\mathcal{C}}(A, B)$ are functions from A to B with special properties. The set $\hom_{\mathcal{C}}(A, B)$ is called a **hom set**. It is common to denote the fact that $f \in \hom_{\mathcal{C}}(A, B)$ by writing $f: A \to B$.

The morphisms in a category must satisfy two conditions: There must be an associative composition and each object must have an identity morphism. Specifically, for $f: A \to B$ and $g: B \to C$, there is a morphism $g \circ f: A \to C$. Moreover, composition is associative:

$$f \circ (g \circ h) = (f \circ g) \circ h$$

whenever the compositions are defined. Also, for each object $A \in \mathbf{Obj}(\mathcal{C})$, there is a morphism $\iota_A: A \to A$ called the **identity morphism** for A, with the property that if $f: A \to B$, then

$$\iota_B \circ f = f \quad \text{and} \quad f \circ \iota_A = f$$

Finally, a morphism $f: A \to B$ is an **isomorphism** if there is a morphism $f^{-1}: B \to A$ for which

$$f^{-1} \circ f = \iota_A \quad \text{and} \quad f \circ f^{-1} = \iota_B$$

Example 10.2 In the category **Set** of sets, the class of objects is the class of all sets and the hom set $\hom(A, B)$ is the set of all functions from A to B. In the category **Grp** of groups, the class of objects is the class of all groups and the hom set $\hom(A, B)$ is the set of all group homomorphisms from A to B.

In the category **FinPoset** of finite posets, the class of objects is the class of all finite posets and the morphisms from A to B are the order-preserving maps from A to B. In the category **FinDLat** of finite distributive lattices, the class of objects is the class of all finite distributive lattices and the morphisms from A to B are the lattice $\{0, 1\}$-homomorphisms from A to B. We give more examples of categories in the appendix.□

A **subcategory** $\mathcal{D}$ of a category $\mathcal{C}$ is a category with the property that every object of $\mathcal{D}$ is an object of $\mathcal{C}$ and every morphism of $\mathcal{D}$ is a morphism of $\mathcal{C}$ and that the composition and identities are the same in $\mathcal{D}$ as in $\mathcal{C}$.

Functors

Structure-preserving "maps" between categories are called *functors*. Since a category consists of both objects and morphisms, a functor must consist of two functions—one to handle the objects and one to handle the morphisms. Also, there are two versions of functors: *covariant* and *contravariant*.

More specifically, let $\mathcal{C}$ and $\mathcal{D}$ be categories. A **functor** $F: \mathcal{C} \Rightarrow \mathcal{D}$ is a pair of functions (as is customary, we use the same symbol F for both functions):

1) An *object function* $F\colon \mathbf{Obj}(\mathcal{C}) \to \mathbf{Obj}(\mathcal{D})$ that maps objects in $\mathcal{C}$ to objects in $\mathcal{D}$.
2) For a **covariant functor**, a *morphism function*

$$F\colon \mathbf{Mor}(\mathcal{C}) \to \mathbf{Mor}(\mathcal{D})$$

that maps each morphism $f\colon A \to B$ in $\mathcal{C}$ to a morphism $Ff\colon FA \to FB$ in $\mathcal{D}$. For a **contravariant functor**, a morphism function

$$F\colon \mathbf{Mor}(\mathcal{C}) \to \mathbf{Mor}(\mathcal{D})$$

that maps each morphism $f\colon A \to B$ in $\mathcal{C}$ to a morphism $Ff\colon FB \to FA$ in $\mathcal{D}$ (note the reversal of direction).
3) Identity and composition are preserved, that is, for a covariant functor,

$$F\iota_A = \iota_{FA} \quad \text{and} \quad F(g \circ f) = Fg \circ Ff$$

and for a contravariant functor,

$$F\iota_A = \iota_{FA} \quad \text{and} \quad F(g \circ f) = Ff \circ Fg$$

whenever all compositions are defined.

The notation $F\colon\mathcal{C} \Rightarrow \mathcal{D}$ is read "F is a functor from $\mathcal{C}$ to $\mathcal{D}$." A functor $F\colon\mathcal{C} \Rightarrow \mathcal{C}$ from $\mathcal{C}$ to itself is referred to as a **functor on** $\mathcal{C}$.

One way to think of a covariant functor $F\colon\mathcal{C} \Rightarrow \mathcal{D}$ is as a mapping of one-arrow *diagrams* in $\mathcal{C}$

$$A \xrightarrow{f} B$$

to one-arrow diagrams in $\mathcal{D}$

$$FA \xrightarrow{Ff} FB$$

with the property that identity and composition are preserved.

Example 10.3 The **power set functor** $\mathcal{P}\colon \mathbf{Set} \Rightarrow \mathbf{Set}$ sends a set A to its power set $\wp(A)$ and each set function $f\colon A \to B$ to the induced function $\mathcal{P}f\colon \wp(A) \to \wp(B)$ that sends a subset $X \subseteq A$ to fX. It is easy to see that this defines a covariant functor.

More to the purpose of this book, the **spectrum functor**

$$F\colon \mathbf{FinDLat} \Rightarrow \mathbf{FinPoset}$$

sends a finite distributive lattice L to its spectrum $\mathrm{Spec}(L)$ and a lattice homomorphism $f\colon L \to M$ to the induced inverse map

$$f^{-1}\colon \mathrm{Spec}(M) \to \mathrm{Spec}(L)$$

restricted to prime ideals. In the opposite direction, the **down-set functor**

$$\mathcal{O}: \textbf{FinPoset} \Rightarrow \textbf{FinDLat}$$

sends a finite poset P to the lattice $\mathcal{O}(P)$ of down-sets of P and a monotone map $g: P \to Q$ between posets to the induced inverse map $g^{-1}: \mathcal{O}(Q) \to \mathcal{O}(P)$, restricted to down-sets. We will show later that these are contravariant functors.□

Functors can be composed in the "obvious" way. Specifically, if $F: \mathcal{C} \Rightarrow \mathcal{D}$ and $G: \mathcal{D} \Rightarrow \mathcal{E}$ are functors, then $G \circ F: \mathcal{C} \Rightarrow \mathcal{E}$ is defined by

$$(G \circ F)(A) = G(FA) \quad \text{and} \quad (G \circ F)(f) = G(Ff)$$

for $A \in \textbf{Obj}(\mathcal{C})$ and $f \in \hom_{\mathcal{C}}(A, B)$.

The morphism portion of a functor $F: \mathcal{C} \Rightarrow \mathcal{D}$ maps hom sets to hom sets, specifically,

$$F: \hom(A, B) \to \hom(FA, FB)$$

or

$$F: \hom(A, B) \to \hom(FB, FA)$$

in the case of a contravariant functor. If each of these hom set maps is surjective, then F is said to be **full** and if each of these hom set maps is injective, then F is said to be **faithful**.

Natural Transformations

A structure-preserving map between functors is called a *natural transformation.* Consider a pair of covariant functors $F, G: \mathcal{C} \Rightarrow \mathcal{D}$, as shown in Figure 10.1.

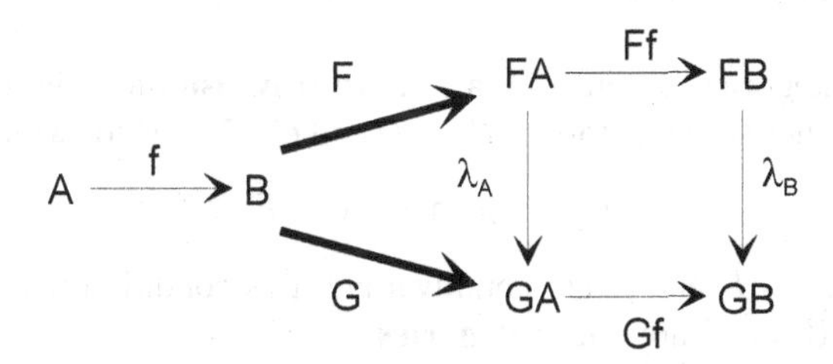

Figure 10.1

We have remarked that the essence of a functor is what it does to the one-arrow diagrams $f: A \to B$. A **natural transformation** $\lambda: F \dashrightarrow G$ is intended to map, in a nice way, the family of one-arrow diagrams that characterize the functor F to the corresponding family of one-arrow diagrams that characterize G. As shown in Figure 10.1, this is accomplished by a *family*

$$\{\lambda_A: FA \to GA \mid A \in \textbf{Obj}(\mathcal{C})\}$$

of morphisms in $\mathcal{D}$—one for each object in $\mathcal{C}$— with the property that the square in Figure 10.1 commutes, that is,

$$Gf \circ \lambda_A = \lambda_B \circ Ff$$

We write $\lambda: F \dashrightarrow G$ or $\{\lambda_A\}: F \dashrightarrow G$ to denote a natural transformation. Each map λ_A is called a **component** of the natural transformation.

If each component λ_A is an isomorphism, then λ is called a **natural isomorphism** and we write $\{\lambda_A\}: F \approx G$. In this case, the family $\{\lambda_A^{-1}\}$ is a natural isomorphism from G to F.

Some very important commonly known maps are natural transformations. We give two examples in the appendix: the determinant and the coordinate map for finite-dimensional vector spaces.

In this chapter, we will show that the family

$$\rho = \{\rho_L : L \approx \mathcal{O}(\mathrm{Spec}(L)) \mid L \text{ is a finite distributive lattice}\}$$

of rho maps is a natural isomorphism.

Dual Categories

It seems reasonable to say that two categories $\mathcal{C}$ and $\mathcal{D}$ are *isomorphic* if there exist functors $F: \mathcal{C} \Rightarrow \mathcal{D}$ and $G: \mathcal{D} \Rightarrow \mathcal{C}$ for which

$$F \circ G = I_{\mathcal{D}} \quad \text{and} \quad G \circ F = I_{\mathcal{C}}$$

where $I_{\mathcal{C}}$ is the identity functor on $\mathcal{C}$ and similarly for $\mathcal{D}$. However, this notion turns out to be too restrictive. A more useful notion comes when we require only that the two compositions $F \circ G$ and $G \circ F$ are *naturally isomorphic* to the respective identity functors.

Thus, two categories $\mathcal{C}$ and $\mathcal{D}$ are **naturally isomorphic** (or **naturally equivalent**) if there are functors $F: \mathcal{C} \Rightarrow \mathcal{D}$ and $G: \mathcal{D} \Rightarrow \mathcal{C}$ for which

$$F \circ G \approx I_{\mathcal{D}} \quad \text{and} \quad G \circ F \approx I_{\mathcal{C}}$$

When the functors F and G are contravariant, this condition is often expressed by saying that $\mathcal{C}$ and $\mathcal{D}$ are **dual categories**.

We will show that the categories **FinDLat** and **FinPoset** are dual, as are the categories **BDLat** of (arbitrary) bounded distributive lattices and **OrdBool** of ordered Boolean spaces (defined later). In fact, these dualities are the subject of this chapter.

Finally, it is not hard to see that if $F: \mathcal{C} \Rightarrow \mathcal{D}$ and $G: \mathcal{D} \Rightarrow \mathcal{C}$ are functors for which

$$F \circ G \approx I_{\mathcal{D}} \quad \text{and} \quad G \circ F \approx I_{\mathcal{C}}$$

then F and G are both full and faithful. This follows from the fact that functors are ordinary functions when restricted to hom sets.

The Duality Between Finite Distributive Lattices and Finite Posets

The duality between finite distributive lattices and finite posets is described in Figure 10.2.

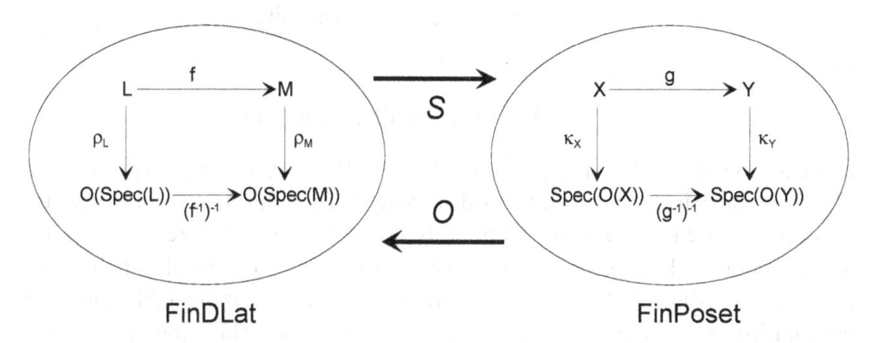

Figure 10.2

The Categories and Functors

Let **FinDLat** be the category of all finite distributive lattices, with lattice $\{0,1\}$-homomorphisms and let **FinPoset** be the category of all finite posets, with monotone functions.

The **spectrum functor** $\mathcal{S}$: **FinDLat** $\Rightarrow$ **FinPoset** is defined by

$$\mathcal{S}(L) = \text{Spec}(L) \quad \text{and} \quad \mathcal{S}(f) = f^{-1}: \text{Spec}(M) \to \text{Spec}(L)$$

where L and M are finite distributive lattices and $f: L \to M$ is a $\{0,1\}$-homomorphism. Note that the map f^{-1} is the induced inverse map, restricted to the prime ideals of M. It is easy to see that $\mathcal{S}$ is a contravariant functor, since

$$\mathcal{S}(\iota) = \iota^{-1} = \iota$$

and

$$\mathcal{S}(g \circ f) = (g \circ f)^{-1} = f^{-1} \circ g^{-1} = \mathcal{S}(f) \circ \mathcal{S}(g)$$

The **down-set functor** $\mathcal{O}$: **FinPoset** $\Rightarrow$ **FinDLat** is defined by

$$\mathcal{O}: P \mapsto \mathcal{O}(P) \quad \text{and} \quad \mathcal{O}(g) = g^{-1}: \mathcal{O}(Q) \to \mathcal{O}(P)$$

where P and Q are finite posets and $g: P \to Q$ is a monotone map. The map

g^{-1} is the induced inverse map, restricted to the down-sets of Q. It is clear that $\mathcal{O}$ is also a contravariant functor.

The Natural Isomorphisms

We have seen that the rho maps:

$$\rho = \{\rho_L: L \approx \mathcal{O}(\text{Spec}(L)) \mid L \text{ is a finite distributive lattice}\}$$

are lattice isomorphisms. On the **FinPoset** side, we define a family of order isomorphisms

$$\kappa = \{\kappa_P: P \approx \text{Spec}(\mathcal{O}(P)) \mid P \text{ is a finite poset}\}$$

by

$$\kappa_P(a) = \mathcal{K}_{\neg a} := \{D \in \mathcal{O}(P) \mid a \notin D\}$$

It is easy to see that $\mathcal{K}_{\neg a}$ is a prime ideal in $\mathcal{O}(P)$. Since P is finite, all filters in $\mathcal{O}(P)$ are principal. Since $\mathcal{O}(P)$ is distributive, Theorem 4.26 implies that the prime filters are precisely the filters of the form $\uparrow_{\mathcal{O}(P)}(D)$ where $D \in \mathcal{O}(P)$ is join-irreducible. Hence, the prime ideals are precisely the ideals of the form $(\uparrow_{\mathcal{O}(P)}(D))^c$ where $D \in \mathcal{O}(P)$ is join-irreducible. But Example 3.20 shows that the join-irreducibles in $\mathcal{O}(P)$ are the prime ideals $\downarrow_P a$. Thus, the prime ideals in $\mathcal{O}(P)$ are precisely the ideals of the form

$$(\uparrow_{\mathcal{O}(P)}(\downarrow_P a))^c = (\mathcal{K}_a)^c = \mathcal{K}_{\neg a}$$

This shows that κ_P is surjective.

To see that κ_P is an order isomorphism, we have

$$a \leq b \quad \Rightarrow \quad \mathcal{K}_{\neg a} \subseteq \mathcal{K}_{\neg b}$$

and

$$\mathcal{K}_{\neg a} \subseteq \mathcal{K}_{\neg b} \quad \Rightarrow \quad \downarrow b \notin \mathcal{K}_{\neg a} \quad \Rightarrow \quad a \in \downarrow b \quad \Rightarrow \quad a \leq b$$

The fact that the families ρ and κ are natural isomorphisms is shown in the diagrams in Figure 10.2. Note that the functors $\mathcal{S}$ and $\mathcal{O}$ are defined in such a way that

$$(f^{-1})^{-1}: \mathcal{O}(\text{Spec}(L)) \to \mathcal{O}(\text{Spec}(M))$$

and

$$(g^{-1})^{-1}: \text{Spec}(\mathcal{O}(P)) \to \text{Spec}(\mathcal{O}(Q))$$

Now, the diagram on the left in Figure 10.2 commutes if and only if

$$\rho_M \circ f = (f^{-1})^{-1} \circ \rho_L$$

that is, for all $a \in L$,

$$\mathcal{P}_{\neg f(a)} = (f^{-1})^{-1}\mathcal{P}_{\neg a}$$

But this holds since

$$\begin{aligned}(f^{-1})^{-1}\mathcal{P}_{\neg a} &= (f^{-1})^{-1}(\{P \in \mathrm{Spec}(L) \mid a \notin P\}) \\ &= \{Q \in \mathrm{Spec}(M) \mid a \notin f^{-1}(Q)\} \\ &= \{Q \in \mathrm{Spec}(M) \mid f(a) \notin Q\} \\ &= \mathcal{P}_{\neg f(a)}\end{aligned}$$

Proof that the second diagram commutes:

$$\kappa_Q \circ g = (g^{-1})^{-1} \circ \kappa_P$$

is similar. Thus, ρ and κ are natural isomorphisms and the categories **FinDLat** and **FinPoset** are dual.

Theorem 10.4 *The categories* **FinDLat** *and* **FinPoset** *are dual. Specifically:*

1) *The following are contravariant functors:*
   a) $\mathcal{S}$: **FinDLat** $\Rightarrow$ **FinPoset** *defined by*

   $$\mathcal{S}: L \mapsto \mathrm{Spec}(L) \quad \textit{and} \quad \mathcal{S}(f) = f^{-1}: \mathrm{Spec}(M) \to \mathrm{Spec}(L)$$

   for $f: L \to M$.
   b) $\mathcal{O}$: **FinPoset** $\Rightarrow$ **FinDLat** *defined by*

   $$\mathcal{O}: P \mapsto \mathcal{O}(P) \quad \textit{and} \quad \mathcal{O}(g) = g^{-1}: \mathcal{O}(Q) \to \mathcal{O}(P)$$

   for $g: P \to Q$.
2) *The family of rho maps*

   $$\rho = \{\rho_L: L \approx \mathcal{O}(\mathrm{Spec}(L)) \mid L \in \textbf{FinDLat}\}$$

   defined by

   $$\rho_L(a) = \mathcal{P}_{\neg a} := \{I \in \mathrm{Spec}(L) \mid a \notin I\}$$

   is a natural isomorphism $\rho: I_{\textbf{FinDLat}} \to \mathcal{O} \circ \mathcal{S}$. *The family of kappa maps*

   $$\kappa = \{\kappa_P: P \approx \mathrm{Spec}(\mathcal{O}(P)) \mid P \in \textbf{FinPoset}\}$$

   defined by

   $$\kappa_P(p) = \mathcal{K}_{\neg p} := \{D \in \mathcal{O}(P) \mid p \notin D\}$$

   is a natural isomorphism $\kappa: I_{\textbf{FinPoset}} \to \mathcal{S} \circ \mathcal{O}$.
3) *The functors* $\mathcal{S}$ *and* $\mathcal{O}$ *are full and faithful. That is:*
   a) $\mathcal{S}$ *defines a bijection between hom sets:*

   $$\mathcal{S}: \hom_{\textbf{FinDLat}}(L, M) \leftrightarrow \hom_{\textbf{FinPoset}}(\mathrm{Spec}(M), \mathrm{Spec}(L))$$

b) *$\mathcal{O}$ defines a bijection between hom sets:*

$$\mathcal{O}: \hom_{\mathbf{FinPoset}}(P,Q) \leftrightarrow \hom_{\mathbf{FinDLat}}(\mathcal{O}(Q),\mathcal{O}(P))$$

Proof. For part 3), see Theorem A2.8.□

Totally Order-Separated Spaces

As mentioned earlier, in order to describe the duality theory for arbitrary bounded distributive lattices, we require a few basic notions from point-set topology, which are summarized in a short appendix. We also require a few more specialized facts, presented here. Proofs will, in general, be left to the reader.

Definition *An* **ordered topological space** *is a triple $(X,\tau,\le)$ where (X,τ) is a topological space and $(X,\le)$ is a poset. We will denote the set of all clopen sets in X by $\wp^{\mathrm{clo}}(X)$ and the set of all clopen down-sets in X by $\mathcal{O}^{\mathrm{clo}}(X)$.*□

Definition *If $(X,\tau,\le)$ and $(Y,\sigma,\le)$ are ordered topological spaces then a map $f:X\to Y$ is an* **order-homeomorphism** *if it is an order isomorphism of posets and a homeomorphism of topological spaces.*□

Definition

1) *A topological space (X,τ) is* **totally separated** *if the clopen sets separate points, that is, if*

$$x\ne y \quad\Rightarrow\quad \exists D\in\wp^{\mathrm{clo}}(X) \text{ with } y\in D \text{ and } x\notin D$$

A compact, totally separated topological space is called a **Boolean space** *or* **Stone space**.

2) *An ordered topological space $(X,\tau,\le)$ is* **totally order-separated** *if the clopen down-sets separate points, that is, if*

$$x\not\le y \quad\Rightarrow\quad \exists D\in\mathcal{O}^{\mathrm{clo}}(X) \text{ with } y\in D \text{ and } x\notin D$$

A compact, totally order-separated ordered topological space is called an **ordered Boolean space** *or a* **Priestley topological space**.□

We note the following:

1) Totally separated and totally order-separated spaces are Hausdorff.
2) In a totally order-separated space, the order is determined by the *clopen* down sets, since

$$x\le y \quad\Leftrightarrow\quad (y\in D\Rightarrow x\in D \text{ for all } D\in\mathcal{O}^{\mathrm{clo}}(X))$$

3) Since a Boolean space or an ordered Boolean space X is compact and Hausdorff, the closed subsets of X are the same as the compact subsets of X. It follows that any clopen set is a *finite* union of basic open sets.

4) Any Boolean space X is an ordered Boolean space under the partial order of equality. For in this case, $x \leq y$ means $x = y$ and so every nonempty subset of X is a down-set. Since $y \not\leq x$ means $y \neq x$ and since X is totally separated, there is a clopen (down-) set K for which $x \in K$ and $y \notin K$. Thus, X is totally order-separated and so is an ordered Boolean space. Moreover,

$$\mathcal{O}^{\mathrm{clo}}(X) = \wp^{\mathrm{clo}}(X)$$

Conversely, if X is an ordered Boolean space under equality, then $\mathcal{O}^{\mathrm{clo}}(X) = \wp^{\mathrm{clo}}(X)$ and X is a Boolean space.

In a totally order-separated space, separation can be extended to certain compact subsets.

Theorem 10.5 *Let X be totally order-separated. If A and B are disjoint compact sets with the property that no element of B is less than any element of A, that is, if*

$$B \cap (\downarrow A) = \emptyset$$

then there is a clopen down-set C such that $A \subseteq C$ and $B \cap C = \emptyset$.□

Of course, the intersection of clopen down-sets is a closed down-set and the union of clopen down-sets is an open down-set. In an ordered Boolean space, the converses are also true.

Theorem 10.6 *Let X be an ordered Boolean space.*

1) The closed down-sets of X are precisely the intersections of clopen down-sets.
2) The open down-sets of X are precisely the unions of clopen down-sets.□

Theorem 10.7 *If C is a closed subset of an ordered Boolean space X, then $\downarrow C$ and $\uparrow C$ are also closed.*□

The Priestley Prime Ideal Space

Let L be a bounded lattice. We gather a few facts about the sets $\mathcal{P}_{\neg a}$, $\mathcal{P}_b$ and

$$\mathcal{P}_{\neg a,b} := \mathcal{P}_{\neg a} \cap \mathcal{P}_b$$

for $a, b \in L$.

Theorem 10.8 *Let L be a nontrivial bounded lattice and let*

$$\mathcal{S} = \{\mathcal{P}_{\neg a} \mid a \in L\} \quad \text{and} \quad \mathcal{B} = \{\mathcal{P}_{\neg a,b} \mid a, b \in L\}$$

Then $\mathcal{S} \subseteq \mathcal{B}$ since

$$\mathcal{P}_{\neg a} = \mathcal{P}_{\neg a,0}$$

for all $a \in L$. Moreover,

1) *$\mathcal{S}$ is a sublattice of the power set lattice $\wp(\mathrm{Spec}(L))$, in particular,*

$$\mathcal{P}_{\neg a} \cap \mathcal{P}_{\neg b} = \mathcal{P}_{\neg(a \wedge b)} \quad \text{and} \quad \mathcal{P}_{\neg a} \cup \mathcal{P}_{\neg b} = \mathcal{P}_{\neg(a \vee b)}$$

If L is complemented, then $\mathcal{S}$ is a subalgebra of $\wp(\mathrm{Spec}(L))$, that is,

$$\mathcal{P}_a = (\mathcal{P}_{\neg a})^c = \mathcal{P}_{\neg(a')}$$

2) *$\mathcal{B}$ has finite intersections, since*

$$\mathcal{P}_{\neg a_1,b_1} \cap \mathcal{P}_{\neg a_2,b_2} = \mathcal{P}_{\neg(a_1 \wedge a_2),(b_1 \vee b_2)}$$

The complement of an element of $\mathcal{B}$ is

$$\mathcal{P}^c_{\neg a,b} = \mathcal{P}_{\neg b,0} \cup \mathcal{P}_{\neg 1,a}$$

3) *If L is complemented, then $\mathcal{P}_b = \mathcal{P}_{\neg b'}$ and so*

$$\mathcal{P}_{\neg a,b} = \mathcal{P}_{\neg(a \wedge b')}$$

Thus, $\mathcal{B} = \mathcal{S}$. □

Since $\mathcal{S}$ and $\mathcal{B}$ each have finite intersections, Theorem A1.2 implies that they can each serve as a basis for a topology on $\mathrm{Spec}(L)$. Moreover, since

$$\mathcal{P}^c_{\neg a,b} = \mathcal{P}_{\neg b,0} \cup \mathcal{P}_{\neg 1,a}$$

each basis element $\mathcal{P}_{\neg a,b}$ is clopen in this topology.

Definition *Let L be a bounded distributive lattice.*

1) *The ordered topological space $(\mathrm{Spec}(L), \tau, \subseteq)$, where τ is the topology with basis consisting of the clopen sets*

$$\mathcal{B} = \{\mathcal{P}_{\neg a,b} \mid a, b \in L\}$$

is called the **Priestley prime ideal space** *or the* **Priestley dual space** *of L. Note that the sets*

$$\{\mathcal{P}_{\neg a} \mid a \in L\} \cup \{\mathcal{P}_b \mid b \in L\}$$

form a subbasis for the Priestley prime ideal space.

2) *The topological space $(\mathrm{Spec}(L), \sigma)$, where σ is the topology with basis*

$$\mathcal{S} = \{\mathcal{P}_{\neg a} \mid a \in L\}$$

is called the **Stone prime ideal space** *or the* **Stone dual space** *of L.* □

Theorem 10.8 implies that if L is a Boolean algebra, then $\mathcal{B} = \mathcal{S}$ so the Stone and Priestley prime ideal topologies coincide. We will show in a moment that the Priestley prime ideal space of a bounded distributive lattice L is an ordered Boolean space. However, we leave it as an exercise to show that for non-Boolean distributive lattices, the Stone prime ideal space is not Hausdorff and so it is not a Boolean space. This is why the Priestley prime ideal topology is more suitable to the study of arbitrary bounded distributive lattices.

Theorem 10.9

1) *The Priestley prime ideal space* $(\mathrm{Spec}(L), \tau, \subseteq)$ *of a bounded distributive lattice L is an ordered Boolean space.*
2) *The Stone prime ideal space* $(\mathrm{Spec}(B), \tau)$ *of a Boolean algebra* B *is a Boolean space.*

Proof. For part 1), to see that τ is totally order-separated, suppose that $Q \not\subseteq P$. Then there is an $a \in Q \setminus P$ and so $P \in \mathcal{P}_{\neg a}$ and $Q \notin \mathcal{P}_{\neg a}$. Since $\mathcal{P}_{\neg a} = \mathcal{P}_{\neg a,0}$ is a clopen down-set, τ is totally order-separated.

To prove compactness, the Alexander subbasis lemma (Theorem A1.4) implies that it is sufficient to show that any cover $\mathcal{U}$ of $\mathrm{Spec}(L)$ by subbasis elements has a finite subcover. Suppose that

$$\mathcal{U} = \{\mathcal{P}_{\neg d} \mid d \in D\} \cup \{\mathcal{P}_e \mid e \in E\}$$

is a cover of $\mathrm{Spec}(L)$ by subbasic elements. To insure that D and E are nonempty, we can assume that $0 \in D$ since $\mathcal{P}_{\neg 0} = \emptyset$ and $1 \in E$ since $\mathcal{P}_1 = \emptyset$.

Thus, if P is a prime ideal, then $d \notin P$ for some $d \in D$ or $e \in P$ for some $e \in E$. Put another way, there is no prime ideal containing D and disjoint from E. Equivalently, there is no prime ideal containing $(D]$ and disjoint from $[E)$. But then the prime separation theorem (Theorem 8.5) implies that $(D]$ and $E)$ cannot be disjoint and so there is an $x \in (D] \cap [E)$. Hence, there exist $d_1, \ldots, d_n \in D$ and $e_1, \ldots, e_m \in E$ for which

$$e_1 \wedge \cdots \wedge e_m \leq x \leq d_1 \vee \cdots \vee d_n$$

If $P \notin \mathcal{P}_{e_i}$ for all i, then $P \in \mathcal{P}_{\neg e_i}$ for all i and so

$$P \in \bigcap \mathcal{P}_{\neg e_i} = \mathcal{P}_{\neg(e_1 \wedge \cdots \wedge e_m)} \subseteq \mathcal{P}_{\neg(d_1 \vee \cdots \vee d_n)} = \bigcup \mathcal{P}_{\neg d_j}$$

and so $P \in \mathcal{P}_{\neg d_j}$ for some j. This shows that the family

$$\{\mathcal{P}_{e_1}, \ldots, \mathcal{P}_{e_m}, \mathcal{P}_{\neg d_1}, \ldots, \mathcal{P}_{\neg d_n}\}$$

is a finite subcover of $\mathcal{U}$.

For part 2), if B is a Boolean algebra, then the Stone prime ideal space $\mathrm{Spec}(B)$ is the same as with the Priestley prime ideal space $\mathrm{Spec}(B)$ and so it is an ordered Boolean space. However, since $\mathrm{Spec}(B)$ is the family of all maximal ideals of B, it is an antichain and so the partial order on $\mathrm{Spec}(B)$ is equality. Hence,

$$\mathcal{O}^{\text{clo}}(\text{Spec}(B)) = \wp^{\text{clo}}(\text{Spec}(B))$$

and $\text{Spec}(B)$ is a Boolean space.□

Now we can prove that the image of the rho map $\rho\colon L \hookrightarrow \mathcal{O}(\text{Spec}(L))$ is the set $\mathcal{O}^{\text{clo}}(\text{Spec}(L))$ of clopen down-sets, that is, all clopen down-sets have the form $\mathcal{P}_{\neg a}$ for some $a \in L$.

Theorem 10.10 *Let L be a bounded distributive lattice. Let*

$$\mathcal{B} = \{\mathcal{P}_{\neg a,b} \mid a, b \in L\}$$

be the basis for the Priestley prime ideal space $\text{Spec}(L)$.

1) **(Clopen sets)** *The clopen sets $\wp^{\text{clo}}(\text{Spec}(L))$ are precisely the finite unions of basis elements $\mathcal{P}_{\neg a,b}$.*

2) **(Clopen down-sets)** *The clopen down-sets are*

$$\mathcal{O}^{\text{clo}}(\text{Spec}(L)) = \mathcal{S} = \{P_{\neg a} \mid a \in L\}$$

and so the rho map

$$\rho\colon L \to \mathcal{O}^{\text{clo}}(\text{Spec}(L))$$

is surjective and therefore a lattice isomorphism.

Proof. For part 1), any finite union of the clopen basic sets is a clopen set. Conversely, if K is clopen, then it is a union of basic sets. However, K is also compact and so it is a finite union of basic sets.

For part 2), let $I \in D$. For each $J \notin D$, we have $J \not\subseteq I$ and so if we take $x_J \in J \setminus I$, then $I \in \mathcal{P}_{\neg x_J}$ and $J \in \mathcal{P}_{x_J}$. The family $\{\mathcal{P}_{x_J} \mid J \notin D\}$ is an open cover of the compact set D^c and so it has a finite subcover $\{\mathcal{P}_{x_1}, \ldots, \mathcal{P}_{x_n}\}$, where we have simplified the subscripts a bit. Hence,

$$D^c \subseteq \mathcal{P}_{x_1} \cup \cdots \cup \mathcal{P}_{x_n}$$

and so

$$I \in \mathcal{P}_{\neg x_1} \cap \cdots \cap \mathcal{P}_{\neg x_n} \subseteq D$$

Thus, D is the union of sets $\mathcal{P}_{\neg x_1} \cap \cdots \cap \mathcal{P}_{\neg x_n} = \mathcal{P}_{\neg(x_1 \wedge \cdots \wedge x_n)}$ in $\mathcal{S}$. Since D is compact, it is the union of finitely many of these sets and so

$$D = \mathcal{P}_{\neg a_1} \cup \cdots \cup \mathcal{P}_{\neg a_n} = \mathcal{P}_{\neg(a_1 \vee \cdots \vee a_n)}$$ □

The Priestley Duality

The **Priestley duality**, shown in Figure 10.3,

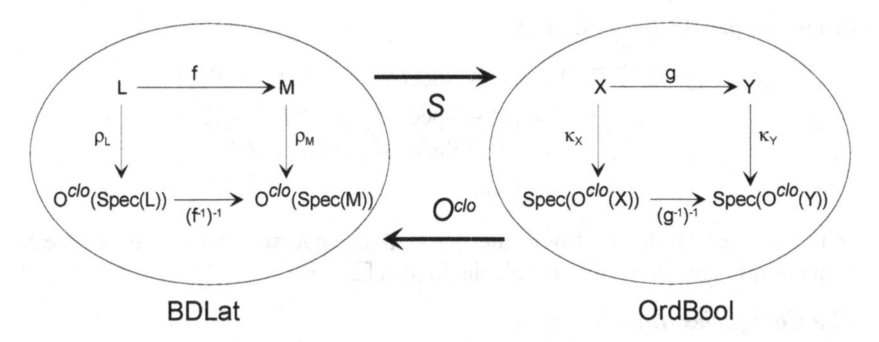

Figure 10.3

is a duality between the category **BDLat** of bounded distributive lattices, together with lattice $\{0,1\}$-homomorphisms and the category **OrdBool** of ordered Boolean spaces with continuous monotone maps. We begin our discussion with a preliminary result. For an ordered Boolean space X, let

$$\kappa(x) = \mathcal{K}_{\neg x} := \{D \in \mathcal{O}^{\text{clo}}(X) \mid x \notin D\}$$

Theorem 10.11

1) *If $f: L \to M$ is a lattice $\{0,1\}$-homomorphism, then the restricted induced inverse map*

$$f^{-1}: \text{Spec}(M) \to \text{Spec}(L)$$

satisfies, for $a, b \in L$,

$$\begin{aligned}(f^{-1})^{-1}(\mathcal{P}_{\neg a}) &= \mathcal{P}_{\neg f(a)} \\ (f^{-1})^{-1}(\mathcal{P}_{a}) &= \mathcal{P}_{f(a)} \\ (f^{-1})^{-1}(\mathcal{P}_{\neg a,b}) &= \mathcal{P}_{\neg f(a),f(b)}\end{aligned} \tag{10.12}$$

2) *If $g: X \to Y$ is a continuous monotone map, then the restricted induced inverse map*

$$g^{-1}: \mathcal{O}^{\text{clo}}(Y) \to \mathcal{O}^{\text{clo}}(X)$$

satisfies

$$\begin{aligned}(g^{-1})^{-1}(\mathcal{K}_{\neg a}) &= \mathcal{K}_{\neg g(a)} \\ (g^{-1})^{-1}(\mathcal{K}_{a}) &= \mathcal{K}_{g(a)} \\ (g^{-1})^{-1}(\mathcal{K}_{\neg a} \cap \mathcal{K}_b) &= \mathcal{K}_{\neg g(a)} \cap \mathcal{K}_{g(b)}\end{aligned} \tag{10.13}$$

Proof. For part 1), if $a \in L$, then

$$\begin{aligned}(f^{-1})^{-1}(\mathcal{P}_{\neg a}) &= \{P \in \operatorname{Spec}(M) \mid f^{-1}(P) \in \mathcal{P}_{\neg a}\} \\ &= \{P \in \operatorname{Spec}(M) \mid a \notin f^{-1}(P)\} \\ &= \{P \in \operatorname{Spec}(M) \mid f(a) \notin P\} \\ &= \mathcal{P}_{\neg f(a)}\end{aligned}$$

and the rest follows from the fact that inverse maps are Boolean homomorphisms. Part 2) is entirely analogous.□

The Categories and Functors

Let **BDLat** be the category of bounded distributive lattices, together with lattice $\{0,1\}$-homomorphisms. Let **OrdBool** be the category of ordered Boolean spaces with continuous monotone maps. The relevant functors are described in the next theorem.

Theorem 10.14 *The following are contravariant functors.*
1) $\mathcal{S}$: **BDLat** $\Rightarrow$ **OrdBool** *defined by*

$$\mathcal{S}(L) = \operatorname{Spec}(L) \quad \textit{and} \quad \mathcal{S}(f) = f^{-1}: \operatorname{Spec}(M) \to \operatorname{Spec}(L)$$

for any $L \in$ **BDLat** *and any* $\{0,1\}$*-homomorphism* $f: L \to M$.
2) $\mathcal{O}^{\text{clo}}$: **OrdBool** $\Rightarrow$ **BDLat** *defined by*

$$\mathcal{O}^{\text{clo}}(X) = \mathcal{O}^{\text{clo}}(X) \quad \textit{and} \quad \mathcal{O}^{\text{clo}}(g) = g^{-1}: \mathcal{O}^{\text{clo}}(Y) \to \mathcal{O}^{\text{clo}}(X)$$

for any $X \in$ **OrdBool** *and any continuous monotone map* $g: X \to Y$.
Proof. For part 1), it is clear that $\mathcal{S}(f) = f^{-1}$ is monotone. Also, f^{-1} is continuous since Theorem 10.11 implies that the inverse image $(f^{-1})^{-1}(\mathcal{P}_{\neg a,b})$ of a basic open set $\mathcal{P}_{\neg a,b}$ is open. Finally, if $f: L \to M$ and $g: M \to N$, then

$$\mathcal{S}(g \circ f) = (g \circ f)^{-1} = f^{-1} \circ g^{-1} = \mathcal{S}(f) \circ \mathcal{S}(g)$$

and $\mathcal{S}(\iota) = \iota$. Hence, $\mathcal{S}$ is a contravariant functor.

For part 2), $\mathcal{O}^{\text{clo}}(f) = f^{-1}$ is a lattice homomorphism and

$$f^{-1}(\emptyset) = \emptyset \quad \text{and} \quad f^{-1}(Y) = X$$

Also, if $f: X \to Y$ and $g: Y \to Z$, then

$$\mathcal{O}^{\text{clo}}(g \circ f) = (g \circ f)^{-1} = f^{-1} \circ g^{-1} = \mathcal{O}^{\text{clo}}(f) \circ \mathcal{O}^{\text{clo}}(g)$$

and $\mathcal{O}^{\text{clo}}(\iota) = \iota$. Hence, $\mathcal{O}^{\text{clo}}$ is a contravariant functor.□

The Priestley and Stone Representations

On the lattice side, Theorem 10.10 implies that the rho map

$$\rho: L \approx \mathcal{O}^{\mathrm{clo}}(\mathrm{Spec}(L))$$

is a lattice isomorphism from a bounded distributive lattice L to the family of all clopen down-sets of a certain ordered Boolean space, namely, the Priestley prime ideal space $\mathrm{Spec}(L)$.

On the ordered Boolean space side, for a given ordered Boolean space X, let

$$\mathcal{K}_{\neg x} = \{D \in \mathcal{O}^{\mathrm{clo}}(X) \mid x \notin D\}$$

It is clear that $\mathcal{K}_{\neg x} \in \mathrm{Spec}(\mathcal{O}^{\mathrm{clo}}(X))$. We define the **kappa function** $\kappa: X \to \mathrm{Spec}(\mathcal{O}^{\mathrm{clo}}(X))$ by

$$\kappa(x) = \mathcal{K}_{\neg x}$$

We would like to show that κ is an order-homeomorphism, where $\mathrm{Spec}(\mathcal{O}^{\mathrm{clo}}(X))$ has the Priestley prime ideal topology, that is, the sets

$$\mathcal{P}_{\neg D,E} = \{\mathbb{I} \in \mathrm{Spec}(\mathcal{O}^{\mathrm{clo}}(X)) \mid D \notin \mathbb{I}, E \in \mathbb{I}\}$$

for $D, E \in \mathcal{O}^{\mathrm{clo}}(X)$, form a basis for the topology. Theorem A1.6 implies that it is sufficient to show that κ is a continuous order isomorphism.

To see that κ is an order embedding, note that the order in X is determined by the clopen down-sets and so

$$x \le y \quad \Leftrightarrow \quad \mathcal{K}_y \subseteq \mathcal{K}_x \quad \Leftrightarrow \quad \mathcal{K}_{\neg x} \subseteq \mathcal{K}_{\neg y} \quad \Leftrightarrow \quad \kappa(x) \subseteq \kappa(y)$$

To show that κ is continuous, it is sufficient to show that the inverse image of any basic open set is open. But

$$\begin{aligned}\kappa^{-1}(\mathcal{P}_{\neg D,E}) &= \{x \in X \mid \kappa(x) \in \mathcal{P}_{\neg D,E}\} \\ &= \{x \in X \mid \mathcal{K}_{\neg x} \in \mathcal{P}_{\neg D,E}\} \\ &= \{x \in X \mid D \notin \mathcal{K}_{\neg x} \text{ and } E \in \mathcal{K}_{\neg x}\} \\ &= \{x \in X \mid x \in D \text{ and } x \notin E\} \\ &= D \setminus E\end{aligned}$$

and since D and E are clopen, so is $D \setminus E$.

Finally, to see that κ is surjective, the image $\kappa(X)$ is compact and therefore closed. We show that any nonempty basic open set meets $\kappa(X)$, which implies that $\kappa(X) = \mathrm{Spec}(\mathcal{O}^{\mathrm{clo}}(X))$. For a nonempty proper basic open set

$$\mathcal{P}_{\neg D,E} = \{\mathbb{I} \in \mathrm{Spec}(\mathcal{O}^{\mathrm{clo}}(X)) \mid D \notin \mathbb{I}, E \in \mathbb{I}\}$$

we have $D \not\subseteq E$ and so there is an $x \in D \setminus E$. But $x \in D$ implies that $D \notin \mathcal{K}_{\neg x}$ and $x \notin E$ implies that $E \in \mathcal{K}_{\neg x}$ and so $\mathcal{K}_{\neg x} \in \mathcal{P}_{\neg D,E}$.

The fact that the individual rho and kappa maps are isomorphisms gives representation theorems for bounded distributive lattices and ordered Boolean spaces.

Theorem 10.15

1) **(Priestley representation theorem for bounded distributive lattices)** *Let L be a bounded distributive lattice. The rho map $\rho: L \to \mathcal{O}^{\text{clo}}(\text{Spec}(L))$ defined by*

$$\rho(a) = \mathcal{P}_{\neg a}$$

is a lattice isomorphism and so

$$\rho: L \approx \mathcal{O}^{\text{clo}}(\text{Spec}(L))$$

Thus, every bounded distributive lattice is isomorphic to the power set sublattice of clopen down-sets of a Priestley prime ideal space, namely, $\text{Spec}(L)$.

2) Let X be an ordered Boolean space. The kappa map $\kappa: X \to \text{Spec}(\mathcal{O}^{\text{clo}}(X))$ defined by

$$\kappa(x) = \mathcal{K}_{\neg x}$$

is an order-homeomorphism and so

$$\kappa: X \approx \text{Spec}(\mathcal{O}^{\text{clo}}(X))$$

Thus, every ordered Boolean space is order-homeomorphic to the Priestley prime ideal space of a bounded distributive lattice, namely, $\mathcal{O}^{\text{clo}}(X)$.□

The Natural Isomorphisms

We have seen that the families

$$\rho = \{\rho_L: L \approx \mathcal{O}^{\text{clo}}(\text{Spec}(L)) \mid L \text{ is a bounded distributive lattice}\}$$

and

$$\kappa = \{\kappa_X: X \approx \text{Spec}(\mathcal{O}^{\text{clo}}(X)) \mid X \text{ is an ordered Boolean space}\}$$

are families of lattice isomorphisms and order-homeomorphisms, respectively. In fact, ρ and κ are natural isomorphisms, in symbols,

$$\rho: I_{\mathbf{BDLat}} \mathrel{\dot{\approx}} \mathcal{O}^{\text{clo}} \circ \mathcal{S} \quad \text{and} \quad \kappa: I_{\mathbf{OrdBool}} \mathrel{\dot{\approx}} \mathcal{S} \circ \mathcal{O}^{\text{clo}}$$

and so the categories **BDLat** and **OrdBool** are dual. Indeed, the first equation in (10.12) can be written in the form

$$(f^{-1})^{-1} \circ \rho_L = \rho_M \circ f$$

and the first equation in (10.13) can be written

$$(g^{-1})^{-1} \circ \kappa_X = \kappa_Y \circ g$$

and these equations are precisely the defining conditions for natural transformations, as shown in Figure 10.3, which we repeat here in Figure 10.4.

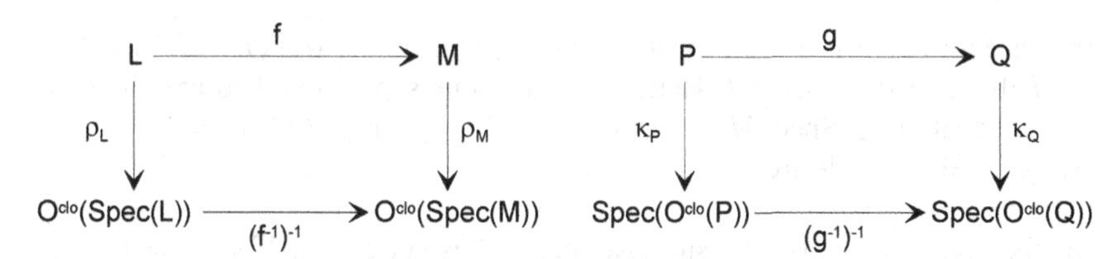

Figure 10.4

Theorem 10.16 (The Priestley duality)

1) *The families*

$$\rho = \{\rho_L\}: I \dot{\rightarrow} \mathcal{O}^{\text{clo}} \circ \mathcal{S} \quad \text{and} \quad \kappa = \{\kappa_X\}: I \dot{\rightarrow} \mathcal{S} \circ \mathcal{O}^{\text{clo}}$$

are natural isomorphisms and so **BDLat** *and* **OrdBool** *are dual categories. In particular,*

$$(f^{-1})^{-1} \circ \rho_L = \rho_M \circ f$$

for any lattice $\{0,1\}$*-homomorphism* $f: L \to M$ *and*

$$(g^{-1})^{-1} \circ \kappa_X = \kappa_Y \circ g$$

for any continuous monotone map $g: X \to Y$.

2) *The functors* $\mathcal{S}$ *and* $\mathcal{O}^{\text{clo}}$ *are full and faithful, that is, the maps*

$$\mathcal{S}: \text{hom}_{\mathbf{BDLat}}(L, M) \to \text{hom}_{\mathbf{OrdBool}}(\text{Spec}(L), \text{Spec}(M))$$

and

$$\mathcal{O}^{\text{clo}}: \text{hom}_{\mathbf{OrdBool}}(X, Y) \to \text{hom}_{\mathbf{BDLat}}(\mathcal{O}^{\text{clo}}(X), \mathcal{O}^{\text{clo}}(Y))$$

are bijections.□

We can say a bit more about the connection between a lattice $\{0,1\}$-homomorphism $f: L \to M$ and its image $\mathcal{S}(f) = f^{-1}$ under the spectrum functor.

Theorem 10.17 *Let* $f: L \to M$ *be a lattice* $\{0,1\}$*-homomorphism, where* L *and* M *are bounded distributive lattices.*

1) f *is injective if and only if* $\mathcal{S}(f) = f^{-1}$ *is surjective.*

2) *f is surjective if and only if $S(f)=f^{-1}$ is an order embedding.*

Proof. For part 1), if $f:L\hookrightarrow M$ is injective, then $f:L\approx f(L)$ is an isomorphism and so $f^{-1}:\mathrm{Spec}(f(L))\to\mathrm{Spec}(L)$ is surjective. Hence, if $P\in\mathrm{Spec}(L)$, then there is an $R\in\mathrm{Spec}(f(L))$ for which $f^{-1}(R)=P$. We need only show that there is a $Q\in\mathrm{Spec}(M)$ for which $Q\cap f(L)=R$, since then

$$f^{-1}(Q)=f^{-1}(Q\cap f(L))=f^{-1}(R)=P$$

To this end, $I=\downarrow_M R$ is a proper ideal of M and $F=\uparrow_M(f(L)\setminus R)$ is a filter of M that is disjoint from I. Hence, by the prime separation theorem, there is a prime ideal $Q\in\mathrm{Spec}(M)$ for which $R\subseteq Q$ and $Q\cap F=\emptyset$, that is, $Q\cap f(L)=R$, as desired.

For the converse to part 1), suppose that f^{-1} is surjective. If $f(a)=f(b)$ for $b\not\leq a$, then the prime separation theorem for points implies that there is a $P\in\mathrm{Spec}(L)$ such that $a\in P$ and $b\notin P$. Since f^{-1} is surjective, there is a $Q\in\mathrm{Spec}(M)$ such that $f^{-1}(Q)=P$. But

$$f(b)=f(a)\in f(P)\subseteq Q$$

implies that $b\in f^{-1}(Q)$, which is false. Hence, f is injective.

For part 2), if f is surjective, then

$$A\subseteq B\quad\Leftrightarrow\quad f^{-1}(A)\subseteq f^{-1}(B)$$

and so f^{-1} is an order embedding. For the converse, if f^{-1} is an order embedding, then for any prime ideals P and Q of M,

$$\begin{aligned}P\subseteq Q\quad&\Leftrightarrow\quad f^{-1}(P)\subseteq f^{-1}(Q)\\&\Leftrightarrow\quad f^{-1}(P\cap f(L))\subseteq f^{-1}(Q)\\&\Leftrightarrow\quad P\cap f(L)\subseteq Q\end{aligned}$$

Thus, assuming that f is not surjective, we can obtain a contradiction by finding prime ideals P and Q for which $P\not\subseteq Q$ but $P\cap f(L)\subseteq Q$.

Since f is not surjective, there is an $a\in M\setminus f(L)$ and we need only show that there are prime ideals P and Q for which

$$a\in P,\quad a\notin Q\quad\text{and}\quad P\cap f(L)\subseteq Q$$

or equivalently,

$$\downarrow a\subseteq P,\quad \uparrow a\cap Q=\emptyset\quad\text{and}\quad P\cap f(L)\subseteq Q$$

Now, if $P\cap f(L)$ and $\uparrow a$ are disjoint, then there is such a prime ideal Q, so it is sufficient to show that there is a prime ideal P for which

$$\downarrow a \subseteq P \quad \text{and} \quad (P \cap f(L)) \cap \uparrow a = \emptyset$$

But $\downarrow a$ is disjoint from $f(L) \cap \uparrow a$, since $a \notin f(L)$ and so P also exists.□

The Case of Boolean Algebras

The Priestley duality between bounded distributive lattices and ordered Boolean spaces can easily be restricted to a duality between the category **BoolAlg** of Boolean algebras with Boolean homomorphisms and the category **Bool** of Boolean algebras with continuous maps, as shown in Figure 10.5.

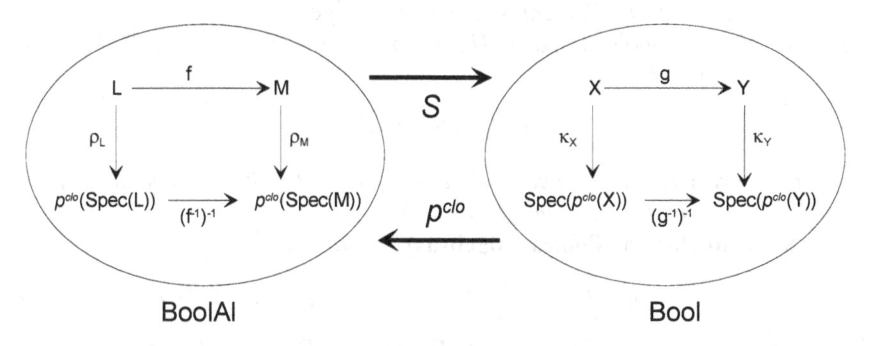

Figure 10.5

First, we note that **BoolAlg** is a subcategory of **BDLat**. Also, we have seen that any Boolean space can be thought of as an ordered Boolean space under equality. Thus, since all maps are monotone in this case, it follows that **Bool** is a subcategory of **OrdBool**.

As to the functors, the spectrum functor sends a Boolean algebra B to the Stone prime ideal space $\text{Spec}(B)$, which is a Boolean space according to Theorem 10.9. Also, $\mathcal{S}(f) = f^{-1}$ is a Boolean homomorphism.

As to the down-set functor, in a Boolean space X the order is equality and so all sets are down-sets, that is, $\mathcal{O}^{\text{clo}}(X) = \wp^{\text{clo}}(X)$ and so the restriction of the functor $\mathcal{O}^{\text{clo}}$ to **Bool** is the functor $\wp^{\text{clo}}$, sending X to $\wp^{\text{clo}}(X)$ and sending a continuous map $g: X \to Y$ to the induced inverse map $g^{-1}: \wp^{\text{clo}}(Y) \to \wp^{\text{clo}}(X)$.

Thus, we have the two contravariant functors

$$\mathcal{S}: \textbf{BoolAlg} \Rightarrow \textbf{Bool} \quad \text{and} \quad \wp^{\text{clo}}: \textbf{Bool} \Rightarrow \textbf{BoolAlg}$$

Moreover, we have seen that the $\{0,1\}$-homomorphisms ρ_L are Boolean isomorphisms when L is a Boolean algebra. Also, the kappa maps κ_X are homeomorphisms of Boolean spaces.

Thus, we arrive at the duality in Figure 10.5.

Theorem 10.18

1) **(Stone representation theorem for Boolean algebras)** *Let B be a Boolean algebra. The rho map $\rho: B \to \wp^{\text{clo}}(\text{Spec}(B))$ defined by*

$$\rho(a) = \mathcal{P}_{\neg a}$$

is a Boolean algebra isomorphism and so

$$\rho: B \approx \wp^{\text{clo}}(\text{Spec}(B))$$

Thus, every Boolean algebra is isomorphic to the Boolean algebra of clopen subsets of a Boolean space, namely, $\text{Spec}(B)$.

2) *Let X be a Boolean space. The kappa map $\kappa: X \to \text{Spec}(\wp^{\text{clo}}(X))$ is a homeomorphism and so*

$$\kappa: X \approx \text{Spec}(\wp^{\text{clo}}(X))$$

Thus, every Boolean space is homeomorphic to the Stone prime ideal space of a Boolean algebra, namely, $\wp^{\text{clo}}(X)$.

3) **(Stone duality for Boolean algebras)** *The families*

$$\rho = \{\rho_L\}: I \dashrightarrow \wp^{\text{clo}} \circ \mathcal{S} \quad \textit{and} \quad \kappa = \{\kappa_X\}: I \dashrightarrow \mathcal{S} \circ \wp^{\text{clo}}$$

are natural isomorphisms and so **BoolAl** *and* **Bool** *are dual categories. In particular,*

$$(f^{-1})^{-1} \circ \rho_L = \rho_M \circ f$$

for any Boolean homomorphism $f: L \to M$ and

$$(g^{-1})^{-1} \circ \kappa_X = \kappa_Y \circ g$$

for any continuous map $g: X \to Y$.

4) *The functors $\mathcal{S}$ and $\wp^{\text{clo}}$ are full and faithful, that is, the maps*

$$\mathcal{S}: \text{hom}_{\textbf{BoolAl}}(L, M) \to \text{hom}_{\textbf{Bool}}(\text{Spec}(L), \text{Spec}(M))$$

and

$$\wp^{\text{clo}}: \text{hom}_{\textbf{Bool}}(X, Y) \to \text{hom}_{\textbf{BoolAl}}(\wp^{\text{clo}}(X), \wp^{\text{clo}}(Y))$$

are bijections. □

Note that in a Boolean prime ideal space $\text{Spec}(B)$, the clopen sets are precisely the sets $\mathcal{P}_{\neg a}$. Also, the open sets are precisely the unions of clopen sets and the closed sets are precisely the intersections of clopen sets.

Applications

According to the Priestley duality, we may think of any bounded distributive lattice L as the family $\mathcal{O}^{\text{clo}}(X)$ of clopen down-sets of a Priestley prime ideal space $X = \text{Spec}(L)$. What is the advantage of doing this? The space X is, in some sense, simpler than the lattice $\mathcal{O}^{\text{clo}}(X)$ of clopen down-sets. Thus, many lattice-theoretic concepts may have simpler descriptions in topological terms than in lattice terms.

For instance, the Priestley representation makes it easy to show that any bounded distributive lattice L is a sublattice of a Boolean algebra and to describe, up to isomorphism, a minimal such Boolean algebra. We first identify L with its faithful representation

$$\rho(L) = \{\mathcal{P}_{\neg a}\} = \mathcal{O}^{\text{clo}}(\text{Spec}(L))$$

Now, $\mathcal{O}^{\text{clo}}(\text{Spec}(L))$ is a sublattice of the Boolean algebra $\wp^{\text{clo}}(\text{Spec}(L))$. Moreover, if $\mathcal{B}$ is a Boolean subalgebra of $\wp^{\text{clo}}(\text{Spec}(L))$ containing $\mathcal{O}^{\text{clo}}(\text{Spec}(L))$, then for any basic clopen set $\mathcal{P}_{\neg a,b}$ of $\text{Spec}(L)$, we have

$$\mathcal{P}_{\neg a,b} = \mathcal{P}_{\neg a} \cap \mathcal{P}_b = \mathcal{P}_{\neg a} \cap (\mathcal{P}_{\neg b})^c \in \mathcal{B}$$

and so $\mathcal{P}_{\neg a,b} \in \mathcal{B}$ for all $a, b \in L$. Then Theorem 10.10 implies that any element of $\wp^{\text{clo}}(\text{Spec}(L))$, being a *finite* union of the basic clopen sets $\mathcal{P}_{\neg a,b}$, belongs to $\mathcal{B}$ and so

$$\mathcal{B} = \wp^{\text{clo}}(\text{Spec}(L))$$

Thus, $\wp^{\text{clo}}(\text{Spec}(L))$ is a minimal Boolean algebra containing $\mathcal{O}^{\text{clo}}(\text{Spec}(L))$ as a sublattice.

Theorem 10.19 *If L is a bounded distributive lattice, then $\wp^{\text{clo}}(\text{Spec}(L))$ is a minimal Boolean algebra containing the isomorphic copy $\mathcal{O}^{\text{clo}}(\text{Spec}(L))$ of L as a sublattice.* □

The Priestley representation can be used to prove certain lattice-theoretic results, such as the result above, and to describe certain lattice-theoretic notions via the topological notions. Here are some additional examples.

Recall that the **pseudocomplement** a^* of $a \in L$ is defined to be the maximum semicomplement of a, that is,

$$a^* = \max\{x \in L \mid a \wedge x = 0\}$$

if the maximum exists.

We will need the following facts, discussed in an earlier exercise. Two lattice $\{0,1\}$-epimorphisms $f\colon L \to M$ and $g\colon L \to N$ with domain L are **kernel equivalent**, written $f \equiv g$, if the corresponding congruence kernels θ_f and θ_g are equal, that is, if

$$f(a) = f(b) \quad \Leftrightarrow \quad g(a) = g(b)$$

This is easily seen to be an equivalence relation on the class $\text{Epi}(L)$ of $\{0,1\}$-epimorphisms with domain L.

Theorem 10.20 *Let* $(X, \tau, \leq)$ *be an ordered Boolean space, with clopen down-sets* $\mathcal{O}^{\text{op}}(X)$.

1) **(Ideals of $\mathcal{O}^{\text{clo}}(X)$)** *The union map*

$$\mathcal{U}: \mathcal{I}(\mathcal{O}^{\text{clo}}(X)) \approx \mathcal{O}^{\text{op}}(X)$$

defined by

$$\mathcal{U}(\mathbb{I}) = \bigcup \mathbb{I}$$

is a lattice isomorphism and so $\mathcal{O}^{\text{op}}(X)$ *is isomorphic to its own ideal lattice. Thus, any bounded distributive lattice is isomorphic to its ideal lattice.*

2) **(Pseudocomplements in $\mathcal{O}^{\text{clo}}(X)$)** *The pseudocomplemented elements* U *in* $\mathcal{O}^{\text{clo}}(X)$ *are precisely the elements* U *for which* $\uparrow U$ *is clopen, in which case* $U^* = X \setminus \uparrow U$.

3) **(Congruence relations on $\mathcal{O}^{\text{clo}}(X)$)** *The lattice* $\text{Con}(\mathcal{O}^{\text{clo}}(X))$ *of congruence relations on* $\mathcal{O}^{\text{clo}}(X)$ *is isomorphic to the lattice of open subsets of* X.

Proof. We prove the third statement, leaving the others for the exercises. Let $L = \mathcal{O}^{\text{clo}}(X)$. Note that $\{x\}$ is a closed set for all $x \in X$, since X is Hausdorff. Also, $\uparrow x$ and $\downarrow x$ are closed by Theorem 10.7.

The plan is to find a bijection

$$\Gamma(X) \leftrightarrow \text{Con}(L)$$

from the family $\Gamma(X)$ of closed sets of X onto the set $\text{Con}(L)$. Then we will show that this bijection is a lattice anti-isomorphism. We can then compose this map with the complement map to get a lattice isomorphism from the open sets onto $\text{Con}(L)$.

To find the bijection mentioned above, we first find a complete system of distinct representatives $\{f_K \mid K \in \Gamma(X)\}$ for the kernel equivalence classes of $\text{Epi}(L)$, indexed by the family $\Gamma(X)$ of closed sets. Then the map

$$K \to f_K \to \theta_{f_K}$$

where θ_g is the congruence kernel of g, is a bijection from $\Gamma(X)$ to $\text{Con}(L)$, since every congruence relation θ is a congruence kernel, namely, the congruence kernel of the projection π_θ.

For each closed set K (with the subspace topology), consider the intersection by K map

$$f_K : \mathcal{O}^{\text{clo}}(X) \to \mathcal{O}^{\text{clo}}(K)$$

defined by

$$f_K(D) = D \cap K$$

for any $D \in \mathcal{O}^{\text{clo}}(X)$. This map is a lattice $\{0,1\}$-homomorphism and we leave it as an exercise to show that f_K is surjective, that is, $f_K \in \text{Epi}(L)$.

We claim that the maps $\{f_K \mid K \in \mathcal{O}^{\text{clo}}(X)\}$ form a complete system of distinct representatives for $\text{Epi}(L)/\equiv$. If $K_1 \neq K_2$ are distinct closed subsets of X, then f_{K_1} and f_{K_2} are not kernel equivalent. To see this, suppose that $K_1 \not\subseteq K_2$ and let $x \in K_1 \setminus K_2$. We show that there are $U, V \in \mathcal{O}^{\text{clo}}(X)$ for which

$$f_{K_2}(U) = f_{K_2}(V) \quad \text{and} \quad f_{K_1}(U) \neq f_{K_1}(V)$$

that is,

$$U \cap K_2 = V \cap K_2 \quad \text{and} \quad U \cap K_1 \neq V \cap K_1$$

If we find U and V for which $x \notin U$ and $x \in V$, then $x \notin U \cap K_1$ and $x \in V \cap K_1$ and so $U \cap K_1 \neq V \cap K_1$. Thus, it is sufficient to show that

$$x \notin U, \quad x \in V \quad \text{and} \quad U \cap K_2 = V \cap K_2$$

or, equivalently,

$$\uparrow x \cap U = \emptyset, \quad \downarrow x \subseteq V \quad \text{and} \quad U \cap K_2 = V \cap K_2 \tag{10.21}$$

Now, about the only thing we can separate is

$$K_2 \cap \downarrow x \quad \text{and} \quad \uparrow x$$

or

$$K_2 \cap \uparrow x \quad \text{and} \quad \downarrow x$$

since $x \notin K_2$. Choosing the former, there is a clopen down-set U for which

$$K_2 \cap \downarrow x \subseteq U \quad \text{and} \quad U \cap \uparrow x = \emptyset$$

If we can find a clopen down-set V that contains U as well as $\downarrow x$, then $U \cap K_2 \subseteq V \cap K_2$ and we only arrange it so that $V \cap K_2 \subseteq U$, or equivalently, that $V \cap (K_2 \setminus U) = \emptyset$. So we want to separate $U \cup \downarrow x$ from $K_2 \setminus U$, which is possible since no element of $K_2 \setminus U$ is less than or equal to any element of $\downarrow x \cup U$. Hence, there is a clopen down-set V for which

$$\downarrow x \cup U \subseteq V \quad \text{and} \quad V \cap (K_2 \setminus U) = \emptyset$$

and so $U \cap K_2 = V \cap K_2$. Thus, (10.21) holds.

It follows that for closed subsets K_1 and K_2 of X,

$$K_1 = K_2 \quad \Leftrightarrow \quad f_{K_1} \equiv f_{K_2}$$

and so each kernel equivalence class has at most one member of the form f_K.

Now suppose that $f: \mathcal{O}^{\text{clo}}(X) \to \mathcal{O}^{\text{clo}}(Z)$ is a $\{0,1\}$-epimorphism, where Z is an ordered Boolean space. Since the functor $\mathcal{O}^{\text{clo}}$ is full, there is a continuous monotone map $h: Z \to X$ for which $h^{-1} = f$. Since f is surjective, Theorem 10.17 implies that h is a continuous order embedding. Now, for $D, E \in \mathcal{O}^{\text{clo}}(X)$, we have

$$\begin{aligned} f(D) = f(E) &\Leftrightarrow h^{-1}(D) = h^{-1}(E) \\ &\Leftrightarrow h^{-1}(D \cap h(Z)) = h^{-1}(E \cap h(Z)) \end{aligned}$$

But $h: Z \to h(Z)$ is an order isomorphism and so the last equation above holds if and only if

$$D \cap h(Z) = E \cap h(Z)$$

Finally, since h is continuous and Z is compact, the image $h(Z)$ is compact and therefore closed in X. Thus, we have shown that

$$f(D) = f(E) \quad \Leftrightarrow \quad f_{h(Z)}(D) = f_{h(Z)}(E)$$

where $h(Z)$ is closed in X. Thus, any $\{0,1\}$-epimorphism f is equivalent to a $\{0,1\}$-epimorphism of the form f_K for some closed set K in X.

This shows that the family $\{f_K \mid K \in \Gamma(X)\}$ is a complete system of distinct representatives for kernel equivalence. Hence, the map $\Pi: \Gamma(X) \to \text{Con}(L)$ defined by

$$\Pi(K) = \theta_{f_K}$$

is a bijection. Moreover, if $K_1 \subseteq K_2$ then

$$U \cap K_2 = V \cap K_2 \quad \Rightarrow \quad U \cap K_1 = V \cap K_1$$

that is,

$$f_{K_2}(U) = f_{K_2}(V) \quad \Rightarrow \quad f_{K_1}(U) = f_{K_1}(V)$$

But we have already shown that if $K_1 \not\subseteq K_2$, then there are clopen down-sets U and V for which

$$U \cap K_2 = V \cap K_2 \quad \text{and} \quad U \cap K_1 \neq V \cap K_1$$

and so

$$f_{K_2}(U) = f_{K_2}(V) \quad \not\Rightarrow \quad f_{K_1}(U) = f_{K_1}(V)$$

Thus,

$$(f_{K_2}(U) = f_{K_2}(V) \Rightarrow f_{K_1}(U) = f_{K_1}(V)) \quad \Leftrightarrow \quad K_1 \subseteq K_2$$

In terms of congruence relations, this is

$$\theta_{f_{K_2}} \subseteq \theta_{f_{K_2}} \quad \Leftrightarrow \quad K_1 \subseteq K_2$$

which shows that the bijection Π is order-reversing and therefore a lattice anti-isomorphism. Now, the complement map $\gamma: A \mapsto A^c$ on X is an latttice anti-isomorphism from the lattice τ of open sets in X onto $\Gamma(X)$ and so the composition $\Pi \circ \gamma: \tau \to \mathrm{Con}(L)$ is a lattice isomorphism.□

Exercises

1. Let X be totally separated. Prove that for any two disjoint compact sets A and B in X, there is a clopen set C such that $A \subseteq C$ and $B \cap C = \emptyset$.
2. Let X be totally order-separated. Prove that if A and B are disjoint compact sets with the property that no element of B is less than any element of A, that is,

$$B \cap (\downarrow A) = \emptyset$$

then there is a clopen down-set C such that $A \subseteq C$ and $B \cap C = \emptyset$.
3. Let X be an ordered Boolean space. Verify the following statements.
   a) The closed down-sets of X are precisely the intersections of clopen down-sets.
   b) The open down-sets of X are precisely the unions of clopen down-sets.
4. Let $(X, \tau, \leq)$ be an ordered Boolean space. Prove that if C is closed in X then $\uparrow C$ and $\downarrow C$ are also closed.
5. Let L be a bounded distributive lattice. Prove that $a \in L$ has a complement in L if and only if $\mathcal{P}_a$ is a down-set.
6. a) Prove that the Cantor set $\mathcal{C}$ is an ordered Boolean space under the order inherited from $\mathbb{R}$.
   b) Prove that $\mathcal{O}^{\mathrm{clo}}(\mathcal{C})$ is a countable Boolean algebra with no atoms.
7. Prove that the intersection maps

$$f_K: \mathcal{O}^{\mathrm{clo}}(X) \to \mathcal{O}^{\mathrm{clo}}(K)$$

defined by

$$f_K(D) = D \cap K$$

for any $D \in \mathcal{O}^{\mathrm{clo}}(X)$ are surjective.
8. **(Characterizing ideals in a bounded distributive lattice)** Let $(X, \tau, \leq)$ be an ordered Boolean space. Let $\mathcal{O}^{\mathrm{op}}(X)$ denote the family of open down-sets of X. Consider the maps

$$\mathcal{U}: \mathcal{I}(\mathcal{O}^{\mathrm{clo}}(X)) \to \mathcal{O}^{\mathrm{op}}(X)$$

defined by

$$\mathcal{U}(\mathbb{I}) = \bigcup \mathbb{I}$$

and

$$\mathcal{V}: \mathcal{O}^{\mathrm{op}}(X) \to \mathcal{I}(\mathcal{O}^{\mathrm{clo}}(X))$$

defined by

$$\mathcal{V}(D) = \{A \in \mathcal{O}^{\mathrm{clo}}(X) \mid A \subseteq D\}$$

In words, $\mathcal{U}(\mathbb{I})$ is the union of all of the elements (clopen down-sets) of $\mathbb{I}$ and $\mathcal{V}(D)$ is the set of all clopen down-sets contained in D.

a) Show that $\mathcal{U}(\mathbb{I})$ is an open down-set in X.
b) Show that $\mathcal{V}(D)$ is an ideal of $\mathcal{O}^{\mathrm{clo}}(X)$.
c) Show that

$$\mathcal{U} \circ \mathcal{V} = \iota \quad \text{and} \quad \mathcal{V} \circ \mathcal{U} = \iota$$

where ι is the appropriate identity function.

d) Show that $\mathcal{U}$ is a lattice isomorphism.

9. **(Characterizing pseudocomplemented elements of a bounded distributive lattice)** Let $L = \mathcal{O}^{\mathrm{clo}}(X)$. Prove that the pseudocomplemented elements $U \in \mathcal{O}^{\mathrm{clo}}(X)$ are precisely the elements U for which $\uparrow U$ is clopen, in which case $U^* = X \setminus \uparrow U$. *Hint*: Suppose that $U \in \mathcal{O}^{\mathrm{clo}}(X)$ has a pseudocomplement U^*, as shown in Figure 10.6.

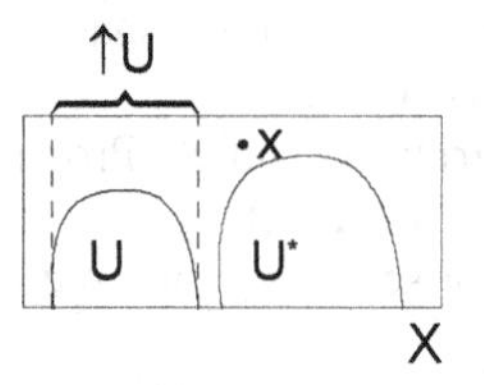

Figure 10.6

What happens if there is an $x \in (\uparrow U)^c \setminus U^*$? Use Exercise 4.

The Stone space of a bounded distributive lattice

Definition *Let L be a bounded distributive lattice. The set* $\mathrm{Spec}(L)$ *together with the topology σ with basis*

$$\mathcal{S} = \{\mathcal{P}_{\neg a} \mid a \in L\}$$

is called the **Stone prime ideal space** *or the* **Stone dual space** *of L.* □

10. Show that for a Boolean algebra, the Stone and Priestley prime ideal spaces are the same.
11. Prove that the Stone prime ideal space is compact.

12. Prove that the open sets of the Stone prime ideal space are the sets

$$\mathcal{P}_{\neg I} := \{P \in \mathrm{Spec}(L) \mid I \not\subseteq P\}$$

for $I \in \mathcal{I}(L)$.

13. Prove that the compact open sets of the Stone prime ideal space are the basis elements

$$\{\mathcal{P}_{\neg a} \mid a \in L\}$$

14. Prove that the Stone prime ideal space is Hausdorff if and only if L is a Boolean algebra. Consequently, the Stone prime ideal space of a non-Boolean bounded distributive lattice is not Hausdorff and so not a Boolean (Stone) space.

Chapter 11
Free Lattices

Lattice Identities

Let us recall a few definitions from earlier in the book.

Definition *Let X be a nonempty set. A* **lattice term** (*or just* **term**), *also called a* **lattice polynomial**, *over X is any expression defined as follows. Let $\sqcup$ and $\sqcap$ be formal symbols.*
1) *The elements of X are terms of* **weight** 1.
2) *If p and q are terms, then so are*

$$(p \sqcap q) \quad \textit{and} \quad (p \sqcup q)$$

whose weights are the sum of the weights of p and q.
3) *An expression in the symbols $X \cup \{\sqcap, \sqcup, (,)\}$ is a term if it can be formed by a finite number of applications of 1) and 2).*

We denote the set of all lattice terms over X by $\mathcal{T}_X$. The elements of X are called **variables**. □

Two terms are equal if and only if they are identical. Thus, for example, $x_1 \sqcup x_2$ is not equal to $x_2 \sqcup x_1$.

It is customary to omit the final pair of parentheses when writing lattice terms. A lattice term involving some or all of the variables $x_{i_1}, \ldots, x_{i_n} \in X$ but no others is denoted by $p(x_{i_1}, \ldots, x_{i_n})$.

We now define two binary operations $\wedge$ and $\vee$, called meet and join respectively, on $\mathcal{T}_X$. For $p, q \in \mathcal{T}_X$, let

$$p \wedge q = p \sqcap q \quad \text{and} \quad p \vee q = p \sqcup q$$

Then since $x_i \sqcap x_j = x_i \wedge x_j$ for all $x_i, x_j \in X$ and similarly for join, we can write any lattice term $p \in \mathcal{T}_X$ using only the variables and the meet and join symbols (and parentheses).

S. Roman (ed.), *Lattices and Ordered Sets*, doi: 10.1007/978-0-387-78901-9_11,

Note that while the set $\mathcal{T}_X$ of all terms over X is not a lattice, it is an Ω-algebra, where Ω consists of the two binary operations of meet and join.

Recall also that given a lattice L and a function $f: X \to A$, where $A \subseteq L$, the **evaluation map** $\epsilon_f: \mathcal{T}_X \to L$ is defined, for each $p(x_{i_1}, \ldots, x_{i_n}) \in \mathcal{T}_X$ by

$$\epsilon_f p(x_{i_1}, \ldots, x_{i_n}) = p(fx_{i_1}, \ldots, fx_{i_n})$$

Definition *A* **lattice identity** *or* **lattice equation** *over a nonempty set X is an equation between lattice terms*

$$p(x_1, \ldots, x_n) = q(x_1, \ldots, x_n)$$

where $x_i \in X$.□

Free and Relatively Free Lattices

Now, given a nonempty set X, we wish to describe the construction of the most general lattice L generated by the elements of X (or a copy of these elements). This is called the *free lattice* generated by X. By most general, we mean that a lattice identity should hold in F if and only if it holds in all lattices. Such a lattice can be characterized, up to isomorphism, by a *universal mapping property.*

However, before giving this property, we want to generalize this notion a bit. A lattice is a set with two algebraic operations (meet and join) that satisfy certain lattice identities, namely, the associative, commutative, idempotent and absorption laws. If we add the modular law, we get modular lattices. If we add the distributive law, we get distributive lattices. More generally, we have the following concept.

Definition *Let Σ be a set of lattice equations. An* **equational class** *or* **variety** *based on Σ is the family $\mathcal{K}$ of all lattices that satisfy the equations in Σ. The lattices in $\mathcal{K}$ are called* **$\mathcal{K}$-lattices**. *The* **trivial variety** *is the variety for which $\Sigma = \{x = y\}$ and so the $\mathcal{K}$-lattices are precisely the one-element lattices.*□

The family of all lattices is an equational class with $\Sigma = \emptyset$ or $\Sigma = \{x = x\}$.. Also, the families of modular lattices and distributive lattices are equational classes.

Definition *Let $\mathcal{K}$ be an equational class. A lattice identity*

$$p(x_1, \ldots, x_n) = q(x_1, \ldots, x_n)$$

holds in all $\mathcal{K}$-lattices *if*

$$\epsilon_f p(x_1, \ldots, x_n) = \epsilon_f q(x_1, \ldots, x_n)$$

that is,

$$p(fx_1, \ldots, fx_n) = q(fx_1, \ldots, fx_n)$$

for all $\mathcal{K}$-lattices L and all functions $f: X \to L$.$\square$

Now, given an equational class $\mathcal{K}$, we wish to describe the most general $\mathcal{K}$-lattice F, in the sense that any lattice equation that holds in F also holds in all $\mathcal{K}$-lattices.

Definition *Let $\mathcal{K}$ be an equational class and let X be a nonempty set, whose elements are called* **variables**. *Referring to Figure 11.1*

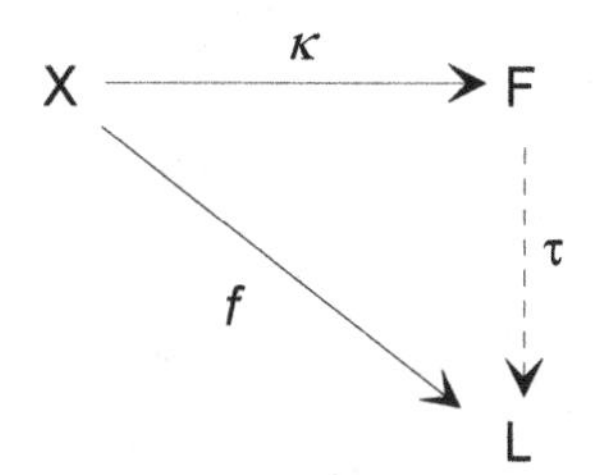

Figure 11.1

we say that the pair (F, κ) where F is a $\mathcal{K}$-lattice and $\kappa: X \to F$ is a function is **$\mathcal{K}$-universal** *if for any function $f: X \to L$, where L is a $\mathcal{K}$-lattice, there is a unique lattice homomorphism $\tau_f: F \to L$ for which*

$$\tau_f \circ \kappa = f$$

We will refer to the map τ_f as the **mediating morphism** *for f. If such a pair exists, then F is called a* **$\mathcal{K}$-free lattice** *and is said to be* **$\mathcal{K}$-freely generated** *by X or have* **basis** *X. The $\mathcal{K}$-free lattice F is denoted by $\mathrm{FL}_{\mathcal{K}}(X)$. A lattice is* **relatively free** *if it is a $\mathcal{K}$-free lattice for some equational class $\mathcal{K}$. If $\Sigma = \emptyset$, then a $\mathcal{K}$-free lattice $\mathrm{FL}_{\mathcal{K}}(X)$ is called a* **free lattice**, *is denoted by $\mathrm{FL}(X)$ and is said to be* **freely generated** *by X.*$\square$

If the equational class $\mathcal{K}$ is nontrivial, then the map $\kappa: X \to \mathrm{FL}_{\mathcal{K}}(X)$ in the definition above must be injective. To see this, let L be a $\mathcal{K}$-lattice with two distinct elements $a, b \in L$. If $x \neq y$ in X, then we can define a function $f: X \to L$ for which $f(x) = a$ and $f(y) = b$. It follows that $\kappa x \neq \kappa y$. Thus, if $\mathcal{K}$ is nontrivial, the image $\kappa(X)$ is a *faithful* copy of X in $\mathrm{FL}_{\mathcal{K}}(X)$.

To see that the universal mapping property captures the spirit of $\mathcal{K}$-free lattices, we have the following.

Theorem 11.1 *Let X be a nonempty set and let $(F, \kappa: X \to F)$ be a $\mathcal{K}$-universal pair. Then a lattice identity*

$$p(x_1, \ldots, x_n) = q(x_1, \ldots, x_n)$$

over X holds in all $\mathcal{K}$-lattices if and only if it holds in the $\mathcal{K}$-free lattice F, that is, if and only if

$$p(\kappa x_1, \ldots, \kappa x_n) = q(\kappa x_1, \ldots, \kappa x_n)$$

in F.

Proof. One direction is clear so assume that the identity holds in the $\mathcal{K}$-free lattice F. Let L be a $\mathcal{K}$-lattice and let $f: X \to L$. If $\tau: F \to L$ is the mediating morphism for f, then

$$\tau p(\kappa x_1, \ldots, \kappa x_n) = \tau q(\kappa x_1, \ldots, \kappa x_n)$$

that is,

$$p(\tau\kappa x_1, \ldots, \tau\kappa x_n) = q(\tau\kappa x_1, \ldots, \tau\kappa x_n)$$

and so

$$p(f x_1, \ldots, f x_n) = q(f x_1, \ldots, f x_n)$$

that is,

$$\epsilon_f p(x_1, \ldots, x_n) = \epsilon_f q(x_1, \ldots, x_n)$$

Since this holds for all $\mathcal{K}$-lattices L and all evaluation maps $\epsilon_f: X \to L$, it follows that $p = q$ holds in all $\mathcal{K}$-lattices.□

Before dealing with the existence of universal pairs, let us consider the issue of uniqueness.

Theorem 11.2 *Let $(F, \kappa: X \to F)$ and $(L, f: X \to L)$ be $\mathcal{K}$-universal pairs. Then there is a lattice isomorphism $\mu: F \approx L$ for which*

$$\mu \circ \kappa = f$$

In particular, F and L are isomorphic. For this reason, one often refers to the *$\mathcal{K}$-free lattice on X.*

Proof. Referring to Figure 11.2, the maps $\tau: F \to L$ and $\sigma: L \to F$ are mediating morphisms and

$$f = \tau \circ \kappa \quad \text{and} \quad \kappa = \sigma \circ f$$

Hence,

$$f = (\tau \circ \sigma) \circ f \quad \text{and} \quad \kappa = (\sigma \circ \tau) \circ \kappa$$

However, the map $\sigma \circ \tau: F \to F$ and the identity map $\iota: F \to F$ are both mediating morphisms as well (as shown in the final diagram in the figure) and so the uniqueness requirement implies that $\sigma \circ \tau = \iota$. Similarly, $\tau \circ \sigma = \iota$ and so $\mu = \tau$ is an isomorphism.□

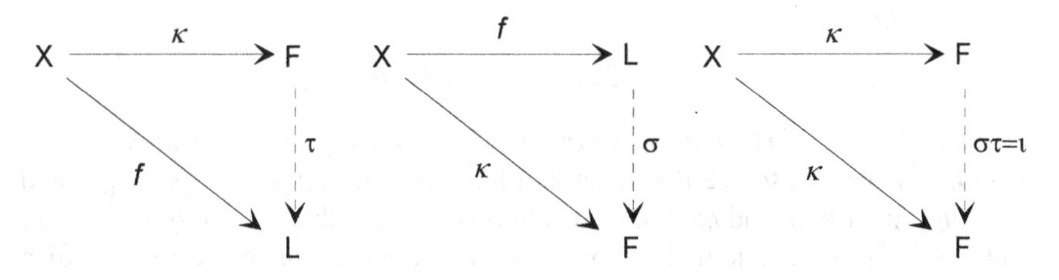

Figure 11.2

Constructing a Relatively Free Lattice

If $\mathcal{T}_X$ is the family of lattice terms over X and $\kappa: X \to \mathcal{T}_X$ is the inclusion map, then the pair $(\mathcal{T}_X, \kappa)$ almost fits the definition of a $\mathcal{K}$-free lattice. In particular, for any $f: X \to L$ and any $x \in X$, we have

$$\epsilon_f \circ \kappa(x) = f(x)$$

where ϵ_f is the evaluation map associated to f. Moreover, ϵ_f preserves meets and joins. The problem is that $\mathcal{T}_X$ is not a $\mathcal{K}$-lattice!

To fix this problem, we pass to equivalence classes. Define a binary relation on $\mathcal{T}_X$ by setting

$$p\theta q \quad \text{if} \quad p = q \text{ holds in all } \mathcal{K}\text{-lattices}$$

It is clear that this is an equivalence relation. Moreover, if $p_1\theta q_1$ and $p_2\theta q_2$, then for any $f: X \to L$, where L is a $\mathcal{K}$-lattice, we have $\epsilon_f p_1 = \epsilon_f q_1$ and $\epsilon_f p_2 = \epsilon_f q_2$ and so

$$\epsilon_f(p_1 \vee p_2) = \epsilon_f p_1 \vee \epsilon_f p_2 = \epsilon_f q_1 \vee \epsilon_f q_2 = \epsilon_f(q_1 \vee q_2)$$

Hence,

$$p_1\theta q_1 \text{ and } p_2\theta q_2 \quad \Rightarrow \quad (p_1 \vee p_2)\theta(q_1 \vee q_2)$$

and similarly,

$$p_1\theta q_1 \text{ and } p_2\theta q_2 \quad \Rightarrow \quad (p_1 \wedge p_2)\theta(q_1 \wedge q_2)$$

These two properties can be expressed by saying that θ is a **congruence relation** on the Ω-algebra $\mathcal{T}_X$.

Theorem 11.3 *Let X be a nonempty set.*

1) *The set $F = \mathcal{T}_X/\theta$ is a $\mathcal{K}$-lattice under meet and join defined by*

$$[p] \wedge [q] = [p \wedge q] \quad \textit{and} \quad [p] \vee [q] = [p \vee q]$$

respectively.

2) *The pair*

$$(\mathcal{T}_X/\theta, \kappa: X \to \mathcal{T}_X/\theta)$$

where $\kappa(x) = [x]$, *is universal and so* $\mathcal{T}_X/\theta$ *is a* $\mathcal{K}$*-free lattice on* X.

Proof. For part 1), to see that meet and join are well defined, if $[p] = [p']$ and $[q] = [q']$, then $p\theta p'$ and $q\theta q'$ and so $(p \wedge q)\theta(p' \wedge q')$, that is, $[p \wedge q] = [p' \wedge q']$ and similarly for the join. Moreover, F is a $\mathcal{K}$-lattice since the axioms of a lattice and the additional lattice equations that define $\mathcal{K}$ hold in all $\mathcal{K}$-lattices. For example, since

$$((p \wedge q) \wedge r)\theta(p \wedge (q \wedge r))$$

we have

$$[(p \wedge q) \wedge r] = [p \wedge (q \wedge r)]$$

and so

$$([p] \wedge [q]) \wedge [r] = [(p \wedge q) \wedge r] = [p \wedge (q \wedge r)] = [p] \wedge ([q] \wedge [r])$$

For part 2), if $f: X \to L$, where L is a $\mathcal{K}$-lattice, then the mediating morphism condition is

$$\tau_f([x]) = fx$$

for all $x \in X$. Now, the map $\tau: F \to L$ defined by

$$\tau([p(x_1, \ldots, x_n)]) = \epsilon_f(p(x_1, \ldots, x_n)) = p(fx_1, \ldots, fx_n)$$

is well defined, since ϵ_f is constant on the congruence classes of F. Also,

$$\tau([p \wedge q]) = \epsilon_f(p \wedge q) = \epsilon_f(p) \wedge \epsilon_f(q) = \tau_f([p]) \wedge \tau([q])$$

and similarly for join. Thus, since $\tau([x]) = fx$, it follows that τ is a mediating morphism for f. As to uniqueness, if $\sigma: F \to L$ is also a lattice homomorphism for which $\sigma \circ \kappa = f$, that is, if $\sigma([x]) = fx$ for all $x \in X$, then

$$\tau([x]) = fx = \sigma([x])$$

and so $\tau p = \sigma p$ for any element $p([x_1], \ldots, [x_n])$ of F, whence $\tau = \sigma$.□

It is customary to drop the congruence class notation and denote the elements of $\mathrm{FL}_{\mathcal{K}}(X) = \mathcal{T}_X/\theta$ as lattice terms and to refer to κ as inclusion. Then we refer to the elements of $\mathcal{T}_X$ as **formal lattice terms**. Of course, it must be kept in mind that distinct formal lattice terms may be the same element of the $\mathcal{K}$-free lattice $\mathrm{FL}_{\mathcal{K}}(X)$. Thus, $x_1 \vee x_2$ and $x_2 \vee x_1$ are distinct elements of $\mathcal{T}_X$, but are the same elements of $\mathrm{FL}_{\mathcal{K}}(X)$ by the commutative law. Also, one must carefully check that functions defined on $\mathcal{T}_X$ are well defined when thought of as functions on $\mathrm{FL}_{\mathcal{K}}(X)$.

Characterizing Equational Classes of Lattices

In 1935, Garrett Birkhoff characterized equational classes of lattices among all families of lattices.

Theorem 11.4 (Birkhoff, 1935 [7]) *Let $\mathcal{K}$ be a family of lattices. The following are equivalent:*

1) *$\mathcal{K}$ is an equational class*
2) *$\mathcal{K}$ is closed under the taking of sublattices, homomorphic images and direct products*
3) *$\mathcal{K}$ is either trivial or else $\mathcal{K}$ is closed under homomorphic images and for every nonempty set X, there is a $\mathcal{K}$-free lattice $\mathrm{FL}_{\mathcal{K}}(X)$.*

Proof. It is not hard to see that 1) implies 2). Suppose that 2) holds and let $\mathcal{K}$ be nontrivial. Let $\mathrm{FL}(X) = \mathcal{T}_X/\theta$ be the free lattice on X, where θ is the congruence relation on $\mathcal{T}_X$ defined by

$$p\theta q \quad \text{if} \quad p = q \text{ holds in all lattices}$$

Now consider the family

$$\mathcal{F} = \{\mu \in \mathrm{Con}(F) \mid F/\mu \in \mathcal{K}\}$$

To see that $\mathcal{F}$ is nonempty, note that since $\mathcal{K}$ is closed under sublattices and homomorphic images, it contains all one-element lattices. Hence, the relation μ defined by $a\mu b$ for all $a, b \in F$ has the property that $F/\mu \in \mathcal{K}$.

The smallest member of $\mathcal{F}$ is the meet

$$\lambda = \bigwedge\{\mu \in \mathrm{Con}(F) \mid F/\mu \in \mathcal{K}\}$$

and Theorem 9.17 implies that F/λ is isomorphic to a sublattice of the direct product

$$\prod\{F/\mu \mid F/\mu \in \mathcal{K}\}$$

and so $F/\lambda \in \mathcal{K}$. To see that F/λ is $\mathcal{K}$-free on X, consider Figure 11.3, where $f: X \to L$ and $L \in \mathcal{K}$ and $j: X \to F$ is the inclusion map.

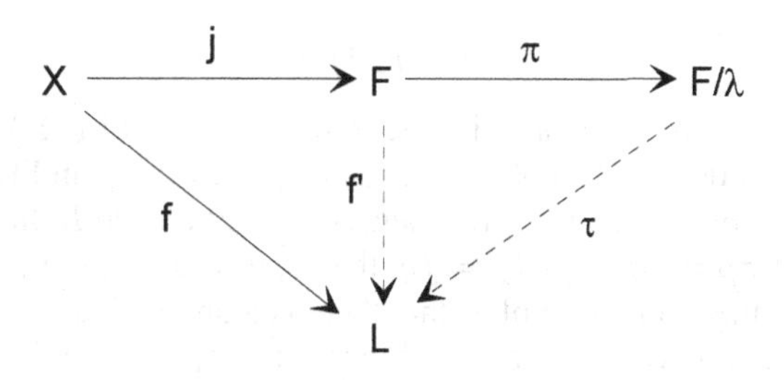

Figure 11.3

The universality of the free lattice implies that there is a unique lattice homomorphism $f'\colon F \to L$. Therefore, by the first isomorphism theorem, $F/\ker(f')$ is isomorphic to a sublattice of L and so $F/\ker(f') \in \mathcal{K}$, which implies that $\lambda \subseteq \ker(f')$ and so the universal mapping property of the quotient implies that there is a unique lattice homomorphism $\tau\colon F/\lambda \to L$ for which

$$\tau \circ \pi \circ j = f$$

Thus, the pair $(F/\lambda, \pi \circ j)$ is $\mathcal{K}$-universal and so F/λ is $\mathcal{K}$-free on X. Thus, 3) holds.

Finally, suppose that 3) holds and continue the notation above with reference to Figure 11.3. Theorem 11.1 implies that a lattice identity $p = q$ holds in a given $\mathcal{K}$-free lattice if and only if $p = q$ holds in all $\mathcal{K}$-lattices. Now, let $X_0 = \{x_1, x_2, \ldots\}$ be a countably infinite set of variables and let Σ be the set of all lattice identities in X_0 that hold in $\mathrm{FL}_{\mathcal{K}}(X_0)$. Let $\mathcal{E}_\Sigma$ be the equational class associated to Σ. We want to show that $\mathcal{K} = \mathcal{E}_\Sigma$. First, the identities in Σ are over X_0 and hold in $\mathrm{FL}_{\mathcal{K}}(X_0)$ and therefore also hold in all $\mathcal{K}$-lattices. Hence, $\mathcal{K} \subseteq \mathcal{E}_\Sigma$.

For the reverse inclusion, let $L \in \mathcal{E}_\Sigma$, that is, all of the identities in Σ hold in L. Referring to Figure 11.4, let Y be a generating set for L. The function $f\colon \mathcal{T}_Y \to L$ is defined by

$$fp(y_1, \ldots, y_n) = p(y_1, \ldots, y_n)$$

where the $p(y_1, \ldots, y_n)$ on the left is a lattice term in $\mathcal{T}_Y$, whereas the $p(y_1, \ldots, y_n)$ on the right is an element of L.

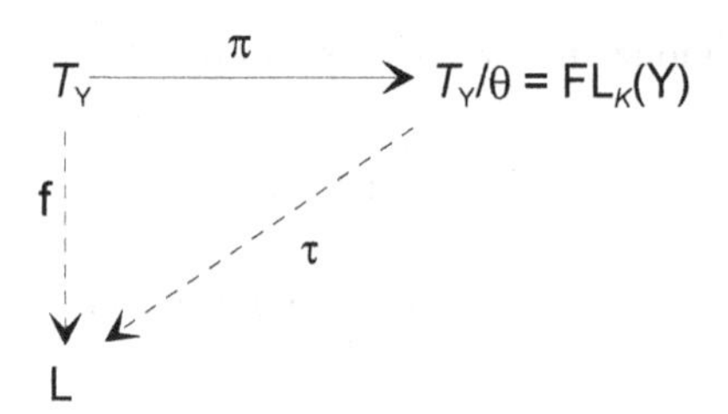

Figure 11.4

If $p(y_1, \ldots, y_n) = q(y_1, \ldots, y_n)$ in the $\mathcal{K}$-free lattice $\mathrm{FL}_{\mathcal{K}}(Y) = \mathcal{T}_Y/\theta$, then $p = q$ in all $\mathcal{K}$-lattices and so $p(x_1, \ldots, x_n) = q(x_1, \ldots, x_n)$ in $\mathrm{FL}_{\mathcal{K}}(X_0)$, that is, $p(x_1, \ldots, x_n) = q(x_1, \ldots, x_n)$ is in Σ and so $p = q$ holds in L. In the notation of Figure 11.4, if $\pi p = \pi q$, then $fp = fq$, that is, $\theta = \ker(\pi) \subseteq \ker(f)$ and so the universality of the quotient implies that there is a unique lattice homomorphism $\tau\colon \mathrm{FL}_{\mathcal{K}}(Y) \to L$ for which $\tau \circ \pi = f$. Since f is surjective, so is τ and so L is a homomorphic image of $\mathrm{FL}_{\mathcal{K}}(Y)$ and is therefore a $\mathcal{K}$-lattice. Hence, $L \in \mathcal{K}$ and so $\mathcal{K} = \mathcal{E}_\Sigma$. Thus, 1) holds.□

The Word Problem for Free Lattices

The **word problem** for $\mathcal{K}$-free lattices is the problem of determining, for any $\mathcal{K}$-free lattice $\mathrm{FL}_{\mathcal{K}}(X)$, when two formal lattice terms in X represent the same element of $\mathrm{FL}_{\mathcal{K}}(X)$. A solution to the word problem is an algorithm that makes this determination. If no such algorithm exists, then the word problem is said to be **unsolvable**.

The word problem for free lattices ($\mathcal{K} = \emptyset$) was first solved by the logician Thoralf Skolem in 1920 [56]. Skolem's solution was apparently not well known. In 1941, Whitman [65] produced a second solution to the word problem that is well known today.

For other equational classes, the word problem is not necessarily solvable. In fact, Ralph Freese [18] proved in 1980 that there is no algorithm for deciding whether $p = q$ in the free modular lattice generated by a set of size 5 and in 1982, Christian Herrmann [28] did the same for the free modular lattice generated by a set of size 4. (Dedekind showed in 1900 [15] that the free modular lattice generated by a set of size 3 has only 28 elements.)

For free lattices, Whitman's idea is to describe a decision procedure for all possible inequalities of the form $p \leq q$ among lattice terms in $\mathrm{FL}(X)$. Then, of course, we have $p = q$ if and only if $p \leq q$ and $q \leq p$.

Theorem 11.5 *Let* $\mathrm{FL}(X)$ *be the free lattice on* X. *Then the following hold, where* $x, y \in X$ *and* $p_i, q_i \in \mathrm{FL}(X)$ *for* $i = 1, 2$*:*

1) **(Variables are incomparable)**

$$x \leq y \quad \Leftrightarrow \quad x = y$$

2) **(Variables are join prime)**

$$x \leq q_1 \vee q_2 \quad \Leftrightarrow \quad x \leq q_1 \textit{ or } x \leq q_2$$

3) **(Variables are meet prime)**

$$p_1 \wedge p_2 \leq x \quad \Leftrightarrow \quad p_1 \leq x \textit{ or } p_2 \leq x$$

4) **(Definition of meet)**

$$p \leq q_1 \wedge q_2 \quad \Leftrightarrow \quad p \leq q_1 \textit{ and } x \leq q_2$$

5) **(Definition of join)**

$$p_1 \vee p_2 \leq q \quad \Leftrightarrow \quad p_1 \leq q \textit{ and } p_2 \leq q$$

6) **(Whitman's condition)**

$$p := p_1 \wedge p_2 \leq q := q_1 \vee q_2 \quad \Leftrightarrow \quad p_1 \leq q \textit{ or } p_2 \leq q \textit{ or } p \leq q_1 \textit{ or } p \leq q_2$$

Proof. The "if" statements hold in all lattices. Also, statements 4) and 5) hold in all lattices. Next, we show that for any $x \in X$ and $y_1, \ldots, y_m \in X \setminus \{x\}$,

$$x \not\leq \bigvee y_i$$

which, in particular, proves 1). To this end, define a function $f: X \to \mathbf{2}$, where $\mathbf{2} = \{0,1\}$ is the lattice with two elements $0 < 1$ by setting $fx = 1$ and $fy = 0$ for all $y \in X \setminus \{x\}$. This can be extended uniquely to a lattice homomorphism $\tau_f: \mathrm{FL}(X) \to \mathbf{2}$. But then

$$x \leq \bigvee y_i \quad \Rightarrow \quad 1 = fx \leq \bigvee fy_i = 0$$

which is false. Hence, $x \not\leq \bigvee y_i$.

Next, we observe that all $w \in \mathrm{FL}(X)$ satisfy

$$x \leq w \quad \text{or} \quad w \leq \bigvee_{i=1}^{n} y_i \text{ for some } y_i \in X \setminus \{x\}$$

For if S is the set of elements of $\mathrm{FL}(X)$ for which this holds, then $X \subseteq S$ and S inherits meets and joins from $\mathrm{FL}(X)$, whence $S = \mathrm{FL}(X)$.

Now we can prove 2). Suppose that $x \leq q_1 \vee q_2$. If $x \not\leq q_1$ and $x \not\leq q_2$, then

$$q_1 \leq \bigvee_{i=1}^{n} y_i \quad \text{and} \quad q_1 \leq \bigvee_{i=1}^{m} z_i$$

for some $y_i, z_i \in X \setminus \{x\}$ and so

$$x \leq q_1 \vee q_2 \leq \left(\bigvee_{i=1}^{n} y_i\right) \vee \left(\bigvee_{i=1}^{m} z_i\right)$$

which we have shown to be false. Statement 3) is the dual of 2).

As to Whitman's condition, Alan Day [14] has provided a simple proof. Suppose that Whitman's condition fails in $F = \mathrm{FL}(X)$ for some elements

$$p := p_1 \wedge p_2 \leq q := q_1 \vee q_2$$

We construct a new lattice in which $p \not\leq q$, by replacing the interval $I = [p,q]$ in F with a pair of intervals that are distinct copies of I:

$$I_0 = I \times \{0\} = \{(u,0) \mid u \in I\}$$

and

$$I_1 = I \times \{1\} = \{(u,1) \mid u \in I\}$$

Let

$$F^* = (F \setminus I) \cup I_1 \cup I_2$$

Extend the order on $F \setminus \{I\}$ to F^* by setting (for $i = 0, 1$)

$$\begin{aligned} v \le (u, i) \text{ in } F^* \quad &\Leftrightarrow \quad v \le u \text{ in } F \\ (u, i) \le v \text{ in } F^* \quad &\Leftrightarrow \quad u \le v \text{ in } F \\ (u, i) \le (v, j) \text{ in } F^* \quad &\Leftrightarrow \quad u \le v \text{ in } F \text{ and } i \le j \end{aligned}$$

We leave it to the reader to check that F^* is a lattice under this order. Also, in F^*, we have

$$p_1 \wedge p_2 = (p, 1) \not\le (q, 0) = q_1 \vee q_2$$

We can associate two maps with this construction. The first map $\pi: F^* \to F$ is projection of F^* onto F defined by

$$\pi u = u \text{ for } u \in F \setminus I \quad \text{and} \quad \pi(u, i) = u \text{ for } i = 0, 1$$

This is a lattice epimorphism. For the second map, the injection $g: X \to F^*$ defined by

$$g(x) = \begin{cases} x & \text{for } x \notin I \\ (x, 0) & \text{for } x \in I \end{cases}$$

can be extended to a lattice homomorphism $\tau_g: F \to F^*$. Note that the composition $\pi \circ \tau_g: F \to F$ satisfies $\pi(\tau_g(x)) = x$ for all $x \in X$ and so $\pi \circ \tau_g = 1$, the identity on F.

Now, the failure of Whitman's condition means that

$$p_1 \not\le q \quad \text{and} \quad p_2 \not\le q \quad \text{and} \quad p \not\le q_1 \quad \text{and} \quad p \not\le q_2$$

and so $p_i, q_i \notin I = [p, q]$. Therefore, if $s \in \{p_1, p_2, q_1, q_2\}$, then $\pi[\tau_g(s)] = s = \pi s$ and since π is injective on $F \setminus I$, it follows that $\tau_g(s) = s$. Therefore, $\tau_g(p) = p$ and $\tau_g(q) = q$ and so in F^*, we have

$$\tau_g(p) = p \not\le q = \tau_g(q)$$

But $p \le q$ in F implies that $\tau_g(p) \le \tau_g(q)$ in F^*, which is a contradiction. Thus, Whitman's condition holds.□

Theorem 11.5 provides a decision procedure for determining whether $p = q$ is true in the free lattice $\mathrm{FL}(X)$. We simply apply the rules in the theorem to the inequalities $p \le q$ and $q \le p$. Each application reduces the weight of the inequality and so must eventually terminate with a yes or no answer.

For example, to decide whether

$$x \vee (y \vee (x \wedge z)) \le x \vee y$$

we apply 5) to get

$$x \le x \vee y \quad \text{and} \quad y \vee (x \wedge z) \le x \vee y$$

The first inequality is true by 2). As to the second, another application of 5) gives

$$y \le x \vee y \quad \text{and} \quad x \wedge z \le x \vee y$$

the first of which is true. As to the second, Whitman's condition says that this holds since $x \wedge z \le x$ by 4). Thus, the decision is yes.

Canonical Forms

The following concept is important in the study of free lattices.

Definition *Let A and B be finite subsets of a lattice L. If each $a \in A$ is contained in some $b \in B$, then A* **refines** *B, denoted by $A \ll B$. The dual notion is that every element of A contains an element of B, written $A \gg B$.*□

The basic properties of refinement are given in the exercises. We will require the following property, whose proof is an exercise.

Lemma 11.6 *If $A \ll B$ and $B \ll A$, then A and B have the same maximal elements.*□

Let X be a nonempty set. Using the equivalence class notation, every element $[t]$ of the free lattice $\mathrm{FL}(X) = \mathcal{T}_X/\theta$ is an equivalence class of terms in $\mathcal{T}_X$ and we refer to each element of $[t]$ as a **representative** of $[t]$. We wish to show that the representatives of $[t]$ of minimum weight are unique up to applications of the associativity and commutativity laws. For example, the following are the same up to associativity and commutativity,

$$x \wedge (y \wedge z) \quad \text{and} \quad (y \wedge x) \wedge z$$

Let us make this idea more precise. Every term $t \in \mathcal{T}_X$ that is not a variable has an outermost operation (meet or join). If the outermost operation for t is the join, then t has the form

$$t = (u_1) \vee (u_2)$$

If the outermost operation of u_1 is join, then we can write

$$t = ((u_{1,1}) \vee (u_{1,2})) \vee (u_2)$$

and use the formal associativity law to remove parentheses:

$$(u_{1,1}) \vee (u_{1,2}) \vee (u_2)$$

If we continue to remove parentheses in this manner, the result is a join

$$(v_1) \vee \cdots \vee (v_n)$$

where each v_i is either a variable or the outermost operation of v_i is meet. We can repeat the process with the terms v_i, resulting in

$$((v_{1,1}) \wedge \cdots \wedge (v_{1,k_1})) \vee \cdots \vee ((v_{n,1}) \wedge \cdots \wedge (v_{n,k_n}))$$

where each $v_{i,j}$ is either a variable or the outermost operation of $v_{i,j}$ is join. This process can be repeated until it cannot be continued further, resulting in a string of one of the following forms, which we refer to as *normal form* (by analogy with the disjunctive/conjunctive normal forms of Boolean algebra).

Definition *Let $A = X \cup \{\wedge, \vee, (,)\}$ and let $\mathcal{W}_A$ be the set of all strings over the alphabet A. An element of X is called a* **variable**. *The* **weight** *$w(s)$ of a string $s \in \mathcal{W}_A$ is the number of occurrences of variables in s.*

1) *A member of X has* **join-normal form** *and* **meet-normal form** *in $\mathcal{W}_A$.*
2) *Assume that join-normal form and meet-normal form have been defined for members of $\mathcal{W}_A$ of weight less than $n \geq 2$.*
   a) *A string over A of weight $n \geq 2$ has* **join-normal form** *if it has the form*

   $$s = (s_1) \vee \cdots \vee (s_k)$$

   where $k \geq 2$ and each s_i has meet-normal form.
   b) *A string over A of weight $n \geq 2$ has* **meet-normal form** *if it has the form*

   $$s = (s_1) \wedge \cdots \wedge (s_k)$$

   where $k \geq 2$ and each s_i has join-normal form.

An element of $\mathcal{W}_X$ has **normal form** *if it has either meet-normal or join-normal form. Variables have* **trivial normal form**; *all other normal-form strings have* **nontrivial normal form**. □

Each term $t \in \mathcal{T}_X$ can be reduced to a unique string $\mathrm{NF}(t)$ in normal form, called the **normal form** of t. Note that $\mathrm{NF}(t)$ is obtained from t by removing only parentheses. Hence, the occurrences and *relative* locations of the variables are the same in both expressions. It is also clear that any string in normal form can be "reverse engineered," generally in more than one way, to produce a term in $\mathcal{T}_X$. Also, the weights of t and $\mathrm{NF}(t)$ are the same, that is,

$$w(t) = w(\mathrm{NF}(t))$$

and their values *as members of the free lattice* $\mathrm{FL}(X)$ are the same, that is,

$$[t] = [\mathrm{NF}(t)]$$

Having the same normal form expresses the fact that two terms are equivalent up to associativity. To deal with commutativity, we define an equivalence relation $\equiv_c$ on the family of normal-form strings by saying that two normal-form strings s and t are equivalent if one can be obtained from the other by reordering the terms within zero or more join substrings or meet substrings

$$(u_1) \vee \cdots \vee (u_k) \quad \text{or} \quad (u_1) \wedge \cdots \wedge (u_k)$$

anywhere within s and t. Then we define an equivalence relation $\equiv$ on $\mathcal{T}_X$ by

$$s \equiv_{ac} t \quad \text{if} \quad \text{NF}(s) \equiv_c \text{NF}(t)$$

When $s \equiv_{ac} t$, we say that s and t are **equivalent up to associativity and commutativity**.

We wish to show that if $u \in [t]$ has minimum weight, then the minimum weight terms in $[t]$ are the terms that are $\equiv_{ac}$-equivalent to u. Since taking the normal form preserves weight, we need only show that any two terms of minimum weight are $\equiv_{ac}$-equivalent. For this, it is sufficient to show that if $\alpha, \beta \in \mathcal{W}_X$ are normal forms of minimum weight for elements of $[t]$, then $\alpha \equiv_c \beta$.

If $u \in [t]$ has minimum weight, we can get some information about $\text{NF}(u)$ from the nature of $[t]$. If $[t] = \{x\}$ for some variable $x \in X$, then $u = t = x$ and $\text{NF}(u) = x$. If

$$[t] = \bigvee [t_i]$$

is a **proper join** in $\text{FL}(X)$, that is, if $[t_i] < [t]$ for all i, then $\text{NF}(u)$ must be a nontrivial *join*-normal form. For if $\text{NF}(u) = x$ is a variable, then

$$[x] = [t] = \bigvee [t_i]$$

and so $[x] \leq [t_k]$ for some k, whence $[t_k] = \bigvee [t_i] = [t]$, contradicting the fact that $\bigvee [t_i]$ is a proper join. Also, if $\text{NF}(u)$ has nontrivial meet-normal form

$$\text{NF}(u) = \bigwedge (u_j)$$

then

$$\bigwedge [u_j] = [\text{NF}(u)] = [u] = [t] = \bigvee [t_i]$$

and Whitman's condition implies that one of the following must hold:

1) $[u_k] \leq \bigvee [t_i]$ for some k. Then $[u_k] = \bigwedge [u_j] = [t]$, which implies that $u_k \in [t]$. But $w(u_k) < w(\text{NF}(u)) = w(u)$, which contradicts the fact that u has minimum weight.
2) $\bigwedge [u_j] \leq [t_k]$ for some k. But then $[t_k] = \bigvee [t_i] = [t]$, contradicting the fact that $\bigvee [t_i]$ is a proper join.

Thus, $\text{NF}(u)$ has nontrivial join-normal form

$$\text{NF}(u) = \bigvee\{(u_i) \mid i \in I\}$$

If u has minimum weight in $[t]$, there are several things we can say about the elements $[u_i]$ of the free lattice $\text{FL}(X)$.

Theorem 11.7 *Let $t \in \mathcal{T}_X$ and let $u \in [t]$ have minimum weight.*
1) If

$$[t] = [t_1] \vee \cdots \vee [t_n]$$

is a proper join in $\text{FL}(X)$*, then u has a nontrivial join-normal form*

$$\text{NF}(u) = (u_1) \vee \cdots \vee (u_m)$$

and the following properties hold in $\text{FL}(X)$*:*
P1) The $[u_i]$ are distinct and $U = \{[u_1], \ldots, [u_m]\}$ is an antichain
P2) If $u_k = a_1 \wedge \cdots \wedge a_m$, then $[a_i] \not\leq [t]$ for all i
P3) Each u_i is a minimum weight representation of $[u_i]$.
P4) $\{u_1, \ldots, u_m\} \ll \{t_1, \ldots, t_n\}$
2) Dually, if

$$[t] = [t_1] \wedge \cdots \wedge [t_n]$$

is a proper meet in $\text{FL}(X)$*, then u has a nontrivial meet-normal form*

$$\text{NF}(u) = (u_1) \wedge \cdots \wedge (u_m)$$

and the following properties hold in $\text{FL}(X)$*:*
P1′) = P1)
P2′) If $u_k = a_1 \vee \cdots \vee a_m$, then $[t] \not\leq [a_i]$ for all i
P3′) = P3)
P4′) $\{u_1, \ldots, u_m\} \gg \{t_1, \ldots, t_n\}$

Proof. If P1) fails, then there is a proper subset I_0 of I for which

$$[u] = \bigvee\{[u_i] \mid i \in I\} = \bigvee\{[u_i] \mid i \in I_0\}$$

and then the string

$$\bigvee\{u_i \mid i \in I_0\}$$

can be reverse engineered to a term $p \in [u] = [t]$ of smaller weight than that of u and so P1) holds.

For P2), if $u_k = a_1 \wedge \cdots \wedge a_m$, then

$$[a_1] \wedge \cdots \wedge [a_m] \leq [t_1] \vee \cdots \vee [t_n]$$

and so Whitman's condition implies that $[a_j] \leq [t]$ for some j or $[u_k] \leq t_j$ for some j. But if $[a_j] \leq [t]$ for some j, then,

$$[t] \leq [a_j] \vee \left(\bigvee_{i \neq k} [t_i]\right) \leq [t]$$

and so

$$[t] = [a_j] \vee \left(\bigvee_{i \neq k} [t_i]\right)$$

which implies that $[t]$ has a shorter representation, obtained by reverse engineering the string

$$a \vee \left(\bigvee_{i \neq k} t_i\right)$$

Thus,

$$[a_i] \not\leq [t] \text{ for all } i$$

which is P2). Note that P4) follows from P2) and Whitman's condition. P3) is clear.□

We can now show that minimum weight representatives are unique up to associativity and commutativity.

Theorem 11.8 *Let* $t \in \mathcal{T}_X$.

1) *If* $u, v \in [t]$ *have minimum weight in* $[t]$, *then*

$$\text{NF}(u) \equiv {}_c\text{NF}(v)$$

that is, u *and* v *are equivalent up to associativity and commutativity.*

2) a) *If* $[t]$ *is a proper join in* $\text{FL}(X)$, *then* $v \in [t]$ *has minimum weight if and only if its join-normal form*

$$\text{NF}(v) = (v_1) \vee \cdots \vee (v_m)$$

satisfies properties P1)–P3), and therefore P4) of Theorem 11.7.

b) *If* $[t]$ *is proper meet in* $\text{FL}(X)$, *then* $v \in [t]$ *has minimum weight if and only if its meet-normal form*

$$\text{NF}(v) = (v_1) \wedge \cdots \wedge (v_m)$$

satisfies properties P1')–P3'), and therefore P4') of Theorem 11.7.

Proof. We have seen that minimum weight representations satisfy properties P1)–P4) or P1')–P4'). For the converses of 2a) and 2b), let

$$\text{NF}(v) = (v_1) \vee \cdots \vee (v_m)$$

satisfy properties P1)–P3), and therefore P4), and let $u \in [t]$ have minimum weight, with

$$\mathrm{NF}(u) = (u_1) \vee \cdots \vee (u_n)$$

Since $U = \{[u_1], \ldots, [u_n]\}$ and $V = \{[v_1], \ldots, [v_m]\}$ both satisfy P4), we have

$$U \ll V \quad \text{and} \quad V \ll U$$

and since U and V are antichains, Lemma 11.6 implies that $U = V$. Thus, by reindexing if necessary, we have $m = n$ and $[v_i] = [u_i]$ for all i. It follows that both v_i and u_i are minimum weight representatives of $[u_i]$ and so $w(v_i) = w(u_i)$ for all i. Hence, $w(v) = w(u)$ is minimum.□

Definition *If $t \in \mathcal{T}_X$, then the terms in $[t]$ of mimimum weight are called* **canonical forms** *for $[t]$.*□

Thus, the canonical forms for a term $t \in \mathcal{T}_X$ are $\equiv_{ac}$-equivalent, that is, equivalent up to associativity and commutativity.

The Free Lattice on Three Generators Is Infinite

We wish to show that the free lattice $\mathrm{FL}(X)$ on three generators $X = \{x, y, z\}$ has an infinite strictly increasing sequence, which shows that $\mathrm{FL}(X)$ is infinite. We begin with an application of Theorem 11.5 to get some cancellation rules.

Lemma 11.9 *Let F_X be a free lattice over X and let u, v, w be distinct members of X. Then the following cancellation rules hold for $p, q \in F_X$:*

1) $u \leq v \vee p \quad \Rightarrow \quad u \leq p$
2) $v \wedge p \leq u \quad \Rightarrow \quad p \leq u$
3) $u \wedge (v \vee p) \leq v \vee (w \wedge q) \quad \Rightarrow \quad p \leq v \vee (w \wedge q)$
4) $u \wedge (v \vee p) \leq w \vee (u \wedge q) \quad \Rightarrow \quad u \wedge (v \vee p) \leq q$

We write these cancellation rules in the following abbreviated form, which shows the cancellations in bold:

1) $u \leq \boldsymbol{v} \vee p$
2) $\boldsymbol{v} \wedge p \leq u$
3) $\boldsymbol{u} \wedge \boldsymbol{(v} \vee p) \leq v \vee (w \wedge q)$
4) $u \wedge (v \vee p) \leq \boldsymbol{w} \vee \boldsymbol{(u \wedge} q)$

Proof. Parts 1) and 2) follow directly from Theorem 11.5. For part 3), Whitman's condition implies that one of the following must hold:

5) $u \leq v \vee (w \wedge q)$
6) $v \vee p \leq v \vee (w \wedge q)$
7) $u \wedge (v \vee p) \leq v$
8) $u \wedge (v \vee p) \leq w \wedge q$

Now, 5) and 1) imply that $u \leq w \wedge q \leq w$, which is false. Also, 8) implies that

$$u \wedge (v \vee p) \leq w$$

and so 2) gives

$$v \le v \vee p \le w$$

which is false. On the other hand, 6) implies that

$$p \le v \vee (w \wedge q)$$

which gives 3). Finally, 7) and 2) imply that $v \vee p \le v$, that is,

$$p \le v \le v \vee (w \wedge q)$$

which is also 3). Thus, 3) holds.

For part 4), Whitman's condition implies that one of the following must hold:

9) $u \le w \vee (u \wedge q)$
10) $v \vee p \le w \vee (u \wedge q)$
11) $u \wedge (v \vee p) \le w$
12) $u \wedge (v \vee p) \le u \wedge q$

Now, 9) and 1) imply that $u \le u \wedge q \le q$ and so 4) holds in this case. Statement 10) implies that

$$v \le w \vee (u \wedge q)$$

and so 1) gives $v \le u \wedge q \le u$, which is false. Statement 11) and 2) give

$$v \le v \vee p \le w$$

which is false. Finally, 12) implies that

$$u \wedge (v \vee p) \le q$$

which is 4). Thus 4) holds in all cases.□

Now we can prove that $\mathrm{FL}(\{x, y, z\})$ is infinite.

Theorem 11.10 *The free lattice* $\mathrm{FL}(X)$ *on the set* $X = \{x, y, z\}$ *has an infinite strictly ascending chain and is therefore an infinite lattice.*
Proof. We use the cancellation rules

1) $u \le \boldsymbol{v} \vee p$
2) $\boldsymbol{v} \wedge p \le u$
3) $\boldsymbol{u} \wedge (\boldsymbol{v} \vee p) \le v \vee (w \wedge q)$
4) $u \wedge (v \vee p) \le \boldsymbol{w} \vee (\boldsymbol{u} \wedge q)$
5) $u \le \boldsymbol{v} \wedge p$
6) $\boldsymbol{v} \vee p \le u$

the last two being obvious. Let $p_0 = x$ and for $n > 0$, let

$$p_n = x \vee (y \wedge (z \vee (x \wedge (y \vee (z \wedge p_{n-1})))))$$

This is the alternating meet and join of z, y and x in that order done twice.

We first explore the relationship between the variables x, y and z and the elements p_n. Clearly, $x \le p_1$. Actually, $x < p_1$, for if $p_1 \le x$, that is, if

$$x \vee (y \wedge (z \vee (x \wedge (y \vee (z \wedge x))))) \le x$$

then cancellation rules 6) and 2) imply that

$$z \vee (x \wedge (y \vee (z \wedge x))) \le x$$

and so $z \le x$, which is false. Hence, $x < p_1$. As a result, since the meet and join operations are monotone, the sequence (p_n) is nondecreasing and so

$$x < p_1 \le p_n$$

for all $n \ge 1$. Our eventual goal is to show that (p_n) is strictly increasing.

If $y \le p_n$, that is, if

$$y \le y \wedge (z \vee (x \wedge (y \vee (z \wedge p_{n-1}))))$$

then cancellation rules 5) and 1) imply that

$$y \le x \wedge (y \vee (z \wedge p_{n-1})) \le x$$

which is false and so $y \not\le p_n$. But $p_n \le y$ implies that $x \le y$, which is also false and so $y \parallel p_n$ for all $n \ge 0$.

If $z \le p_n$, that is, if

$$z \le y \wedge (z \vee (x \wedge (y \vee (z \wedge p_{n-1}))))$$

then $z \le y$, which is false. Also, $p_n \le z$ implies that $x \le z$, which is false. Thus,

$$x < p_1 \le p_n \text{ for } n \ge 1 \quad \text{and} \quad y, z \parallel p_n \text{ for } n \ge 0$$

Now suppose that n is the smallest integer for which $p_{n+1} \le p_n$, that is,

$$x \vee (y \wedge (z \vee (x \wedge (y \vee (z \wedge p_n))))) \le p_n$$

then 6) gives

$$y \wedge (z \vee (x \wedge (y \vee (z \wedge p_n)))) \le p_n$$

that is,

$$y \wedge (z \vee \underbrace{(x \wedge (y \vee (z \wedge p_n)))}_{p}) \leq x \vee (y \wedge \underbrace{(z \vee (x \wedge (y \vee (z \wedge p_{n-1}))))}_{q})$$

Making the substitutions indicated by the underbraces above gives

$$y \wedge (z \vee p) \leq x \vee (y \wedge q)$$

and so cancellation rule 4) gives

$$y \wedge (z \vee p) \leq q$$

that is,

$$y \wedge (z \vee \underbrace{(x \wedge (y \vee (z \wedge p_n)))}_{p}) \leq z \vee (x \wedge \underbrace{(y \vee (z \wedge p_{n-1}))}_{q})$$

Again making the substitutions indicated by the underbraces above gives

$$y \wedge (z \vee p) \leq z \vee (x \wedge q)$$

and so cancellation rule 3) gives

$$p \leq z \vee (x \wedge q)$$

that is,

$$x \wedge (y \vee \underbrace{(z \wedge p_n)}_{p}) \leq z \vee (x \wedge \underbrace{(y \vee (z \wedge p_{n-1}))}_{q})$$

Again making the substitutions indicated by the underbraces above gives

$$x \wedge (y \vee p) \leq z \vee (x \wedge q)$$

and so cancellation rule 4) gives

$$x \wedge (y \vee p) \leq q$$

that is,

$$x \wedge (y \vee \underbrace{(z \wedge p_n)}_{p}) \leq y \vee (z \wedge \underbrace{p_{n-1}}_{q})$$

Making the substitutions indicated by the underbraces above gives

$$x \wedge (y \vee p) \leq y \vee (z \wedge q)$$

and so cancellation rule 3) gives

$$p \leq y \vee (z \wedge q)$$

that is,

$$z \wedge p_n \leq y \vee (z \wedge p_{n-1})$$

An application of Whitman's condition implies that one of the following must hold:

7) $z \leq y \vee (z \wedge p_{n-1})$
8) $p_n \leq y \vee (z \wedge p_{n-1})$
9) $z \wedge p_n \leq y$
10) $z \wedge p_n \leq z \wedge p_{n-1}$

But 7) and 1) imply that

$$z \leq z \wedge p_{n-1} \leq p_{n-1}$$

which is false. Also, 8) implies that

$$x \leq p_n \leq y \vee (z \wedge p_{n-1}) \leq y \vee z$$

which is false. Statement 9) implies that $p_n \leq y$, which is false. Finally, 10) implies that

$$z \wedge p_n \leq p_{n-1}$$

By Whitman's condition, since p_{n-1} is a meet, we have four possibilities to consider:

1) $z \leq p_{n-1}$
2) $p_n \leq p_{n-1}$
3) $z \wedge p_n \leq x$
4) $z \wedge p_n \leq y \wedge (z \vee (x \wedge (y \vee (z \wedge p_{n-2}))))$

Now, 1) is false and 2) is false by the minimality of n. Also, 3) implies that $p_n \leq x$, which is false. Finally, 4) implies that

$$z \wedge p_n \leq y$$

which is also false. This contradiction implies that there is no integer n for which $p_{n+1} \leq p_n$.□

Free lattices can be quite complicated, as witnessed by the fact that a free lattice with 3 generators is infinite. Moreover, it can be shown that a free lattice on 3 generators contains a sublattice that is free on a countably infinite number of generators! We note also that the free lattice $\mathrm{FL}(X)$ is not complete for $|X| \geq 3$.

The reader interested in more information about free lattices should consult the 1995 book *Free Lattices*, by Freese, Jezek and Nation [19].

Exercises

1. If $A \ll B$ then does it follow that $B \gg A$?
2. Let A and B be finite subsets of a lattice L. Prove the following:
   a) $A \ll B$ implies $\bigvee A \leq \bigvee B$.

b) $\ll$ is a quasiorder on L.
c) $A \subseteq B$ implies $A \ll B$.
d) If A is an antichain and $A \ll B$ and $B \ll A$, then $A \subseteq B$.
e) If A and B are antichains and $A \ll B$ and $B \ll A$, then $A = B$.
f) If $A \ll B$ and $B \ll A$, then A and B have the same maximal elements.

3. Describe the free lattice on the set $X = \{x, y\}$. (The free lattice on three elements is infinite.)
4. Prove that the inequality

$$a \wedge (b \vee c) \leq (a \wedge b) \vee (a \wedge c)$$

does not hold in a free lattice. What conclusions do you draw about distributivity?

5. Let $\mathrm{FL}(X)$ be the free lattice on X.
   a) Prove that every element of $\mathrm{FL}(X)$ is either meet-irreducible or join-irreducible.
   b) Prove that the elements of $\mathrm{FL}(X)$ that are both meet-irreducible and join-irreducible are the elements of X.
   c) Prove that $\mathrm{FL}(X) = \mathrm{FL}(Y)$ implies $X = Y$.
6. Suppose that L is a lattice generated by a set X. Suppose also that X satisfies statements 1)–6) in Theorem 11.5. Show that L is free on X. *Hint*: Apply the universal mapping property to the inclusion map $f: X \to L$. Show that the mediating morphism is an isomorphism.
7. Let L be a lattice with generating set X and assume that L satisfies Whitman's condition. Show that L is isomorphic to the free lattice $\mathrm{FL}(X)$ if and only if the following hold for all $x \in X$ and finite subsets $Y \subseteq X$:

$$x \leq \bigvee Y \quad \Rightarrow \quad x \in Y$$

and

$$\bigwedge Y \leq x \quad \Rightarrow \quad x \in Y$$

Hint: You may use an exercise from an earlier chapter, to wit: Let L be a lattice and let $a \in L$. Let X be a generating set for L. For any subset S of L, let $P(S)$ be the property that for any finite subset $S_0 \subseteq S$,

$$a \leq \bigvee S_0 \quad \Rightarrow \quad a < s \text{ for some } s \in S_0$$

Show that if $P(X)$ holds then $P(L)$ holds. Use this exercise to show that X satisfies the conditions 1)–6) of Theorem 11.5. Then use a previous exercise from this chapter to finish this exercise.

8. Show that the following hold in any free lattice $L = \mathrm{FL}(X)$, for u, a, b:
SD$_\vee$) $[u] = [a] \vee [b]$ and $[u] = [a] \vee [c] \quad \Rightarrow \quad [u] = [a] \vee ([b] \wedge [c])$
SD$_\wedge$) $[u] = [a] \wedge [b]$ and $[u] = [a] \wedge [c] \quad \Rightarrow \quad [u] = [a] \wedge ([b] \vee [c])$
These are referred to as the **semidistributive laws**.

9. Let $[t] \in \mathrm{FL}(X)$ and let $[u] \in \mathrm{FL}(X)$ be join irreducible. Prove that $[u]$ has a canonical form of the form $[t] = [u] \vee [a_1] \vee \cdots \vee [a_n]$ if and only if there is an element $[a] \in \mathrm{FL}(X)$ for which $[t] = [u] \vee [a]$ and if $[v] < [u]$, then $[v] \vee [a] < [t]$.
10. Develop an algorithm for computing a canonical form for any $[t] \in \mathrm{FL}(X)$.

Chapter 12
Fixed-Point Theorems

In this chapter, we give a brief introduction to the theory of fixed points for functions defined on a poset P. Let us begin with a few definitions.

Definition *Let P be a poset. A* **fixed point** *for a function $f: P \to P$ is an element $a \in P$ for which*

$$f(a) = a \qquad \square$$

Let us also recall an earlier definition.

Definition *Let P be a poset. A function $f: P \to P$ is* **inflationary** *if*

$$x \le f(x)$$

for all $x \in P$.$\square$

The term *increasing* is also used for the previous concept but, unfortunately, some authors use *increasing* to mean monotone, so we have decided to avoid the term altogether.

We also recall the following definition.

Definition *Let P and Q be posets that have directed joins. A function $f: P \to Q$ is* **(join) continuous** *if for any directed subset D in P, the set $f(D)$ is also directed and*

$$f\left(\overrightarrow{\bigvee}_{d\in D} d\right) = \overrightarrow{\bigvee}_{d\in D} f(d) \qquad \square$$

If the function $f: P \to P$ and the poset P are sufficiently well behaved, then it is not hard to "construct" a fixed point for f. Take $x \in P$ and consider the sequence

$$x, f(x), f^2(x), \ldots$$

S. Roman (ed.), *Lattices and Ordered Sets*, doi: 10.1007/978-0-387-78901-9_12,

If f is inflationary, then this is a chain in P. If P is chain-complete, then we may set $\alpha = \bigvee f^n(x)$. If f is also continuous, then

$$\alpha \le f(\alpha) = f\left(\bigvee f^n(x)\right) = \bigvee f^{n+1}(x) \le \alpha$$

and so α is a fixed point of f. Thus, if P is chain-complete and f is both inflationary and continuous, then f has a fixed point. Our goal in this chapter is to sharpen this result.

Recall that a poset P is **complete** (also called a **CPO**) if it has a smallest element and has directed joins. Theorem 2.19 implies that a poset is complete if and only if it is chain-complete.

Definition *A* **sub-CPO** *of a CPO P is a subset $S \subseteq P$ for which*
1) $0_P \in S$
2) S inherits directed joins from P. □

Fixed Point Terminology

Here are some terms associated with fixed points.

Definition *Let P be a poset and let $f: P \to P$ be a function. Let $a \in P$. Then (see Figure 12.1)*
1) a is a **fixed point** *of f if*

$$f(a) = a$$

2) a is a **pre-fixed point** *of f if*

$$f(a) \le a$$

3) a is a **post-fixed point** *of f if*

$$f(a) \ge a$$

The corresponding sets of fixed points are denoted by $\mathrm{Fix}(f)$, $\mathrm{Pre}(f)$ *and* $\mathrm{Post}(f)$. *The smallest elements of these sets, if they exist, are denoted by* $\mathrm{MinFix}(f)$, $\mathrm{MinPre}(f)$ *and* $\mathrm{MinPost}(f)$, *respectively. The largest elements are denoted by* $\mathrm{MaxFix}(f)$, $\mathrm{MaxPre}(f)$ *and* $\mathrm{MaxPost}(f)$. □

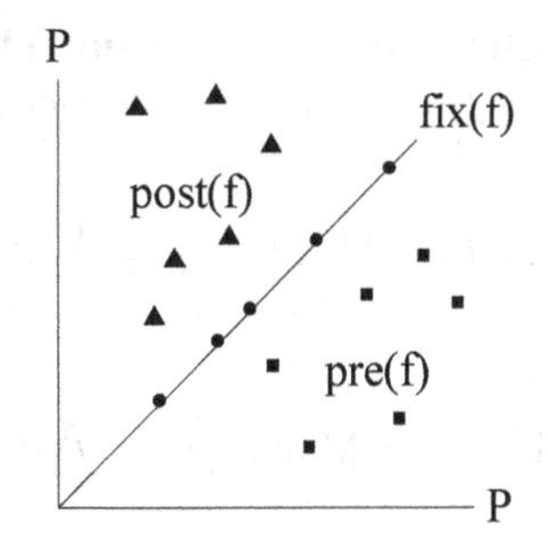

Figure 12.1

Recall that if $f: S \to S$ is a function, then a subset $T \subseteq S$ is said to be ***f*-invariant** if $f(T) \subseteq T$.

Theorem 12.1 *Let $f: P \to P$ be a monotone map.*

1) a) If P is a CPO, then $\mathrm{Pre}(f)$ *is f-invariant and inherits arbitrary nonempty existing P-meets. Moreover,*

$$\mathrm{MinPre}(f) \textit{ exists} \quad \Rightarrow \quad \mathrm{MinPre}(f) = \mathrm{MinFix}(f)$$

b) If P is a complete lattice, then $\mathrm{Pre}(f)$ *is a meet-structure in L and so a complete lattice.*

2) a) If P is a CPO, then $\mathrm{Post}(f)$ *is f-invariant and inherits arbitrary nonempty existing P-joins. Moreover,*

$$\mathrm{MaxPost}(f) \textit{ exists} \quad \Rightarrow \quad \mathrm{MaxPost}(f) = \mathrm{MaxFix}(f)$$

Since $0_P \in$ $\mathrm{Post}(f)$*, it follows that* $\mathrm{Post}(f)$ *is a sub-CPO of P.*

b) If P is a complete lattice, then $\mathrm{Post}(f)$ *is a join-structure in L and so a complete lattice.*

Proof. For part 1), the set $\mathrm{Pre}(f)$ is f-invariant since

$$f(x) \leq x \quad \Rightarrow \quad f(f(x)) \leq f(x)$$

If $\bigwedge x_i$ exists for $x_i \in \mathrm{Pre}(P)$, then

$$f\left(\bigwedge x_i\right) \leq f(x_i) \leq x_i \quad \Rightarrow \quad f\left(\bigwedge x_i\right) \leq \bigwedge x_i$$

and so $\bigwedge x_i \in \mathrm{Pre}(f)$. Finally, if $h = \mathrm{MinPre}(f)$ then $f(h) \in \mathrm{Pre}(f)$ implies that $f(h) \leq h \leq f(h)$ and so $h \in \mathrm{Fix}(f)$. Since $\mathrm{Fix}(f) \subseteq \mathrm{Pre}(f)$, it follows that $h = \mathrm{MinFix}(f)$. Part 2) is dual.□

Fixed-Point Theorems: Complete Lattices

We are now ready to discuss theorems asserting the existence of fixed points, beginning with monotone maps on complete lattices.

Theorem 12.2 (The Knaster–Tarski Fixed Point Theorem, [39], [60]) *Let L be a complete lattice and let $f\colon L \to L$ be a monotone map on L. Then* $\operatorname{Fix}(L)$ *is a complete lattice, with bounds*

$$\operatorname{MaxFix}(f) = \operatorname{MaxPost}(f) = \bigvee \operatorname{Post}(f)$$

and

$$\operatorname{MinFix}(f) = \operatorname{MinPre}(f) = \bigwedge \operatorname{Pre}(f)$$

Proof. Note that

$$\operatorname{Fix}(f) = \operatorname{Pre}(f) \cap \operatorname{Post}(f) = \operatorname{Post}(f|_{\operatorname{Pre}(f)})$$

But $f|_{\operatorname{Pre}(f)}$ is a monotone map on the complete lattice $\operatorname{Pre}(f)$ and so $\operatorname{Post}(f|_{\operatorname{Pre}(f)})$ is also a complete lattice.□

The previous theorem has a converse, proved by Davis [13] in 1955. In particular, a lattice is complete if and only if every monotone function has a fixed point. Since the full proof requires the notions of ordinal and transfinite sequence, we give the proof first in the countable case.

Theorem 12.3 *A countable lattice L is complete if and only if every monotone function on L has a fixed point.*

Proof. We need only prove that if L is incomplete, then there is a monotone function with no fixed point. Let us recall Theorem 3.22: A countably infinite lattice L is incomplete if and only if it has a coalescing pair (B, C) of sequences.

If L has an unbounded strictly increasing infinite sequence

$$B = (b_1 < b_2 < \cdots)$$

then for any $x \in L$, there is a $b_i \in B$ for which $b_i \not\leq x$. Let

$$f(x) = \min\{b_i \mid b_i \not\leq x\}$$

Then $f(x) \neq x$ for all $x \in L$, that is, f has no fixed point. Moreover,

$$x \leq y \quad \Rightarrow \quad \{b_i \mid b_i \not\leq y\} \subseteq \{b_i \mid b_i \not\leq x\} \quad \Rightarrow \quad f(x) \leq f(y)$$

and so f is monotone. The case where L has a strictly decreasing infinite sequence

$$C = (c_1 > c_2 > \cdots)$$

with no lower bound is similar. In particular, for each $x \in L$ let

$$f(x) = \max\{c_i \mid x \not\leq c_i\}$$

Then $f(x) \neq x$ for all $x \in L$ and

$$x \leq y \quad \Rightarrow \quad \{c_i \mid x \not\leq c_i\} \subseteq \{c_i \mid y \not\leq c_i\} \quad \Rightarrow \quad f(x) \leq f(y)$$

and so f is monotone.

Finally, suppose that L has two strictly monotone coalescing infinite sequences

$$B = (b_1 < b_2 < \cdots) \quad \text{and} \quad C = (\cdots < c_2 < c_1)$$

for which $B < C$ and

$$B^u \cap C^\ell = \emptyset$$

If $x \in C^\ell$, then $x \notin B^u$ and we define $f(x)$ by

$$f(x) = \min\{b_i \mid b_i \not\leq x\}$$

On the other hand, if $x \notin C^\ell$, then we let

$$f(x) = \max\{c_i \mid x \not\leq c_i\}$$

In either case, $f(x) \neq x$ and we are left with showing that f is monotone.

Let $x < y$. If $x, y \in C^\ell$ or if $x, y \notin C^\ell$, then the same argument as above shows that $f(x) \leq f(y)$. The only other possibility is that $x \in C^\ell$ and $y \notin C^\ell$. Then

$$f(x) = \min\{b_i \mid b_i \not\leq x\}$$

and

$$f(y) = \max\{c_i \mid y \not\leq c_i\}$$

and since $B < C$, it follows that $f(x) \leq f(y)$. Thus, f is monotone.□

Now we turn to the general case, whose proof is a direct generalization of the previous proof to the transfinite case.

Theorem 12.4 (Davis, 1955 [13]) *A lattice L is complete if and only if every monotone function on L has a fixed point.*
Proof. We need only prove that if L is incomplete, then there is a monotone function with no fixed point. Again, the proof uses Theorem 3.23. If L has a strictly increasing transfinite sequence

$$\mathcal{B} = \langle b_\alpha \mid \alpha < \delta \rangle$$

with no upper bound, then for any $x \in L$, the set of ordinals

$$E_x = \{\alpha < \delta \mid b_\alpha \not\leq x\}$$

is nonempty and so has a least element λ_x. Let

$$f(x) = b_{\lambda_x}$$

Then $\lambda_x \in E_x$ implies that $f(x) \neq x$ and so f has no fixed points. Moreover, if $x \leq y$, then

$$x \leq y \quad \Rightarrow \quad E_y \subseteq E_x \quad \Rightarrow \quad \lambda_x \leq \lambda_y \quad \Rightarrow \quad f(x) \leq f(y)$$

and so f is monotone. The case where L has a strictly decreasing transfinite sequence

$$\mathcal{C} = \langle c_\alpha \mid \alpha < \kappa \rangle$$

with no lower bound is similar. In particular, for each $x \in L$, the set of ordinals

$$F_x = \{\alpha < \delta \mid x \not\leq c_\alpha\}$$

is nonempty and so has a least element μ_x. Let

$$f(x) = c_{\mu_x}$$

Then $f(x) \neq x$ and

$$x \leq y \quad \Rightarrow \quad F_x \subseteq F_y \quad \Rightarrow \quad \mu_y \leq \mu_x \quad \Rightarrow \quad f(x) \leq f(y)$$

and so f is monotone.

Finally, suppose that L has two strictly monotone coalescing infinite sequences

$$\mathcal{B} = \langle b_\alpha \mid \alpha < \delta \rangle \quad \text{and} \quad \mathcal{C} = \langle c_\alpha \mid \alpha < \kappa \rangle$$

for which $\mathcal{B} < \mathcal{C}$ and

$$\mathcal{B}^u \cap \mathcal{C}^\ell = \emptyset$$

If $x \in \mathcal{C}^\ell$, then $x \notin \mathcal{B}^u$ and we define $f(x)$ by

$$f(x) = b_{\lambda_x}$$

as above. On the other hand, if $x \notin \mathcal{C}^\ell$, then we let

$$f(x) = c_{\mu_x}$$

as above. In either case, $f(x) \neq x$ and we are left with showing that f is monotone.

Let $x < y$. If $x, y \in \mathcal{C}^\ell$ or if $x, y \notin \mathcal{C}^\ell$, then the same argument as above shows that $f(x) \leq f(y)$. The only other possibility is that $x \in \mathcal{C}^\ell$ and $y \notin \mathcal{C}^\ell$. Then

$$f(x) = b_{\lambda_x} \in \mathcal{B}$$

and

$$f(x) = c_{\mu_x} \in \mathcal{C}$$

However, since $\mathcal{B} < \mathcal{C}$, it follows that $f(x) \leq f(y)$. Thus, f is monotone.□

Fixed–Point Theorems: Complete Posets

We now turn to the question of the existence of fixed points for functions on a complete poset. As we will see, any inflationary or monotone function on a CPO P has a fixed point. However, we begin by considering functions that are both monotone and inflationary.

Theorem 12.5 *The set $\mathcal{F}$ of all monotone, inflationary functions on a complete poset P has a common fixed point, that is, there is an $a \in P$ for which $f(a) = a$ for all $f \in \mathcal{F}$.*
Proof. The set P^P of all functions on P is partially ordered by pointwise order:

$$f \leq g \quad \text{if} \quad f(x) \leq g(x) \text{ for all } x \in P$$

If $f, g \in \mathcal{F}$, then $f \circ g \in \mathcal{F}$, since

$$x \leq y \quad \Rightarrow \quad g(x) \leq g(y) \quad \Rightarrow \quad f(g(x)) \leq f(g(y))$$

and so $f \circ g$ is monotone. Also, $x \leq g(x)$ implies that $x \leq f(x) \leq f(g(x))$ and so $f \circ g$ is inflationary. Thus, $f \circ g$ is an upper bound for f and g in $\mathcal{F}$ and so $\mathcal{F}$ is a directed family in P^P.

Hence, we can define a function $h: P \to P$ by

$$h(x) = \bigvee\{f(x) \mid f \in \mathcal{F}\}$$

To see that $h \in \mathcal{F}$, if $x \leq y$, then $f(x) \leq f(y)$ for all $f \in \mathcal{F}$ and so $h(x) \leq h(y)$, which shows that h is monotone. Also, since $x \leq f(x)$ for all $f \in \mathcal{F}$, it follows that $x \leq h(x)$ and so h is inflationary. Thus, h is the largest element of $\mathcal{F}$.

Finally, if $f \in \mathcal{F}$, then $h \leq f \circ h \leq h$, that is, $f \circ h = h$. It follows that for any $a \in P$, we have $f(h(a)) = h(a)$, that is, $h(a)$ is a fixed point for every $f \in \mathcal{F}$.□

To improve upon Theorem 12.5, the key is to consider the smallest f-invariant sub-CPO of P. In particular, let $f: P \to P$ be a function on a complete poset P and let

$$\mathcal{F} = \{S \subseteq P \mid S \text{ is an } f\text{-invariant sub-CPO of } P\}$$

which is not empty since $P \in \mathcal{F}$. Moreover, the intersection $P_0 = \bigcap\mathcal{F}$ is nonempty since $0 \in P_0$ and $P_0 \in \mathcal{F}$. Hence, P_0 is the smallest f-invariant sub-CPO of P.

Monotone Functions on a CPO

We can now show that any monotone function on a CPO has a fixed point.

Theorem 12.6 *Let P be a CPO and let $f: P \to P$ be a monotone function. Then f has a least fixed point* $\operatorname{MinFix}(f)$.
Proof. Since f is monotone, $\operatorname{Post}(f) \in \mathcal{F}$ and since $P_0 \subseteq \operatorname{Post}(f)$, it follows that f is inflationary on P_0. Thus, the restriction of f to P_0 is an inflationary monotone function on P_0 and so Theorem 12.5 implies that f has a fixed point $\alpha \in P_0$. Moreover, if $a \in \operatorname{Fix}(f)$, then $\downarrow a \in \mathcal{F}$ since

$$x \in \downarrow a \Rightarrow x \leq a \Rightarrow f(x) \leq f(a) = a \Rightarrow f(x) \in \downarrow a$$

and since $\downarrow a$ inherits directed joins from P. Hence, $P_0 \subseteq \downarrow a$, that is, $P_0 \leq a$. It follows that there is exactly one fixed point $\alpha \in P_0$ and $\alpha = \operatorname{MinFix}(f)$.□

Inflationary Functions on a CPO

We now turn to inflationary functions.

Theorem 12.7 *Let P be a CPO and let $f: P \to P$ be an inflationary function. Then f has a fixed point.*
Proof. We again consider the smallest f-invariant sub-CPO P_0 of P. Our goal is to show that P_0 is directed, since then the join $\alpha = \bigvee P_0$ exists and so

$$\alpha \leq f(\alpha) \leq \bigvee P_0 = \alpha$$

which shows that $\alpha \in \operatorname{Fix}(f)$. If we show that for all $a \in P_0$, every element $b \in P_0$ satisfies $b \leq a$ or $b \geq f(a)$, then P_0 is actually a chain, for in this case, we have either $b \leq a$ or $b \geq f(a) \geq a$ and so a and b are comparable. Thus, we wish to show that

$$P_0 = Z_a := [0, a] \cup [f(a), \infty)$$

where the intervals are taken with respect to P_0. Since $Z_a \subseteq P_0$, we need only show that Z_a is an f-invariant sub-CPO of P, since then $P_0 \subseteq Z_a$.

It is easy to see that Z_a is a sub-CPO of P. Clearly, $0 \in Z_a$. Also, if $D = \{d_i\} \subseteq Z_a$ is directed, let $d = \bigvee_P D \in P_0$. If $d_i \leq a$ for all i, then $d \leq a$ and so $d \in Z_a$ and if $f(a) \leq d_i$ for some i, then $f(a) \leq d_i \leq d$ and so $d \in Z_a$. Thus Z_a is a sub-CPO of P for any $a \in P_0$.

As to the f-invariance of Z_a, since f is inflationary, f maps the interval $[f(a), \infty)$ into itself. Also, f maps a to $f(a) \in Z_a$. Finally, we show that f sends $[0, a)$ into itself, that is,

$$b \in P_0, b < a \quad \Rightarrow \quad f(b) \leq a$$

Let $\mathcal{R}$ be the set of all $a \in P_0$ for which this is true, that is,

$$\mathcal{R} = \{a \in P_0 \mid b \in P_0, b < a \Rightarrow f(b) \leq a\}$$

Thus, $a \in \mathcal{R}$ implies that $Z_a = P_0$. To see that $\mathcal{R} = P_0$, it is sufficient to show that $\mathcal{R}$ is an f-invariant sub-CPO of P.

It is clear that $0 \in \mathcal{R}$. For f-invariance, we must show that $a \in \mathcal{R}$ implies

$$b \in P_0 = Z_a, b < f(a) \quad \Rightarrow \quad f(b) \leq f(a)$$

If $b < a$, then $a \in \mathcal{R}$ implies that $f(b) \leq a \leq f(a)$; if $b = a$, then clearly $f(b) \leq f(a)$; the case $b \geq f(a)$ is not relevant.

To show that $\mathcal{R}$ inherits directed joins, let $D = \{a_i\} \subseteq \mathcal{R}$ be directed. We must show that

$$b < \bigvee a_i \quad \Rightarrow \quad f(b) \leq \bigvee a_i$$

Since $b \in Z_{a_i}$ for each i, it follows that

$$b < a_i \quad \text{or} \quad b = a_i \quad \text{or} \quad b \geq f(a_i) \geq a_i$$

If $b < a_k$ for some k, then $f(b) \leq a_k \leq \bigvee a_i$, as desired, so we may assume that $b \not< a_i$ for all i. If $b \geq f(a_i) \geq a_i$ for all i, then $b \geq \bigvee a_i$ which is not the case. Hence, there is an index k for which $b = a_k$.

Since $b \not< a_i$ for all i, it follows that $a_k = b$ is maximal in D. But if $a_i \parallel a_k = b$ for some i, then since D is directed, there is an $a_j > a_k$, which is false. Hence, $a_k = b$ is the greatest element of D and so $\bigvee a_i = a_k = b$, which is also false. These contradictions imply that $b < a_k$ for some k and complete the proof.□

Constructing a Fixed Point

A first attempt to construct a fixed point might be to look at the sequence

$$0, f(0), f^2(0), \ldots$$

Indeed, if a function $f: P \to P$ on a complete poset P is either inflationary or monotone, then this sequence is increasing. This is clear when f is inflationary and when f is monotone, we have $0 \leq f(0)$ and applying f^n gives $f^n(0) \leq f^{n+1}(0)$. Hence, the join

$$\alpha = \overrightarrow{\bigvee_{n \geq 0}} f^n(0)$$

exists. Now, if f is continuous, then

$$f(\alpha) = f\left(\bigvee_{n \geq 0} f^n(0)\right) = \bigvee_{n \geq 0} f^{n+1}(0) = \bigvee_{n \geq 0} f^n(0) = \alpha$$

and so α is a fixed point. In addition, if f is monotone and a is any fixed point of f, then $0 \leq a$ implies that $f^n(0) \leq f^n(a) = a$ and so $\alpha \leq a$, that is,

$$\alpha = \text{MinFix}(f)$$

Theorem 12.8 *Let $f: P \to P$ be a continuous function on a complete poset P. If f is either monotone or inflationary, then f has a fixed point*

$$\alpha = \overrightarrow{\bigvee_{n \geq 0}} f^n(0)$$

and if f is monotone, then

$$\alpha = \text{MinFix}(f)$$ □

With the help of ordinals, we can generalize Theorem 12.8 to perform a transfinite "construction" of a fixed point that works for both monotone and inflationary functions on a CPO, without requiring continuity. A transfinite sequence $\langle a_\alpha \mid \alpha \in \text{ord} \rangle$ is **increasing** if $\alpha < \beta$ implies $a_\alpha \leq a_\beta$ and **strictly increasing** if $\alpha < \beta$ implies $a_\alpha < a_\beta$.

Let P be a CPO and let $f: P \to P$ be either a monotone or an inflationary function. Consider the following proposition $P(\alpha)$ for $\alpha \in \text{ord}$:

1) If $\alpha = 0$ and $a_\alpha = 0$, then the sequence $\mathcal{A}_\alpha = \langle a_\gamma \mid \gamma \leq \alpha \rangle$ is increasing.
2) If $\alpha = \beta + 1$ for $\beta \in \text{ord}$ and $a_\alpha = f(a_\beta)$, then the sequence $\mathcal{A}_\alpha = \langle a_\gamma \mid \gamma \leq \alpha \rangle$ is increasing.
3) If α is a limit ordinal, the sequence

$$\mathcal{B}_\alpha = \langle a_\gamma \mid \gamma < \alpha \rangle$$

is increasing and if $a_\alpha = \bigvee \mathcal{B}_\alpha$, then the sequence $\mathcal{A}_\alpha = \langle a_\gamma \mid \gamma \leq \alpha \rangle$ is also increasing.

We prove that $P(\alpha)$ holds by induction on α. If $\alpha = 0$, the result is clear. Suppose the result holds for all $\gamma < \alpha$. If $\alpha = \beta + 1$, then $P(\beta)$ holds and so the sequence

$$\mathcal{A}_\beta = \langle a_\gamma \mid \gamma \leq \beta \rangle = \langle a_\gamma \mid \gamma < \alpha \rangle$$

is increasing and we want to show that

$$\mathcal{A}_\alpha = \langle a_\gamma \mid \gamma \leq \alpha \rangle = \mathcal{A}_\beta \cup \{a_\alpha\}$$

is increasing and so it is sufficient to show that $a_\beta \leq a_\alpha$, since a_β is the largest element of $\mathcal{A}_\beta$. If f is inflationary, then

$$a_\beta \leq f(a_\beta) = a_{\beta+1} = a_\alpha$$

If f is monotone, we proceed as follows. If $\beta = \mu + 1$, then $a_\mu, a_\beta \in \mathcal{A}_\beta$ and so $a_\mu \leq a_\beta$. Applying f gives

$$a_\beta = a_{\mu+1} = f(a_\mu) \leq f(a_\beta) = a_{\beta+1} = a_\alpha$$

If β is a limit ordinal, then for any $\gamma < \beta$, we have $a_\gamma \leq a_\beta$ and so

$$a_{\gamma+1} = f(a_\gamma) \leq f(a_\beta) = a_{\beta+1}$$

whence

$$\bigvee\{a_{\gamma+1} \mid \gamma < \beta\} \leq a_{\beta+1} = a_\alpha$$

and since $a_0 = 0$, the join on the left is a_β. Thus, $a_\beta \leq a_\alpha$ in both cases and so $\mathcal{A}_\alpha$ is increasing.

Finally, suppose that α is a limit ordinal. Then $P(\beta)$ holds for all $\beta < \alpha$. To see that $\mathcal{B}_\alpha$ is increasing, let $\gamma < \beta < \alpha$. Then $P(\beta)$ implies that the sequence $\mathcal{A}_\beta$ is increasing and so $a_\gamma \leq a_\beta$. Hence, $\mathcal{B}_\alpha$ is increasing. To see that $\mathcal{A}_\alpha$ is increasing, if $\beta < \alpha$, then

$$a_\alpha = \bigvee \mathcal{B}_\alpha \geq a_\beta$$

Thus, we have shown that $P(\alpha)$ holds for all $\alpha \in \text{ord}$.

Theorem 12.9 *Let P be a CPO and let $f: P \to P$ be either a monotone or an inflationary function. Then f has a fixed point. Moreover, the transfinite sequence $\mathcal{A} = \langle a_\alpha \mid \alpha \in \text{ord}\rangle$ defined by*

$$\begin{aligned} a_0 &= 0 \\ a_{\alpha+1} &= f(a_\alpha) \textit{ for all } \alpha \in \text{ord} \\ a_\lambda &= \sup\{a_\beta \mid \beta < \lambda\} \textit{ for } \lambda \textit{ a limit ordinal} \end{aligned}$$

is well defined and increasing.

1) *The sequence $\mathcal{A}$ eventually becomes constant, that is, there is an ordinal δ for which $a_\delta = a_\alpha$ for all $\alpha > \delta$. Moreover, a_δ is the only fixed point of f contained in $\mathcal{A}$.*
2) *If f is monotone, then a_δ is the least fixed point* $\text{MinFix}(f)$ *of f.*

Proof. We have seen that the sequence $\mathcal{A}$ is well defined, the only issue being that the set $\mathcal{B}_\lambda = \{a_\beta \mid \beta < \lambda\}$ is directed, and that $\mathcal{A}$ is increasing, since if $\beta < \alpha$, then a_β and a_α both lie in the increasing sequence $\mathcal{A}_\alpha \subseteq \mathcal{A}$.

However, $\mathcal{A}$ cannot be strictly increasing, since then $|P| > |\beta|$ for all ordinals β, which is not possible. Hence, there are ordinals $\alpha < \beta$ for which $a_\alpha = a_\beta$ and so the values of a_γ are equal for all $\alpha \leq \gamma \leq \beta$. It follows that $\mathcal{A}$ contains a fixed point of f, for if β is a limit ordinal, then $\alpha < \alpha + 1 < \beta$ and so

$$a_\alpha = a_{\alpha+1} = f(a_\alpha)$$

and if $\beta = \gamma + 1$, then

$$a_\gamma = a_{\gamma+1} = f(a_\gamma)$$

In either case, f has a fixed point a_δ. Moreover, an induction argument shows that $a_\alpha = a_\delta$ for all $\alpha > \delta$. Assume that $a_\beta = a_\delta$ for all $\delta \leq \beta < \alpha$. If

$\alpha = \beta + 1$, then $a_\beta = a_\delta$ and so

$$a_\alpha = f(a_\beta) = f(a_\delta) = a_\delta$$

and if α is a limit ordinal, then

$$a_\alpha = \bigvee \{a_\beta \mid \beta < \alpha\} = a_\delta$$

It follows that $\mathcal{A}$ contains exactly one fixed point of f and 1) holds.

For part 2), if f is monotone, we show by induction that any fixed point x of f satisfies $\mathcal{A} \le x$. Clearly, $0 \le x$. If $a_\alpha \le x$, then

$$a_{\alpha+1} = f(a_\alpha) \le f(x) = x$$

and so $a_{\alpha+1} \le x$. Also, if λ is a limit ordinal and $a_\alpha \le x$ for all $\alpha < \lambda$, then

$$a_\lambda = \bigvee \{a_\alpha \mid \alpha < \lambda\} \le x$$

Thus, $a_\delta \in \mathcal{A}$ is the least fixed point of f.$\square$

Exercises

1. Let P be a CPO. Consider the sets

$$\text{infmon}(P) \subseteq \text{mon}(P) \subseteq \text{func}(P)$$

of inflationary, monotone functions on P, monotone functions on P and all functions on P, respectively, under the pointwise partial order

$$f \le g \quad \text{if} \quad f(x) \le g(x) \text{ for all } x \in P$$

a) Prove that $\text{func}(P)$ is a CPO, where

$$\left(\overrightarrow{\bigvee} f_i \right)(x) = \overrightarrow{\bigvee} f_i(x)$$

b) Prove that $\text{mon}(P)$ is a sub-CPO of $\text{func}(P)$.

2. Show that a nonidentity inflationary function f on a CPO P need not have a least fixed point.
3. Show that an inflationary, monotone function on a poset P need not have a fixed point.
4. Prove that any poset with 0 and with the ACC is a CPO.
5. Let P and Q be CPOs and let $f\colon P \to Q$ be monotone. Prove that if P has the ACC then f is continuous.
6. Find an example of a CPO P and a nonempty subset $S \subseteq P$ that is a CPO under the same order but is not a sub-CPO of P. *Hint*: Look at a special family of subsets of the natural numbers $\mathbb{N}$.
7. Let X be a nonempty set and let X^{**} be the set of all finite or infinite strings over the alphabet X. Order X^{**} by saying that the empty word is the smallest element and that $w \le v$ if w is a prefix of v, that is, if $v = wu$ for

some string u. Show that X^{**} is a CPO under this order. What about the set X^* of finite strings over X?

8. Let P be a countable poset. Prove that P is complete if and only if it is completely inductive.
9. Prove that the underlying set of the sequence $\mathcal{A}$ from Theorem 12.9 is the set P_0. Hence, there is a unique fixed point of f in the set P_0, even when f is inflationary. However, show that it need not be the least fixed point of f in P.
10. A subset S of a poset P is **consistent** if every pair of elements of S has an upper bound in P (but not necessarily in S). Let P be a CPO. Prove that the following are equivalent:
    a) P is **consistently complete**, that is, every consistent subset of P has a join in P.
    b) All subsets that are bounded from above (have an upper bound in P) have a join in P.
    c) All nonempty subsets of P have a meet in P.
    d) $P^{\top}$ (P with a largest element adjoined) is a complete lattice.
    e) $\downarrow x$ is a complete lattice for every $x \in P$.
11. Let P be a poset and let $\{a_{i,j} \mid i \in I, j \in J\}$ be a doubly-indexed subset of P.
    a) Show that if for all $u \in I$ and $v \in J$, the joins

    $$\bigvee_j a_{u,j}, \quad \bigvee_i a_{i,v} \quad \text{and} \quad \bigvee_i \Big(\bigvee_j a_{i,j}\Big)$$

    exist then $\bigvee_{i,j} a_{i,j}$ and $\bigvee_j (\bigvee_i a_{i,j})$ exist and

    $$\bigvee_{i,j} a_{i,j} = \bigvee_i \Big(\bigvee_j a_{i,j}\Big) = \bigvee_j \Big(\bigvee_i a_{i,j}\Big)$$

    b) Let P and Q be posets and let $\mathcal{F}(P,Q)$ be the poset of functions from P to Q, with pointwise ordering: $f \le g$ if $f(x) \le g(x)$ for all $x \in P$. Show that if Q is a CPO then so is $\mathcal{F}(P,Q)$.
    c) Let P be a poset and let Q be a CPO. Let $\mathcal{C}(P,Q)$ be the poset of continuous functions from P to Q, with pointwise ordering. Show that $\mathcal{C}(P,Q)$ is a sub-CPO of $\mathcal{F}(P,Q)$.
12. Show that the KnasterTarski theorem can be used to prove the Schröder-Bernstein theorem: If A and B are sets and $|A| \le |B|$ and $|B| \le |A|$ then $|A| = |B|$. *Hint*: To prove the SchröderBernstein theorem, show that if $f\colon A \to B$ and $g\colon B \to A$ are injective maps then there is a bijection between A and B. Prove that it is sufficient to find subsets $S \subseteq A$ and $T \subseteq B$ such that $\{S, g(T)\}$ is a partition of A and $\{T, f(S)\}$ is a partition of B. Show that this is equivalent to the existence of a subset $S \subseteq A$ such that

    $$S = A \setminus g(B \setminus f(S))$$

Appendix A1
A Bit of Topology

We present the topology needed for the discussion of the representation theorem for distributive lattices.

Topological Spaces

Definition *Let X be a nonempty set. A* **topology** *on X is a nonempty family τ of subsets of X with the following properties:*
1) $\emptyset, X \in \tau$
2) τ *is closed under arbitrary unions*
3) τ *is closed under finite intersections.*
If τ is a topology on X, then (X,τ) is a **topological space**. *One often says that X is a topological space when the topology is understood (or does not need explicit mention). The elements of τ are called* **open sets**. *The complement of an open set is a* **closed set**. *A subset that is both open and closed is* **clopen**. □

Definition *If X is a nonempty set, then the power set $\wp(X)$ is a topology on X, called the* **discrete topology**. *In a topology, every subset of X is open as well as closed and clopen.* □

Subspaces

Theorem A1.1 *Let (X,τ) be a topological space. Let Y be a nonempty subset of X. The family*

$$\sigma = \{A \cap Y \mid A \in \tau\}$$

is a topology on Y, called the **induced topology**, *and the topological space (Y,σ) is called a* **subspace** *of (X,τ). We usually say that Y is a subspace of (X,τ).* □

Bases and Subbases

One way to describe the open sets of a topological space is to give a basis or subbasis for it, as described in the following definition.

Definition *Let (X,τ) be a topological space.*
1) *A family $\mathcal{B}$ of open sets in τ is said to be a* **basis** *(or* **base***) for τ if every open set in τ is the union of elements of $\mathcal{B}$.*
2) *A family $\mathcal{S}$ of open sets in τ is said to be a* **subbasis** *(or* **subbase***) for τ if every open set in τ is a union of finite intersections of elements of $\mathcal{S}$.*□

Note that the union of an empty family is the empty set and the intersection of an empty family is the entire space.

Any nonempty family $\mathcal{F}$ of subsets of a set X qualifies to be a subbasis for a topology: Just take the collection of all unions of finite intersections of members of $\mathcal{F}$. On the other hand, not every nonempty family of subsets of X qualifies as a basis for a topology. The following theorem characterizes those families of subsets that do qualify.

Theorem A1.2 *Let X be a nonempty set and let $\mathcal{F}$ be a nonempty family of subsets of X. Then $\mathcal{F} \cup \{\emptyset, X\}$ is a basis for a topology on X if and only if the intersection of any two members of $\mathcal{F}$ is a union of members of $\mathcal{F}$.*

Connectedness and Separation

Definition *Let (X,τ) be a topological space. A* **separation** *in (X,τ) is a partition $\{U,V\}$ of X by open sets. (Hence, U and V are nonempty.) If $u \in U$ and $v \in V$ we will say that (U,V) is a* **separation of** (u,v).□

Note that if $\{U,V\}$ is a separation, then $V = U^c$ and so U is a nonempty proper clopen set. Conversely, a nonempty proper clopen set C defines a separation $\{C, C^c\}$.

Definition *Let (X,τ) be a topological space.*
1) *(X,τ) is* **Hausdorff** *if for any pair $x \neq y \in X$ there are disjoint open sets U and V such that $x \in U$ and $y \in V$.*
2) *(X,τ) is* **connected** *if X does not have a separation or equivalently, if the only clopen sets in X are $\emptyset$ and X.*
3) *(X,τ) is* **totally separated** *if every pair of distinct elements has a separation or equivalently, if for every $x \neq y \in X$ there is a clopen set C such that $x \in C$ and $y \notin C$.*□

Compactness

Definition *Let (X,τ) be a topological space.*
1) *An* **open cover** *of X is a family $\mathcal{U}$ of open sets in τ whose union is X. If $\mathcal{V} \subseteq \mathcal{U}$ is also an open cover of X, then $\mathcal{V}$ is a* **subcover** *of $\mathcal{U}$.*
2) *(X,τ) is* **compact** *if every open cover of X has a finite subcover.*□

Theorem A1.3 *Let (X,τ) be a topological space.*

1) (X,τ) *is compact if and only if any open cover of* X *by basis elements has a finite subcover.*
2) *A closed subset of a compact space is compact* (*in the induced topology*). $\square$

We will have occasion to use the following result, whose proof seems a bit scarce, so we present it here.

Theorem A1.4 (Alexander subbasis lemma) *Let* (X,τ) *be a topological space with subbasis* $\mathcal{S}$. *If every cover of* X *by elements of* $\mathcal{S}$ *has a finite subcover then* X *is compact.*
Proof. Let $\mathbb{F}$ be the collection of all open covers of X that do not have finite subcovers. For the purposes of contradiction, assume that $\mathbb{F}$ is nonempty and order $\mathbb{F}$ by set inclusion. We wish to show that $\mathbb{F}$ has a maximal element. Let $\mathbb{C}$ be a chain in $\mathbb{F}$ and let $\mathcal{U} = \bigcup \mathbb{C}$. If $\mathcal{U}$ has a finite subcover $\mathcal{K} = \{O_1, \ldots, O_n\}$, then there is a $\mathcal{C} \in \mathbb{C}$ that contains each O_i and so $\mathcal{C}$ has the finite subcover $\mathcal{K}$ as well, which is a contradiction. Hence, $\mathcal{U}$ has no finite subcover and so $\mathcal{U} \in \mathbb{F}$ is an upper bound for the chain $\mathcal{C}$. Thus, Zorn's lemma implies that there is a maximal open cover $\mathcal{M}$ of X having no finite subcover.

The cover $\mathcal{M}$ has some special properties. First, if $O \notin \mathcal{M}$ is open, then the maximality of $\mathcal{M}$ implies that the open cover $\mathcal{M} \cup \{O\}$ has a finite subcover and so

$$X = O \cup M_1 \cup \cdots \cup M_k$$

for some $M_i \in \mathcal{M}$. It follows that no open superset of O can be in $\mathcal{M}$. Thus, if $A \in \mathcal{M}$, then all open subsets O of A are in $\mathcal{M}$ as well.

Second, if A and B are open and $A \notin \mathcal{M}$ and $B \notin \mathcal{M}$, then

$$A \cup M_1 \cup \cdots \cup M_k = X = B \cup N_1 \cup \cdots \cup N_r$$

for some $N_i, M_i \in \mathcal{M}$ and so

$$X = (A \cap B) \cup M_1 \cup \cdots \cup M_k \cup N_1 \cup \cdots \cup N_r$$

which implies that $A \cap B \notin \mathcal{M}$. In other words, $A \cap B \in \mathcal{M}$ implies that $A \in \mathcal{M}$ or $B \in \mathcal{M}$. This extends to any finite intersection of open sets as well.

Now consider the intersection $\mathcal{N} = \mathcal{M} \cap \mathcal{S}$, that is, the elements of $\mathcal{M}$ that are also subbasis elements. We claim that $\mathcal{N}$ is also a cover for X, which cannot be, since $\mathcal{N}$ has a finite subcover, which would also be a finite subcover consisting of elements of $\mathcal{M}$.

If $x \in X$, then there is an $M \in \mathcal{M}$ with $x \in M$. Since M is open, there is a basis element B for which $x \in B \subseteq M$. Thus, we have seen that $B \in \mathcal{M}$. Also, $B = S_1 \cap \cdots \cap S_p$, for some $S_i \in \mathcal{S}$ and so

$$x \in B = S_1 \cap \cdots \cap S_p$$

Hence, by the preceding discussion, $S_i \in \mathcal{M}$ for some i and so $S_i \in \mathcal{N}$. Thus, $\mathcal{N}$ covers X.□

Continuity

Definition *Let (X, τ) and (Y, σ) be topological spaces. A function $f: X \to Y$ is* **continuous** *if the inverse image of every open set in Y is open in X, that is,*

$$U \in \sigma \quad \Rightarrow \quad f^{-1}U \in \tau$$

A continuous bijection f whose inverse f^{-1} is also continuous is called a **homeomorphism.**□

Theorem A1.5 *Let (X, τ) and (Y, σ) be topological spaces. The following are equivalent for a function $f: X \to Y$:*
1) *f is continuous.*
2) *The inverse image of any closed set is closed.*
3) *If $\mathcal{B}$ is a basis for Y, then the inverse image of any element of $\mathcal{B}$ is open in X.*
4) *If $\mathcal{S}$ is a subbasis for Y, then the inverse image of any element of $\mathcal{S}$ is open in X.*□

Theorem A1.6 *Let (X, τ) and (Y, σ) be topological spaces and let $f: X \to Y$ be continuous.*
1) *If C is compact in X then $f(C)$ is compact in Y.*
2) *If C is connected in X then $f(C)$ is connected in Y.*
3) *If X is compact Hausdorff and Y is Hausdorff, then f is a homeomorphism if and only if it is a bijection.*□

We have remarked that closed subsets of a compact space are compact. In a compact Hausdorff space, the converse is also true.

Theorem A1.7 *In a compact Hausdorff space, the compact subsets are the same as the closed subsets.*□

The Product Topology

If $\mathcal{X} = \{(X_i, \tau_i) \mid i \in I\}$ is a family of topological spaces, then the cartesian product

$$X = \prod_{i \in I} X_i = \{f: I \to \bigcup_{i \in I} X_i \mid f(i) \in X_i\}$$

can be given a topology τ called the **product topology** as follows. For each $i \in I$, let $\pi_i: X \to X_i$ be the projection map, defined by $\pi_i(f) = f(i)$ for all $f \in X$. For $B_i \subseteq X_i$, the set

$$\pi_i^{-1}(B_i) = \{f \in X \mid f(i) \in B_i\}$$

is called a **cylinder** with **base** B_i. It is the set of all functions that map the index i into the set B_i. A cylinder is called an **open cylinder** if its base is open and a **closed cylinder** if its base is closed. Note that the entire space X is a cylinder. All other cylinders are **proper cylinders**.

A subbasis for the product topology τ consists of all open cylinders $\pi_i^{-1}(B)$. Thus, a base for τ consists of $\emptyset$, X and all sets of the form

$$\begin{aligned} U_{i_1,\ldots,i_n} &= \pi_{i_1}^{-1}(B_{i_1}) \cap \cdots \cap \pi_{i_n}^{-1}(B_{i_n}) \\ &= \{f \in X \mid f(i_k) \in B_{i_k} \text{ for } k = 1, \ldots, n\} \end{aligned}$$

where B_{i_k} is open in X_{i_k} and $i_k \in I$. In words, the basic open sets are the functions whose values on a finite set $\{i_1, \ldots, i_n\}$ of indices must lie in the prescribed open sets $B_{i_1}, \ldots, B_{i_n}$. It follows that a proper subset O of X is open in the product topology on X if and only if O is the union of sets from $U_{i_1,\ldots,i_n}$.

It is easy to see that the projection maps are continuous and surjective. Moreover, π_i is also an **open map**, that is, π_i maps open sets in X to open sets in X_i. To see this, note that π_i sends any basic open set $U_{i_1,\ldots,i_n}$ in X to an open set in X_i, namely, X_i if $i \notin \{i_1, \ldots, i_n\}$ or B_{i_k} if $i = i_k$. Also, π_i preserves union.

It follows that a cylinder is open in the product topology if and only if it has an open base. Moreover, since π_i^{-1} is a Boolean homomorphism, a cylinder is closed in the product topology if and only if its base is closed. Thus, the terms *open cylinder* and *closed cylinder* are unambiguous.

Appendix A2
A Bit of Category Theory

Here we present the category theory that we will need in the text.

Categories

Informally speaking, a category consists simply of a class of mathematical objects and their structure-preserving maps.

Definition *A* **category** $\mathcal{C}$ *consists of the following:*

1) **(Objects)** *A class* $\mathbf{Obj}(\mathcal{C})$ *whose elements are called the* **objects** *of the category.*
2) **(Morphisms)** *For each pair* (A,B) *of objects of* $\mathcal{C}$*, a set* $\hom_{\mathcal{C}}(A,B)$ *whose elements are called* **morphisms, maps** *or* **arrows** *from* A *to* B*. If* $f\in\hom_{\mathcal{C}}(A,B)$ *we write* $f\colon A\to B$.
3) *The sets* $\hom_{\mathcal{C}}(A,B)$ *are disjoint for distinct pairs* (A,B).
4) **(Composition)** *For* $f\colon A\to B$ *and* $g\colon B\to C$*, there is a morphism* $g\circ f\colon A\to C$*, called the* **composition** *of* g *with* f*. Moreover, composition is associative:*
$$f\circ(g\circ h)=(f\circ g)\circ h$$
whenever the compositions are defined.
5) **(Identity morphisms)** *For each object* $A\in\mathbf{Obj}(\mathcal{C})$ *there is a morphism* $\iota_A\colon A\to A$ *called the* **identity morphism** *for* A*, with the property that if* $f\colon A\to B$*, then*
$$\iota_B\circ f=f \quad\textit{and}\quad f\circ\iota_A=f$$
□

One often writes $A\in\mathcal{C}$ as a shorthand for $A\in\mathbf{Obj}(\mathcal{C})$.

Definition *A morphism* $f\colon A\to B$ *is an* **isomorphism** *if there is a morphism* $f^{-1}\colon B\to A$*, called the* **(two-sided) inverse** *of* f*, for which*
$$f^{-1}\circ f=\iota_A \quad\textit{and}\quad f\circ f^{-1}=\iota_B$$
In this case, the objects A *and* B *are* **isomorphic** *and we write* $A\approx B$.□

Example A2.1 Here are some examples of common categories. In most cases, composition is the "obvious" one.

The Category **Set** of Sets
Obj is the class of all sets.
$\hom(A, B)$ is the set of all functions from A to B.

The Category **Grp** of Groups
Obj is the class of all groups.
$\hom(A, B)$ is the set of all group homomorphisms from A to B.

The Category **Ab** of Abelian Groups
Obj is the class of all abelian groups.
$\hom(A, B)$ is the set of all group homomorphisms from A to B.

The Category $\mathbf{Vect}_F$ of Vector Spaces over a Field F
Obj is the class of all vector spaces over F.
$\hom(A, B)$ is the set of all linear transformations from A to B.

The Category **Rng** of Rings
Obj is the class of all rings (with unit).
$\hom(A, B)$ is the set of all ring homomorphisms from A to B.

The Category **Poset** of Partially Ordered Sets
Obj is the class of all partially ordered sets.
$\hom(A, B)$ is the set of all monotone functions from A to B.

The Category **FinDLat** of all finite distributive lattices
Obj is the class of all finite distributive lattices.
$\hom(A, B)$ is the set of all lattice $\{0, 1\}$-homomorphisms from A to B.

The Category **Top** of Topological Spaces
Obj is the class of all topological spaces.
$\hom(A, B)$ is the set of all continuous functions from A to B.

Here are a couple of examples of more unusual categories.

The Category $\mathbf{Matr}_F$ of Matrices
Obj is the set of all positive integers.
$\hom(c, r)$ is the set of all $r \times c$ matrices over F, composition being matrix multiplication (which accounts for the order of r and c in $\hom(c, r)$).

The Category $\mathbf{Poset}(P, \leq)$, where $(P, \leq)$ is a partially ordered set.
Obj is the set of elements of $\mathcal{P}$.

hom(A, B) is empty unless $A \leq B$, in which case hom(A, B) contains a single element, denoted by AB. Note that the hom-sets specify the relation $\leq$ on P.

As to composition, there is really only one choice: If $AB: A \to B$ and $BC: B \to C$ then it follows that $A \leq B \leq C$ and so $A \leq C$, which implies that hom$(A, C) \neq \emptyset$. Thus, we set $BC \circ AB = AC$. The hom-set hom(A, A) is nonempty and contains only the identity morphism for A.□

Subcategories are defined as follows.

Definition *A category $\mathcal{D}$ is a* **subcategory** *of a category $\mathcal{C}$ if*

1) $\mathbf{Obj}(\mathcal{D})$ *is a subclass of* $\mathbf{Obj}(\mathcal{C})$
2) *For every* $A, B \in \mathbf{Obj}(\mathcal{D})$

$$\hom_{\mathcal{D}}(A, B) \subseteq \hom_{\mathcal{C}}(A, B)$$

3) *If $f: A \to B$ and $g: B \to C$ are morphisms in $\mathcal{D}$, then the composition $g \circ f$ taken in $\mathcal{C}$ is the composition of $\mathcal{D}$.*
4) *The identity maps in $\mathcal{C}$ are in morphisms in $\mathcal{D}$.*□

Functors

Structure-preserving maps between categories are called *functors*. Since a category consists of both objects and morphisms, a functor must consist of two functions—one to handle the objects and one to handle the morphisms. Also, there are two versions of functors: *covariant* and *contravariant*.

Definition *Let $\mathcal{C}$ and $\mathcal{D}$ be categories. A* **functor** $F: \mathcal{C} \Rightarrow \mathcal{D}$ *is a pair of functions (as is customary, we use the same symbol F for both functions):*

1) *An* **object function** $F: \mathbf{Obj}(\mathcal{C}) \to \mathbf{Obj}(\mathcal{D})$ *that maps objects in $\mathcal{C}$ to objects in $\mathcal{D}$.*
2) *For a* **covariant functor**, *a* **morphism function**

$$F: \mathbf{Mor}(\mathcal{C}) \to \mathbf{Mor}(\mathcal{D})$$

that maps each morphism $f: A \to B$ in $\mathcal{C}$ to a morphism $Ff: FA \to FB$ in $\mathcal{D}$. For a **contravariant functor**, *a* **morphism function**

$$F: \mathbf{Mor}(\mathcal{C}) \to \mathbf{Mor}(\mathcal{D})$$

that maps each morphism $f: A \to B$ in $\mathcal{C}$ to a morphism $Ff: FB \to FA$ in $\mathcal{D}$ (note the reversal of direction).

3) *Moreover, identity and composition are preserved, that is, for a covariant functor,*

$$F\iota_A = \iota_{FA} \quad \text{and} \quad F(g \circ f) = Fg \circ Ff$$

and for a contravariant functor,

$$F\iota_A = \iota_{FA} \quad and \quad F(g \circ f) = Ff \circ Fg$$

whenever all compositions are defined.□

The notation $F\colon \mathcal{C} \Rightarrow \mathcal{D}$ is read "F is a functor from $\mathcal{C}$ to $\mathcal{D}$." A functor $F\colon \mathcal{C} \Rightarrow \mathcal{C}$ from $\mathcal{C}$ to itself is referred to as a **functor on** $\mathcal{C}$.

One informal way to think of a covariant functor $F\colon \mathcal{C} \Rightarrow \mathcal{D}$ is as a mapping of one-arrow *diagrams* in $\mathcal{C}$

$$A \xrightarrow{f} B$$

to one-arrow diagrams in $\mathcal{D}$

$$FA \xrightarrow{Ff} FB$$

with the property that loops and triangles are preserved, as shown in Figure A2.1.

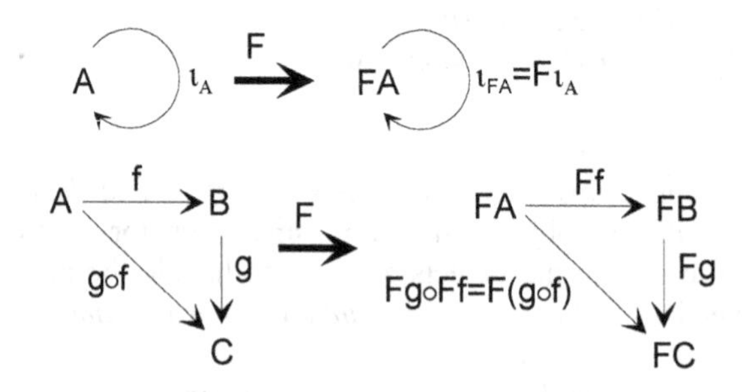

Figure A2.1

Similar pictures can be drawn for contravariant functors.

Functors can be composed in the "obvious" way. Specifically, if $F\colon \mathcal{C} \Rightarrow \mathcal{D}$ and $G\colon \mathcal{D} \Rightarrow \mathcal{E}$ are functors, then $G \circ F\colon \mathcal{C} \Rightarrow \mathcal{E}$ is defined by

$$(G \circ F)(A) = G(FA)$$

for $A \in \mathbf{Obj}(\mathcal{C})$ and

$$(G \circ F)(f) = G(Ff)$$

for $f \in \hom_{\mathcal{C}}(A, B)$.

Definition *Let $F\colon \mathcal{C} \Rightarrow \mathcal{D}$ be a functor.*

1) *F is **full** if for every pair A, B of objects in $\mathcal{C}$, the morphism map is surjective.*
2) *F is **faithful** if for every pair A, B of objects in $\mathcal{C}$, the morphism map is injective.*□

Examples of Functors

We give two simple examples of functors.

Example A2.2 The **power set functor** $\mathcal{P}$: **Set** $\Rightarrow$ **Set** sends a set A to its power set $\wp(A)$ and each set function $f: A \to B$ to the induced function $\mathcal{P}f: \wp(A) \to \wp(B)$ that sends a subset $X \subseteq A$ to fX. It is easy to see that this defines a covariant functor.□

Example A2.3 The following situation is quite common. Let $\mathcal{C}$ be a category. Suppose that $\mathcal{D}$ is another category with the property that every object in $\mathcal{C}$ can be thought of as an object in $\mathcal{D}$ and every arrow $f: A \to B$ of $\mathcal{C}$ can be thought of as an arrow $f: A \to B$ of $\mathcal{D}$.

For instance, every object in **Grp** can be thought of as an object in **Set** simply by ignoring the group operation and every group homomorphism between groups can be thought of as a set function between the corresponding sets. Similarly, every ring can be thought of as an abelian group by ignoring the ring multiplication.

Functors such as these that "forget" some structure are called **forgetful functors**. For categories whose objects are algebraic structures, the "most forgetful" functor is the one that forgets all algebraic structure and thinks of an object (group, ring, field, poset, lattice, etc.) simply as a set.□

Example A2.4 The **spectrum functor** F: **FinDLat** $\Rightarrow$ **FinPoset** sends a finite distributive lattice L to its spectrum $\mathrm{Spec}(L)$ and a lattice homomorphism $f: L \to M$ to the induced inverse map $f^{-1}: \mathrm{Spec}(M) \to \mathrm{Spec}(L)$, restricted to prime ideals. In the opposite direction, the **down-set functor** $\mathcal{O}$: **FinPoset** $\Rightarrow$ **FinDLat** sends a finite poset P to the lattice $\mathcal{O}(P)$ of down-sets of P and a monotone map $g: P \to Q$ between posets to the induced inverse map $g^{-1}: \mathcal{O}(Q) \to \mathcal{O}(P)$, restricted to down-sets.□

Natural Transformations

A structure-preserving map between functors is called a *natural transformation*. Consider a pair of covariant functors $F, G: \mathcal{C} \Rightarrow \mathcal{D}$, as shown in Figure A2.2.

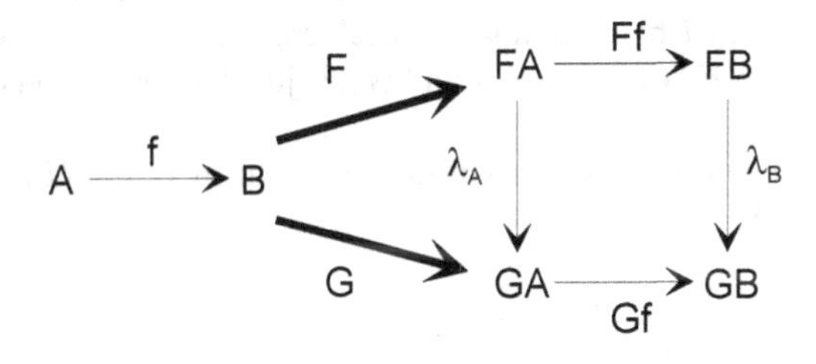

Figure A2.2

Intuitively speaking, the essence of a functor F is how it maps one-arrow diagrams $f: A \to B$ and a natural transformation $\lambda: F \dashrightarrow G$ is intended to map, in a nice way, the family of one-arrow diagrams that characterize the functor F to the corresponding family of one-arrow diagrams that characterize G. As shown in Figure A2.2, this is accomplished by a *family*

$$\{\lambda_A: FA \to GA \mid A \in \mathbf{Obj}(\mathcal{C})\}$$

of morphisms in $\mathcal{D}$—one for each object in $\mathcal{C}$— with the property that the square in Figure A2.2 commutes, that is,

$$Gf \circ \lambda_A = \lambda_B \circ Ff$$

(We also use the notation $\lambda(A)$ in place of λ_A when that is more convenient.)

Let us now give the formal definition of natural transformation.

Definition *Let $F, G: \mathcal{C} \Rightarrow \mathcal{D}$ be covariant functors. A* **natural transformation** *λ from F to G is a family*

$$\lambda = \{\lambda_A: FA \to GA \mid A \in \mathbf{Obj}(\mathcal{C})\}$$

of morphisms in $\mathcal{D}$, indexed by the objects in $\mathcal{C}$, for which the square in Figure A2.2 commutes, that is,

$$\lambda_B \circ Ff = Gf \circ \lambda_A$$

for any $f: A \to B$. This is denoted by $\lambda: F \dashrightarrow G$ or $\{\lambda_A\}: F \dashrightarrow G$. Each morphism λ_A is referred to as a **component** *of λ. If each component λ_A is an isomorphism, then λ is called a* **natural isomorphism** *and we write $\{\lambda_A\}: F \approx G$.*$\square$

If $\lambda: F \approx G$ is a natural isomorphism from F to G, then the family $\{\lambda_A^{-1}\}$ is a natural isomorphism from G to F.

Let us consider a few examples of natural transformations.

Example A2.5 (The determinant) Define functors $F, G: \mathbf{CRng} \Rightarrow \mathbf{Grp}$ as follows. The functor F maps a ring R to the group $GL_n(R)$ of *nonsingular* matrices over R. If $f: R \to S$ is a ring homomorphism, then $(Ff)M$ is the matrix formed from M by applying f to each entry of M. The functor G maps a ring R to the group R^* of units of R and Gf is just the restriction of f to R^*.

Now, the deteminant $\det_R$ is a group homomorphism from $GL_n(R)$ to R^*. As shown in Figure A2.3,

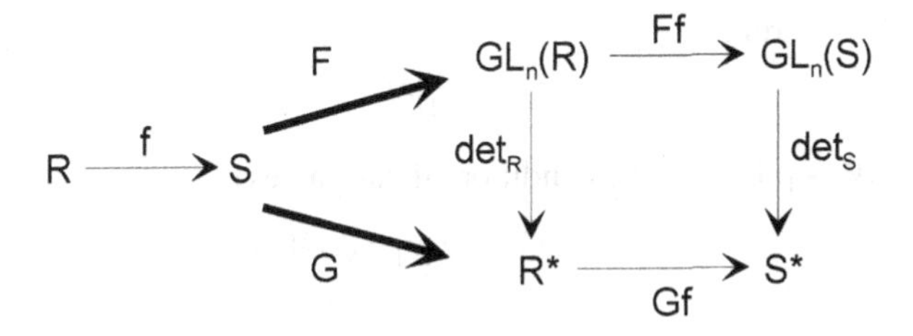

Figure A2.3

the condition that the determinant be natural is

$$\lambda(S) \circ Ff = Gf \circ \lambda(R)$$

that is,

$$\det\begin{bmatrix} fa_{1,1} & \cdots & fa_{1,n} \\ & \vdots & \\ fa_{n,1} & \cdots & fa_{n,n} \end{bmatrix} = f\left(\det\begin{bmatrix} a_{1,1} & \cdots & a_{1,n} \\ & \vdots & \\ a_{n,1} & \cdots & a_{n,n} \end{bmatrix}\right)$$

for any matrix $A = (a_{i,j})$ in $GL_n(R)$. Since this holds, the determinant is a natural transformation from F to G.□

Example A2.6 (The coordinate map) Let $\mathbf{Vect}_k^*$ be the category of *nonzero* vector spaces over the field k. To denote the fact that a vector space V has dimension n, we write V_n. Fix an ordered basis for each vector space V.

Let $\Gamma\colon \mathbf{Vect}_k^* \Rightarrow \mathbf{Vect}_k^*$ be the **matrix representation functor** that sends V_n to the vector space k^n and a linear map $\tau\colon V_n \to W_m$ to multiplication $\mu_{[\tau]}$ by the matrix representation $[\tau]$ of τ with respect to the chosen bases for V and W. Thus,

$$\mu_{[\tau]}(x) = \mu_{[\tau]}x$$

for all $x \in k^n$. The coordinate map $\phi_V\colon V_n \to k^n$, which sends a vector $v \in V$ to its coordinate matrix $[v]$ with respect to the chosen ordered basis for V_n, is an isomorphism of vector spaces. Moreover, the family $\{\phi_V\}$ is a natural isomorphism from the identity functor I to Γ. In fact, as shown in Figure A2.4,

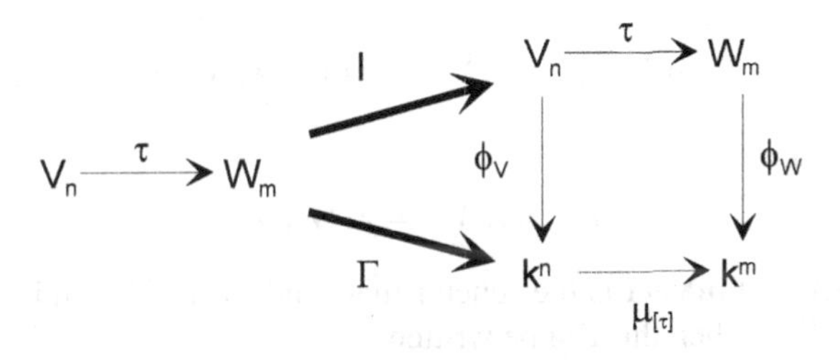

Figure A2.4

the well-known formula

$$[\tau v] = [\tau][v]$$

for $\tau: V \to W$ is precisely the condition of naturalness

$$(\phi_W \circ \tau)v = ([\tau] \circ \phi_V)(v)$$

that is,

$$\phi_W \circ \tau = [\tau] \circ \phi_V$$

□

Example A2.7 (The double-dual map) Let **FinVect** be the category of finite-dimensional vector spaces over a field k. Recall that the *natural map* $\lambda(V): V \to V^{**}$ from a vector space V to its second dual V^{**} is defined by

$$\lambda(V)(v) = \widehat{v}$$

where $\widehat{v}$ is evaluation at v, that is, $\widehat{v}(f) = f(v)$. The natural map is always injective but it is surjective if and only if V is finite-dimensional.

Now, the operator adjoint $\tau^{\times}: W^* \to V^*$ of a linear map $\tau: V \to W$ is defined by

$$\tau^{\times}(f) = f \circ \tau$$

Hence, the second operator adjoint $\tau^{\times\times}: V^{**} \to W^{**}$ is defined by

$$\tau^{\times\times}(\alpha) = \alpha \circ \tau^{\times}$$

for $\alpha \in V^{**}$. Since every $\alpha \in V^{**}$ has the form $\alpha = \widehat{v}$ for some $v \in V$, this can be written

$$\tau^{\times\times}(\widehat{v}) = \widehat{v} \circ \tau^{\times}$$

Now, for $f \in W^*$, we have

$$(\widehat{v} \circ \tau^{\times})f = \widehat{v}(\tau^{\times} f) = \widehat{v}(f \circ \tau) = f(\tau v) = \widehat{\tau v}(f)$$

and so

$$\tau^{\times\times}(\widehat{v}) = \widehat{\tau v}$$

Thus,

$$\tau^{\times\times} \circ \lambda(V)(v) = \lambda(W)(\tau v) = \lambda(W) \circ \tau(v)$$

and so

$$\tau^{\times\times} \circ \lambda(V) = \lambda(W) \circ \tau$$

If F: **FinVect** $\Rightarrow$ **FinVect** is the functor that sends V to V^{**} and $\tau: V \to W$ to $\tau^{\times\times}: V^{**} \to W^{**}$, then this can be written

$$F\tau \circ \lambda(V) = \lambda(W) \circ I\tau$$

where I is the identity functor on **FinVect**. But this says that the family of natural maps $\{\lambda(V)\}$ is a natural transformation from the identity functor I to the double-dual functor F. This accounts for the term *natural map* as it is applied to the map $v \mapsto \widehat{v}$.□

The only categorical result we will need is the following, which says essentially that functors act like ordinary functions with respect to surjectivity and injectivity.

Theorem A2.8

1) *Let $F, G: \mathcal{C} \Rightarrow \mathcal{D}$ be naturally isomorphic functors.*
   a) *F is faithful if and only if G is faithful.*
   b) *F is full if and only if G is full.*

   In particular, if $F \approx I_{\mathcal{C}}$, then F is full and faithful and so the morphism maps

$$F: \hom_{\mathcal{C}}(A, B) \to \hom_{\mathcal{C}}(FA, FB)$$

   are bijections.
2) *Let $F: \mathcal{C} \Rightarrow \mathcal{D}$ and $G: \mathcal{D} \Rightarrow \mathcal{C}$ be functors.*
   a) *If $G \circ F$ is faithful then F is faithful.*
   b) *If $G \circ F$ is full then G is full.*

   In particular, if

$$G \circ F \approx I_{\mathcal{C}} \quad \text{and} \quad F \circ G \approx I_{\mathcal{D}}$$

   then F and G are both full and faithful.

Proof. For part 1), let $\lambda: F \approx G$ be a natural isomorphism. Assume that G is faithful. To see that F is faithful, if $f, g \in \hom_{\mathcal{C}}(A, B)$, then

$$\begin{aligned} Ff = Fg &\Rightarrow \lambda_B \circ Ff \circ \lambda_A^{-1} = \lambda_B \circ Fg \circ \lambda_A^{-1} \\ &\Rightarrow Gf = Gg \\ &\Rightarrow f = g \end{aligned}$$

To see that F is full, let $h \in \hom_{\mathcal{C}}(FA, FB)$. Let

$$k = \lambda_B \circ h \circ \lambda_A^{-1}: GA \to GB$$

Since G is full, there is an $f: A \to B$ such that $Gf = k$ and so

$$h = \lambda_B^{-1} \circ k \circ \lambda_A = \lambda_B^{-1} \circ Gf \circ \lambda_A = Ff$$

Hence F is full.

For part 2), if $G \circ F$ is faithful, then F is faithful, for if $f, g \in \hom_{\mathcal{C}}(A, B)$, we have

$$Ff = Fg \Rightarrow GFf = GFg \Rightarrow f = g$$

Also, if $G \circ F$ is full, then G is full, for if $h \in \hom_{\mathcal{C}}(A, B)$, then there is an $f: A \to B$ such that $h = GFf = G(Ff)$.□

Natural Isomorphism of Categories

It seems reasonable to say that two categories $\mathcal{C}$ and $\mathcal{D}$ are *isomorphic* if there exist functors $F: \mathcal{C} \Rightarrow \mathcal{D}$ and $G: \mathcal{D} \Rightarrow \mathcal{C}$ for which

$$F \circ G = I_{\mathcal{D}} \quad \text{and} \quad G \circ F = I_{\mathcal{C}}$$

However, this notion turns out to be too restrictive. A more useful notion comes when we require only that the two compositions $F \circ G$ and $G \circ F$ are *naturally isomorphic* to the respective identity functors.

Definition *Two categories $\mathcal{C}$ and $\mathcal{D}$ are* **naturally isomorphic** *if there are functors $F: \mathcal{C} \Rightarrow \mathcal{D}$ and $G: \mathcal{D} \Rightarrow \mathcal{C}$ for which*

$$F \circ G \approx I_{\mathcal{D}} \quad \text{and} \quad G \circ F \approx I_{\mathcal{C}}$$

where $I_{\mathcal{D}}$ and $I_{\mathcal{C}}$ are identity functors, as shown in Figure A2.5. When the functors F and G are contravariant, this condition is often expressed by saying that $\mathcal{C}$ and $\mathcal{D}$ are **dual categories.**□

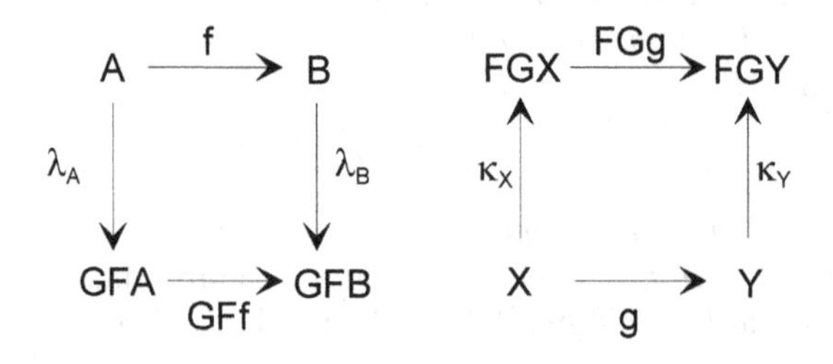

Figure A2.5

Example A2.9 As we will see in the text proper, the category **FinDLat** of finite distributive lattices is dual to the category **FinPoset** of finite posets.□

References

[1] Balbes, R. and Dwinger, Ph., Distributive Lattices, University of Missouri Press, 1974.

[2] Banaschewski, B., On some theorems equivalent with the axiom of choice, Z. Math. Logik Grundlagen Math. 7, (1961), 279–282.

[3] Bell, J. L. and Fremlin, D. H., The maximal ideal theorem for lattices of sets, Bull. London Math. Soc. 4, (1972), 1–2.

[4] Bergman, G., Zur Axiomatic der Elementargeometrie, Monatshefte für Mathematik, Vol. 36 (1929), 269–284.

[5] Birkhoff, G., *Lattice Theory*, Third Edition, AMS Colloquium Publications Vol. 25, 1991.

[6] Birkhoff, G., On the combination of subalgebras, Proc. Camb. Philos. Soc., 29 (1933), 441–464.

[7] Birkhoff, G., On the structure of abstract algebras, Proc. Cambridge Phil. Soc., Vol. 31, (1935), 433–454.

[8] Birkhoff, G. and Frink, O., Representations of Lattices by Sets, Transactions of the American Mathematical Society, Vol. 64, No. 2 (1948), 299–316.

[9] Birkhoff, G. and Ward, M., A characterization of Boolean algebras, Annals of Mathematics, Vol. 40, No. 3 (1939), 609–610.

[10] Blyth, T., *Lattices and Ordered Algebraic Structures*, Springer, 2005.

[11] Crawley, P. and Dilworth, R.P., *Algebraic Theory of Lattices*, Prentice-Hall, 1973.

[12] Davey, B., Priestley, H., *Introduction to Lattices and Order*, Second Edition, Cambridge University Press, 2002.

[13] Davis, A., A characterization of complete lattices, Pacific J. Math., 5 (1955), 311–319.

[14] Day, A., A simple solution of the word problem for lattices, Canad. Math. Bull. 13 (1970), 253–254.

[15] Dedekind, R., Über die drei Moduln erzeugte Dualgruppe, Math. Annalen 53 (1900), 371–403.

[16] Dilworth, R., Lattices with unique complements, Trans. Amer. Math. Soc., Vol. 57, No. 1 (1945), 123–154.

[17] Dilworth, R., The structure of relatively complemented lattices, Ann. Math., 51 (1950), 348–359.

[18] Freese, R., Free modular lattices, Trans. Amer. Math. Soc. 261, (1980), 81–91.

[19] Freese, R., Jezek, J. and Nation, J., *Free Lattices*, Mathematical Surveys and Monographs, AMS, 1995.

[20] Freese, R., Lampe, W.A. and Taylor, W., Congruence lattices of algebras of fixed similarity type I, Pacific J. Math., 82 (1979), 59–68.

[21] Funayama, N., On the completion by cuts of a distributive lattice, Proceedings of the Imperial Academy of Japan, 20 (1944), 1–2.

[22] Gehrke, M., Harding, J. ad Venema, Y., MacNeille completions and canonical extensions, Transactions of the AMS, Vol. 358, No. 2 (2005), 573–590.

[23] Gierz, G., et al., Continuous Lattices and Domains, Cambridge University Press, 2003.

[24] Grätzer, G., *General Lattice Theory*, Second Edition, Birkhäuser, 1998.

[25] Grätzer, G., *The Congruences of a Finite Lattice*: A Proof-by-Picture Approach, Birkhäuser, 2006.

[26] Grätzer, G., Two problems that shaped a century of lattice theory, Notices of the American Mathematical Society, Vol. 54, No. 6 (2007), 696–707.

[27] Grätzer, G. and Schmidt, E.T., Characterizations of congruence lattices of abstract algebras, Acta Sci. Math. (Szeged) 24 (1963), 34–59.

[28] Herrmann, C., Uber die von vier Moduln erzeugte Dualgruppe, Abh. Braunschweig. Wiss. Ges., 33 (1982), 157–159.

[29] Haiman, M., Arguesian lattices which are not linear, Bull. Amer. Math. Soc. 16 (1987), 121–124.

[30] Haiman, M., Arguesian lattices which are not type-1, Algebra Universalis 28 (1991), 128–137.

[31] Hall, M., *The Theory of Groups*, AMS Chelsea, 1976.

[32] Harzheim, E., *Ordered Sets*, Springer, 2005.

[33] Herrlich, H., The axiom of choice holds iff maximal closed filters exist, Math. Log. Quart. 49, No. 3 (2003), 323–324.

[34] Huntington, E. V. Sets of independent postulates for the algebra of logic, Trans. Amer. Math. Soc. 5 (1904), 288–309.

[35] Iwamura, T., A lemma on directed sets (in Japanese), Zenkoku Shijo Sugaku Danwakai 262 (1944), 107–111.

[36] Johnstone, P., *Stone Spaces*, Cambridge University Press, 1982.

[37] Jónsson, B., On the representation of lattices, Math. Scand., 1 (1953), 193–206.

[38] Klimowsky, G., El teorema de Zorn y la existencia de filtros e ideales maximales en los reticulados distributivos, Revista Unión Matemática Argentina 18 (1958), 160–164.

[39] Knaster, B., Une théorème sur les fonctions d'ensembles, Ann. Soc. Pol. Math., 6 (1927), 133–134.

[40] Kurosh, A.G., Durchschnittsdarstellungen mit irreduziblen Komponenten in Ringen und in sogenannten Dualgruppen, Math. Sbornik, 42 (1935), 613–616.

[41] Maeda, F., *Kontinuierliche Geometrien*, Springer-Verlag, 1958.

[42] Maeda, F., Maeda, S., *Theory of Symmetric Lattices*, Springer, 1970.

[43] Markowsky, G., Chain-complete posets and directed sets with applications, Algebra Univ., 6 (1976), 53–68.

[44] Mayer-Kalkschmidt, J. and Steiner, E., Some theorems in set theory and applications in the ideal theory of partially ordered sets, Duke Math J., 31 (1964), 287–289.

[45] McKenzie, R., Equational Bases for Lattice Theories, Math. Scand. 27 (1970), 24–38.

[46] Mrówka, S., On the ideal's extension theorem and its equivalence to the axiom of choice, Fund. Math. 43 (1955), 46–49.

[47] Ore, O., On the foundations of abstract algebra I, Ann. of Math., 36 (1935), 406–437.

[48] Ore, O., On the foundations of abstract algebra II, Ann. of Math., 37 (1936), 265–292.

[49] Padmanabhan, R., Two identities for lattices, Proc. Amer. Math. Soc., 20 (1969), 409–412.

[50] Pudlák, P. and Tůma, J, Every finite lattice can be embedded in a finite partition lattice, Algebra Universalis, 10 (1980), 74–95.

[51] Priestley, H., Ordered Topological Spaces and the Representation of Distributive Lattices, Proc. London Math. Soc., 2, 24 (1972), 507–530.

[52] Priestley, H., Representation of Distributive Lattices By Means of Ordered Stone Spaces, Bull. London Math. Soc., 2 (1970), 186–190.

[53] Rubin, H. and Rubin, J. E., Equivalents of the Axiom of Choice II, North Holland , 1985.

[54] Rubin, H. and Scott, D., Some topological theorems equivalent to the Boolean prime ideal theorem, Bull. Amer. Math. Soc. 60 (1954), 389.

[55] Scott, D.S., The theorem on maximal ideals in lattices and the axiom of choice, Bull. Amer. Math. Soc., 60 (1954), 83.

[56] Skolem, T., Logisch-kombinatorische Untersuchungen über die Erfüllbarkeit und Beweisbarkeit mathematischen Sätze nebst einem Theoreme über dichte Mengen, Videnskapsselskapets skrifter I, Matematisk-naturvidenskabelig klasse, Videnskabsakademiet i Kristiania 4 (1920), 1–36.

[57] Skornyakov, L., *Complemented Modular Lattices and Regular Rings*, Oliver and Boyd, 1964.

[58] Stern, M., *Semimodular Lattices: Theory and Applications*, Cambridge University Press, 1999.

[59] Szász, G., *Introduction to Lattice Theory*, Academic Press, 1963.

[60] Tarski, A., A lattice-theoretical fixpoint theorem and its applications, Pacific Journal of Mathematics, Vol. 5, No. 2 (1955), 285–309.

[61] Tarski, A., Prime ideal theorems for Boolean algebras and the axiom of choice., Bull. Amer. Math. Soc., 60 (1954), 390–391.

[62] Tarski A., Zur Grundlegung der Booleschen Algebra. I, Fundamenta Mathematicae, 24 (1935), 177–198.

[63] Theunissen, M. and Venema, Y., MacNeille completions of lattice expansions, http://staff.science.uva.nl/~yde/recentwork.html.

[64] Wehrung, F., A solution to Dilworth's congruence lattice problem, Adv. Math., to appear.
[65] Whitman, P., Free Lattices, Ann. Math., Vol. 42, No. 1 (1941), 325–330.
[66] Whitman, P., Free Lattices II, Ann. Math., Vol. 43, No. 1 (1942), 104–115.
[67] Whitman, P., Lattices, equivalence relations and subgroups, Bull. Amer. Math. Soc. 52 (1946), 507–522.
[68] Wolk, E., Dedekind completeness and a fixed-point theorem, Canad. J. *Math.*, 9 (1957), 400–405.

Index of Symbols

$\downarrow a = \{x \in P \mid x \leq a\}$
$\uparrow a = \{x \in P \mid x \geq a\}$
$\vee$: join
$\wedge$: meet
$\sqcup$: disjoint union
$\|$: incomparable or parallel
$\sqsubset$: covers
$\sqsubseteq$: covers or is equal to
$a_{[\alpha,\beta]}$: the set of relative complements of a with respect to $[\alpha, \beta]$
A^u: the set of all upper bounds for A
A^ℓ: the set of all lower bounds for A
$B_\downarrow(a) := B \cap \downarrow a$
Con(L): the set of all congruence classes on L
Epi(L): the set of all epimorphisms with domain L
[Epi(L)]: the set of all equivalence classes of epimorphisms with domain L
$\mathcal{F}_0$: proper filters
FL(X): the free lattice over X
ι: the identity function
$\mathcal{I}(L)$: ideals of L
$\mathcal{I}_0(L)$: proper ideals of L
$\mathcal{I}_\emptyset(L) = \mathcal{I}(L) \cup \{\emptyset\}$
$\mathcal{J}(L)$: the set of join-irreducible elements in L
$\mathcal{J}_\downarrow(a) := \mathcal{J}(L) \cap \downarrow a$
$\mathcal{K}(L)$: the set of compact elements of L
$\mathcal{K}_\downarrow(a) := \mathcal{K}(L) \cap \downarrow a$
κ: the kappa function defined by $\kappa(x) = \mathcal{K}_\downarrow(x)$
$^\ell$: the lower bound operator
$\mathcal{M}(L)$: the set of meet-irreducible elements in L
M_3: see Figure 3.4
N_5: see Figure 3.4
NF(t): the normal form of t
$\mathcal{O}(P)$: the down-sets of P
$\mathcal{O}^*(W) := \mathcal{O}(W) \setminus \{W\}$

$\mathcal{P}_{\neg a} = \{P \in \text{Spec}(L) \mid a \notin P\}$
$P[a]$: initial segment of P at a
$\wp(S)$: power set of S
$\wp_0$: proper subsets
StCon(L): the set of all standard congruence relations on L
$\mathcal{T}_X$: the lattice terms over X
$\phi_\downarrow$: the down map defined by $\phi_\downarrow a = \downarrow a$
$\phi_{\neg\uparrow}$: the not-up map defined by $\phi_{\neg\uparrow} a = \{x \in P \mid a \not\leq x\}$
ρ: the rho function defined by $\rho(a) = \mathcal{P}_{\neg a}$
$\mathcal{U}(P)$: the up-sets of P
^{u}: the upper bound operator
$Y \hookrightarrow X$: there is an injection from Y into X
$Y \leftrightarrow X$: there is a bijection from X to Y

Index